Springer Geology

For further volumes:
http://www.springer.com/series/10172

Jens Götze · Robert Möckel
Editors

# Quartz: Deposits, Mineralogy and Analytics

Springer

*Editors*
Jens Götze
TU Bergakademie Freiberg
Institute of Mineralogy
Brennhausgasse 14
09596 Freiberg
Germany

Robert Möckel
TU Bergakademie Freiberg
Institute of Mineralogy
Brennhausgasse 14
09596 Freiberg
Germany

ISBN 978-3-642-22160-6 ISBN 978-3-642-22161-3 (electronic)
DOI 10.1007/978-3-642-22161-3
Springer Heidelberg New York Dordrecht London

Library of Congress Control Number: 2012933435



Printed on acid-free paper

Springer is part of Springer Science+Business Media (www.springer.com)

# Preface

The contributions of this book arose from a conference on "Mineralogy and analytics of highpurity $SiO_2$ raw materials" held in June 2011 in Freiberg, Germany. The 15 chapters are written by international experts and reflect the state of the art in the field of quartz and silica raw materials. The book covers the broad field of $SiO_2$ minerals and rocks and shows the progress made during the last decades in the evaluation of deposits and the application of advanced characterization methods to study the genesis and properties of these materials.

In the broad sense the subject of this book belongs to the properties and specifics of both natural and synthetic materials of the $SiO_2$ system. The chapters are dealing mainly with the mineral quartz, which is one of the most frequent minerals of the Earth's crust and also an important raw material in the industry. Therefore, the book is addressed to graduate students and scientists in mineralogy, geology, chemistry, physics, materials science and engineering, as well as to people from the industry, who are working with the processing and technical application of quartz and $SiO_2$ material.

The 15 chapters of the book should be divided into at least three parts. The first chapter provides a general overview about the classification, mineralogy and industrial potential of $SiO_2$-minerals and rocks. The next five chapters deal with aspects of exploration and qualification of potential deposits of quartz raw materials in terms of certain technical applications. Examples are given for potential deposits of $SiO_2$ raw materials in Argentina, Brazil, Germany and Norway. The third part of the book is focused on modern analytical methods, which are being used for a thorough investigation of quartz and $SiO_2$ materials in geosciences and in the industry. Each chapter presents fundamentals of chemical and structural characterization methods such as microprobe analysis, LA-ICP-MS trace-element analysis, fluid inclusion studies, electron paramagnetic resonance spectroscopy, and cathodoluminescence microscopy and spectroscopy. All these methods are recently used for the microcharacterization of quartz and $SiO_2$ material in science and industry.

The editors of this book are highly appreciated because this book represents a fruitful international collaboration between scientists from Australia, Brazil, Canada, Germany, Norway, Switzerland and the USA.

Jens Götze
Robert Möckel

# Contents

# Contributors

**Sandra B. Barreto** Departamento de Geologia, Universidade Federal do Pernambuco (UFPE), Recife, PE 50740-530, Brazil

**Mario L. S. C. Chaves** Departamento de Geologia, IGC, Universidade Federal de Minas Gerais (UFMG), Belo Horizonte, MG 30123-970, Brazil

**Florian Eichinger** Hydroisotop GmbH, Woelkestrasse 9, 85301 Schweitenkirchen, Germany, e-mail: fe@hydroisotop.de

**Enchbat Dombon** Institute of Mineralogy, TU Bergakademie Freiberg, Brennhausgasse 14, 09596 Freiberg, Germany; Department of Sciences, Technology, and Innovation, Mongolian University of Sciences and Technology, P.O. Box 520, Ulaanbatar 210646, Mongolia

**Belinda Flem** Geological Survey of Norway, Leiv Eirikssons vei 39, 7491 Trondheim, Norway, e-mail: Belinda.Flem@NGU.NO

**Thomas Götte** Institute of Geosciences, Goethe-University Frankfurt, Altenhöferallee 1, 60438 Frankfurt, Germany, e-mail: goette@em.uni-frankfurt.de

**Jens Götze** Institute of Mineralogy, TU Bergakademie Freiberg, Brennhausgasse 14, 09596 Freiberg, Germany, e-mail: goetze@mineral.tu-freiberg.de

**Reiner Haus** Dorfner Analysenzentrum und Anlagenplanungsgesellschaft mbH (ANZAPLAN), Scharhof 1, 92242 Hirschau Germany, e-mail: reiner.haus@dorfner.com

**Peter M. Ihlen** Geological Survey of Norway, Postboks 6315 Sluppen, 7491 Trondheim, Norway

**Ulf Kempe** Institute of Mineralogy, TU Bergakademie Freiberg, Brennhausgasse 14, 09596 Freiberg, Germany, e-mail: kempe@mineral.tu.freiberg.de

**Klaus Krambrock** Departamento de Física, ICEx, Universidade Federal de Minas Gerais (UFMG), CP 702, Belo Horizonte, MG 30123-970, Brazil

**Matthias R. Krbetschek** Institute for Applied Physics, TU Bergakademie Freiberg, Leipziger Str. 23, 09596 Freiberg, Germany, e-mail: quatmi@mailserver.tu-freiberg.de

**Andreas Kronz** Geowissenschaftliches Zentrum, Universität Göttingen, Göttingen, Germany, e-mail: akronz@gwdg.de

**Zucheng Li** Department of Geological Sciences, University of Saskatchewan, Saskatoon, SK S7N5E2, Canada

**Messias G. de Menezes** Departamento de Geologia, Escola de Minas, Universidade Federal de Ouro Preto (UFOP), Ouro Preto, MG 31400-000, Brazil

**Thomas Monecke** Institute of Mineralogy, TU Bergakademie Freiberg, Brennhausgasse 14, 09596 Freiberg, Germany; Department of Geology and Geological Engineering, Colorado School of Mines, 1516 Illinois Street, Golden, CO 80401, USA

**Giulio Morteani** Hydroisotop GmbH, Woelkestrasse 9, 85301 Schweitenkirchen, Germany; Departamento de Geologia, IGC, Universidade Federal de Minas Gerais (UFMG), Belo Horizonte, MG 30123-970, Brazil, e-mail: gmorteani@gmx.de

**Axel Müller** Geological Survey of Norway, Leiv Eirikssons vei 39, Sluppen, 6315, 7491 Trondheim, Norway, e-mail: axel.muller@ngu.no

**Yuanming Pan** Department of Geological Sciences, University of Saskatchewan, Saskatoon, SK S7N5E2, Canada, e-mail: yuanming.pan@usask.ca

**Maurício V. B. Pinheiro** Departamento de Física, ICEx, Universidade Federal de Minas Gerais (UFMG), CP 702, Belo Horizonte, MG 30123-970, Brazil

**Michael Plötze** ClayLab, Institute for Geotechnical Engineering, ETH Zurich, 8093 Zurich, Switzerland, e-mail: michael.ploetze@igt.baug.ethz.ch

**Mikhail Poutivtsev** Institute of Mineralogy, TU Bergakademie Freiberg, Brennhausgasse 14, 09596 Freiberg, Germany; Faculty of Physics, TU Munich, Maier-Leibnitz-Laboratory, James-Franck-Strasse, 85748 Garching, Germany

**Christoph Priess** Dorfner Analysenzentrum und Anlagenplanungsgesellschaft mbH (ANZAPLAN), Scharhof 1, 92242 Hirschau, Germany

**Sebastian Prinz** Dorfner Analysenzentrum und Anlagenplanungsgesellschaft mbH (ANZAPLAN), Scharhof 1, 92242 Hirschau, Germany

**Karl Ramseyer** Institute of Geological Sciences, University Bern, Baltzerstr. 1+3, 3012 Bern, Switzerland

**Brian Rusk** Economic Geology Research Unit, School of Earth and Environmental Sciences, James Cook University, Townsville, QLD 4811, Australia, e-mail: brain.rusk@jcu.edu.au

**Ricardo Scholz** Departamento de Geologia, Escola de Minas, Universidade Federal de Ouro Preto (UFOP), Ouro Preto, MG 31400-000, Brazil, e-mail: r_scholz_dr@yahoo.com

**Marion A. Stevens-Kalceff** School of Physics and Electron Microscope Unit, University of New South Wales, Sydney, NSW 2052, Australia, e-mail: Marion.Stevens-Kalceff@unsw.edu.au

**Alexandre Tarantola** G2R-CNRS, Lorraine University, B.P. 70239 54506 Vandoeuvre-lès-Nancy, France, e-mail: alexandre.tarantola@univ-lorraine.fr

**Alfons M. van den Kerkhof** Geowissenschaftliches Zentrum, Universität Göttingen, Göttingen, Germany, e-mail: akerkho@gwdg.de

**Hartmut B. Walther** Xella Technologie- und Forschungsgesellschaft mbH, Brück, Kloster Lehnin, Germany, e-mail: Hartmut.Walther@xella.com

**Jan Egil Wanvik** Geological Survey of Norway, Postboks 6315 Sluppen, 7491 Trondheim, Norway; Geological Survey of Norway, 6315, 7491 Trondheim, Norway

**Dieter Wolf** Institute of Mineralogy, TU Bergakademie Freiberg, Brennhausgasse 14, 09596 Freiberg, Germany

**Additionally, These External Reviewers are Acknowledged:**

Torsten Graupner (Hannover, Germany)
Anne-Lise Jourdan (London, England)
Jacque L. Kelly (Madison, Wisconsin, USA)
Kitty L. Milliken (Austin, Texas, USA)
Robert Möckel (Freiberg, Germany)
Axel Renno (Freiberg, Germany)
Linards Skuja (Riga, Latvia)

# Chapter 1
# Classification, Mineralogy and Industrial Potential of $SiO_2$ Minerals and Rocks

**Jens Götze**

**Abstract** The various modifications of silica, especially quartz, play a central role in the composition of geological materials. Owing to their abundance and properties, $SiO_2$ minerals and rocks have been used since the beginning of human being in different applications such as tools, weaponries, jewelleries or building materials. In modern industries, silica minerals are widely used as raw materials in high-tech applications (e.g. piezo quartz, optical devices, solar industry) or as mass products (e.g. glass sands, refractory industry, foundry industry, etc.). The occurrence of different silica minerals and the similarity in mineral composition of $SiO_2$ rocks require a clear terminology and nomenclature of silica polymorphs and varieties as well as the different types of $SiO_2$ rocks and their genesis. The knowledge about the development of specific properties (typomorphic properties) in dependence on the specific conditions of formation can be used both for the reconstruction of geological processes and for specific technical applications. The present work gives an overview about the state of the art of the mineralogical and petrographical nomenclature of $SiO_2$ minerals and rocks, the analytical approach for the identification and classification of these materials, and their industrial applications.

J. Götze (✉)
Institute of Mineralogy, TU Bergakademie Freiberg,
Brennhausgasse 14, 09596 Freiberg, Germany
e-mail: goetze@mineral.tu-freiberg.de

J. Götze and R. Möckel (eds.), *Quartz: Deposits, Mineralogy and Analytics*,
Springer Geology, DOI: 10.1007/978-3-642-22161-3_1,

## 1.1 Introduction

$SiO_2$ minerals and rocks play an important role in geological processes and industrial applications. Quartz is not only the most important silica mineral, it is abundant in the Earth crust in igneous, metamorphic and sedimentary rocks. $SiO_2$ minerals and rocks have been formed by primary and secondary magmatic, hydrothermal or sedimentary processes or during diagenesis and metamorphosis (Heaney et al. 1994). Results of these processes are pegmatite or hydrothermal quartz, quartz sands and sandstones, chert, flint or quartzite. The diagenesis of organic silica skeletons, e.g., diatoms, radiolaria, siliceous sponges, may result in the formation of siliceous rocks such as porcellanites, diatomites or radiolarites (Füchtbauer 1988).

Owing to their abundance and properties, $SiO_2$ minerals and rocks have been used in different applications (e.g. tools, weaponries, jewellery, etc.) since early human being and as traditional building materials (e.g. sandstones) worldwide for centuries. Natural silica raw materials represent a wide group of industrial rocks and minerals with interesting chemical and physical properties. Both single crystals of quartz and polycrystalline material or silica rocks are being used in the industry. The application of $SiO_2$ materials is widespread including the use of idiomorphic natural quartz crystals or mass products of $SiO_2$ rocks, e.g. as high-purity quartz sands, refractory materials or silicon ore. Because of the increasing requirements concerning the quality of silica raw materials, synthetic quartz crystals or $SiO_2$ materials are necessary for certain highly advanced applications (Blankenburg et al. 1994).

Silica materials, in particular quartz, are characterized by specific properties (e.g., crystal shape, colour, trace element and isotopic composition, luminescence properties)—ranging from point defects to macroscopic appearance—which are dependent on the geological history and specific conditions of formation. The knowledge of the interrelation between genetic conditions and such properties can be used both for the reconstruction of geological processes and for specific technical applications (Götze 2009a).

The aim of the present work is to provide an overview of the mineralogical and petrographical nomenclature of $SiO_2$ minerals and rocks, the analytical techniques for the characterization of these materials, and their geological importance and industrial applications.

## 1.2 Classification and Properties of $SiO_2$ Minerals

### *1.2.1 $SiO_2$ Modifications and Varieties*

Silica ($SiO_2$) in crystalline and non-crystalline (amorphous) form makes up 12.6 weight% of the Earth's crust. At least 15 ***modifications*** are known (Table 1.1), i.e. mineral phases with the formula $SiO_2$ but a different crystal structure. Quartz

**Table 1.1** The $SiO_2$ system (modified after Strunz and Tennyson 1982)

| | | |
|---|---|---|
| **Quartz-tridymite-cristobalite group** | Low ($\alpha$)-quartz | Trigonal |
| (atmospheric and low pressure) | High ($\beta$)-quartz | Hexagonal |
| | Tridymite | Monoclinic |
| | High-tridymite | Hexagonal |
| | Cristobalite | Tetragonal |
| | High-cristobalite | Cubic |
| | Melanophlogite | Cubic |
| | Fibrous $SiO_2$ (syn.) | Orthorhombic |
| | Moganite | Monoclinic |
| **Keatite-coesite-stishovite group** | Keatite (syn.) | Tetragonal |
| (high and ultra-high pressure) | Coesite | Monoclinic |
| | Stishovite | Tetragonal |
| | Seifertite | Orthorhombic |
| **Lechatelierite-opal group** | Lechatelierite | Natural silica glass |
| (amorphous phases) | Opal | $H_2O$-bearing, solid $SiO_2$ gel |

(trigonal low-temperature alpha-quartz) is the most important silica modification in nature and also most frequently used in technical applications.

Many chemical and physical properties of quartz and the other silica polymorphs such as trace element content, isotopic composition, luminescence behaviour, etc. are determined by their real structure. The close relation between specific pTx-conditions of formation and different types of defects are often reflected in varying properties (so called typomorphic properties) (e.g., Blankenburg et al. 1994; Götze and Zimmerle 2000; Götze 2009a). These varying properties of quartz and other silica minerals result in the existence of numerous ***varieties***, i.e. mineral phases with the same chemical composition of $SiO_2$ and the same crystal structure, but different appearance in shape, colour or varying physical properties (Fig. 1.1). Among these varieties are colour varieties such as in the case of amethyst, smoky quartz, rose quartz and agate or certain growth phenomena such as fibre- and sceptre quartz or chalcedony and quartzine (Rykart 1995; Blankenburg et al. 1994).

The classification of certain non- and microcrystalline silica polymorphs and varieties is often more complicated (Table 1.2), since these minerals are mostly fine-grained and intimately intergrown. These non- and microcrystalline mineral phases play an important role as constituents of sedimentary rocks and, therefore, have to be considered for the characterisation and nomenclature of $SiO_2$-bearing rocks.

## *1.2.2 Real Structure and Properties of Quartz*

Low quartz is the most frequent silica mineral and therefore, the specific properties of quartz do not only determine the quality of $SiO_2$ raw materials but also important properties of $SiO_2$ rocks. Certain point defects, dislocations and

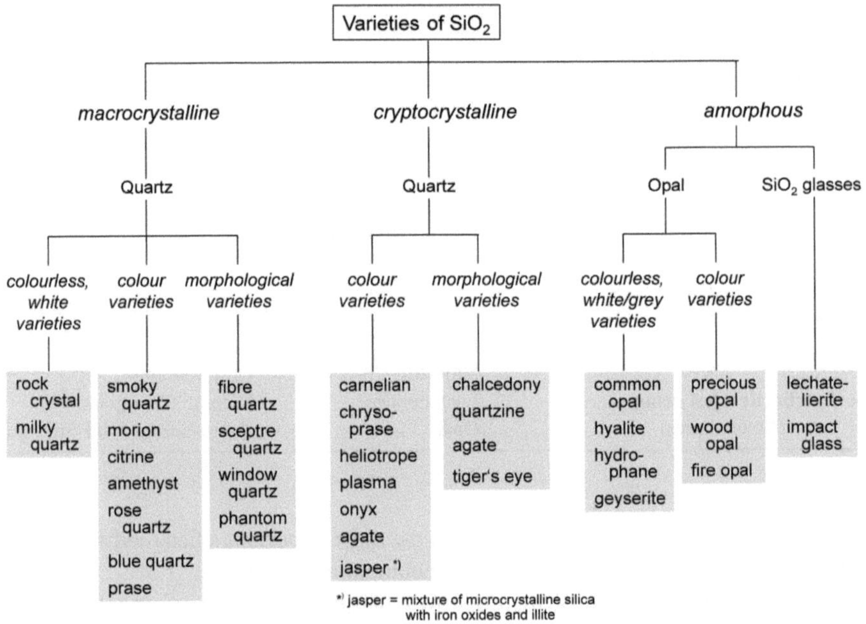


**Fig. 1.1** Varieties of quartz and other silica phases

microinclusions can be incorporated into quartz during crystallization under varying thermodynamic conditions and by secondary processes, such as alteration, irradiation, diagenesis or metamorphosis. The resulting real structure is thus a fingerprint of the specific physico–chemical environment of quartz formation and inherits a lot of genetic information. In addition, the knowledge about the interrelation between conditions of formation—real structure—specific chemical and physical properties of quartz can be useful for several technical applications.

### 1.2.2.1 Point Defects

Point defects in quartz are in general related to the incorporation of foreign ions in lattice and interstitial positions (so called extrinsic defects), to different types of displaced atoms, and/or to defects associated with silicon or oxygen vacancies (intrinsic defects) (Fig. 1.2). More than 20 different types of point defects have been detected in quartz by electron spin resonance (ESR) and luminescence spectroscopy as well as absorption measurements (e.g., Kostov and Bershov 1987; Weil 1984, 1993; Plötze 1995; Stevens-Kalceff et al. 2000; Stevens-Kalceff 2009; Pan et al. 2009).

The most common trace-element related defect centre in quartz is the $[AlO_4]^0$ centre which is caused by substitution of $Si^{4+}$ by $Al^{3+}$ with an electron hole at one of the four nearest $O^{2-}$ ions (Griffiths et al. 1954). The precursor state for this

**Table 1.2** Classification of non- and microcrystalline silica phases (modified after Graetsch 1994)

| Crystal structure or phase | Variety | Subvariety/ synonymous name | Microstructure | Optical character | Water content (weight-%) |
|---|---|---|---|---|---|
| Quartz | Microquartz | | Granular | Positive | <0.4 |
| Disordered quartz (often with moganite) | Chalcedony | | Fibrous [11$\bar{2}$0] | Length-fast | 0.5–2 |
| | Quartzine | | Fibrous [0001] | Length-slow | 0.5–1 |
| Moganite | | | Platy (110), lepidospheric | Length-slow | 1.5–3 |
| Disordered cristobalite | Opal-C | Lussatine | Platy (111)* | Length-fast | 1–3 |
| Cristobalite/ tridymite | Opal-CT | Lussatite | Fibrous [110]* | Length-slow | 3–8 |
| | | Common structureless opal | Platy, lepidospheric | Nearly isotropic | 3–10 |
| Non-crystalline | Opal-AG | Precious opal | Close packing of homometric spheres | Play of colour, anomalous birefringence | 4–8 |
| | | Potch opal | Close packing of heterometric spheres | Isotropic | 4–8 |
| | Opal-AN | Hyalite | Botryoidal crusts and lumps | Strain birefringence | 3–7 |
| | Lechatelierite | Fulgurites | Vitrified tubes | Isotropic | <0.3 |
| | | Impact glass | Meteoritic silica glass | Isotropic | <0.3 |

C—cristobalite, T—tridymite, A—amorphous, G—gel-like, N—network (glass)-like
* Indices refer to cubic setting of cristobalite

paramagnetic centre is the diamagnetic $[AlO_4/M^+]^0$ centre with an adjacent charge compensating cation ($M^+ = H^+, Li^+, Na^+$) in interstitial position. The conversion of the $[AlO_4/M^+]^0$ centre into the optically active $[AlO_4]^0$ centre by gamma-irradiation or X-rays results in the formation of smoky quartz (O'Brien 1955; Cohen 1956). Other common trace-element centres in quartz are associated with the substitutional incorporation of Ti, Ge and Fe (Mackey 1963; Wright et al. 1963; Rinneberg and Weil 1972; Weil 1984, 1993; Rakov et al. 1985; Stegger and Lehmann 1989; Mineeva et al. 1991; Plötze 1995).

Several point defects can be attributed to oxygen and silicon vacancies or oxygen excess. The group of oxygen-deficiency centres (ODC) comprises the neutral oxygen vacancy ($\equiv$Si–Si$\equiv$) and the paramagnetic E′ centres (Griscom 1985; Nishikawa et al. 1994). $O^-$ centres represent different types of defect electrons on oxygen in tetrahedra with silicon vacancies such as the non-bridging oxygen hole centre (NBOHC: $\equiv$Si–O·) (Griscom 1985; Stevens-Kalceff et al. 2000). Oxygen excess centres in quartz include the peroxy linkage and the peroxy

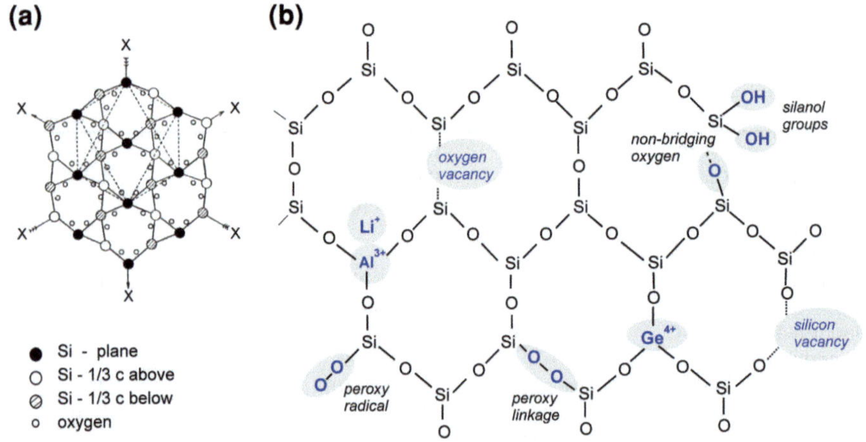


**Fig. 1.2** Schematic structure of α-quartz projected along [001] showing piezoelectric axes XX (**a** modified after Beall 1994) and scheme of most common point defects (**b** modified after Götze 2009a)

radical (Friebele et al. 1979; Baker and Robinson 1983; Pan et al. 2009), whereas hydrogen excess from the $H_2O$ crystallization medium can result in the formation of $OH^-$ centres (silanol groups) in quartz (Weil 1984).

The analysis of point defects is essential for the use of quartz and $SiO_2$ materials in several highly-advanced technical applications. All kinds of point defects can strongly alter the structural, electrical and optical properties resulting in electrical instabilities in $SiO_2$ insulator layers, absorption effects in Piezo quartz or lowering of the quality of optical materials due to loss by dispersion and absorption effects (e.g. formation of smoky quartz) (Blankenburg et al. 1994). Furthermore, condensation of point defects during crystal growth or by mechanical and thermal treatment can result in the formation of dislocations (line defects). Such line defects can be visualized using etching techniques or X-ray topography.

#### 1.2.2.2 Cathodoluminescence Characteristics of $SiO_2$

The various CL colours and emission bands in quartz and other $SiO_2$ minerals, respectively, can be related to different defects. According to Stevens-Kalceff (2009) these are:

- a neutral relaxed oxygen vacancy with an emission band at 290 nm (4.3 eV),
- charge-compensated substitutional $Al^{3+}$ impurity centre with a transient emission at 385 nm (3.3 eV),

- recombination of the self-trapped exciton (STE) with an E′ centre in quartz (emission at 460 nm; 2.7 eV) and in electron-irradiation-amorphized quartz (515 nm; 2.3 eV), respectively,
- a Ti impurity (440–480 nm; 2.6–2.8 eV),
- several types of non-bridging oxygen hole centres (NBOHC) with different precursors causing emissions at 635 (1.95 eV) and 650 nm (1.9 eV), respectively,
- substitutional $Fe^{3+}$ (emission at 750 nm; 1.65 eV),
- interstitial molecular oxygen with a CL emission at 1280 nm (0.968 eV).

In natural quartz and chalcedony, additional emission bands have been observed, which can probably be related to

- $[AlO_4/Li^+]$ or $[TiO_4/Li^+]$ centres (330–340 nm; 3.75–3.6 eV—Demars et al. 1996; Plötze and Wolf 1996),
- interstitial cations (505 nm; 2.45 eV—Ramseyer et al. 1988; Ramseyer and Mullis 1990; Götze et al. 2005) or
- electron defects related to an oxygen vacancy (580 nm; 2.15 eV—Rink et al. 1993; Götze et al. 1999).

The relation of specific luminescence emission bands to different defect centres causes a diversity of CL characteristics both in natural and synthetic quartz and $SiO_2$ minerals (e.g., tridymite, cristobalite, opal) in dependence on the processes of mineral formation or alteration. Using CL microscopy it is possible to visualize the defect structure of quartz and to reveal internal structures, which are not discernable by other analytical methods (Fig. 1.3).

Several studies showed that natural quartz in different rocks develops typical CL properties (e.g. Ramseyer and Mullis 1990; Götze et al. 2001; Richter et al. 2003; Götze 2009a). For instance, the typical CL of igneous and metamorphic quartz consists of two emission bands with maxima at ca. 450 and 650 nm, which results in different visible luminescence colours in dependence on the relative intensities of these two dominant emission bands. Most igneous quartz grains show luminescence colours in different shades of blue caused by the predominance of the 450 nm emission. In volcanic and metamorphic quartz, the intensity of the 650 nm emission band is often higher than in granitic quartz resulting in visible reddish-brown CL. Quartz from pegmatites often exhibits a transient bluish-green CL, which is caused by a typical 500 nm emission. In contrast, hydrothermal quartz crystals may show a variety of luminescence colours, sometimes arranged in distinct zoning. Several studies showed that the CL characteristics of hydrothermal quartz may also depend on the crystallization temperature (Ioannou et al. 2003; Rusk et al. 2008; Jourdan 2008). At high temperatures (>400°C), a homogeneous blue luminescence predominates, whereas at lower temperatures (<300°C) strong zoning is common and yellow CL can be observed. The typical decrease of the blue CL emission at ca. 390 nm under the electron beam in both natural and synthetic hydrothermal quartz is often accompanied by an increase of the 650 nm emission, which is visible by a change in CL colour from initial blue to

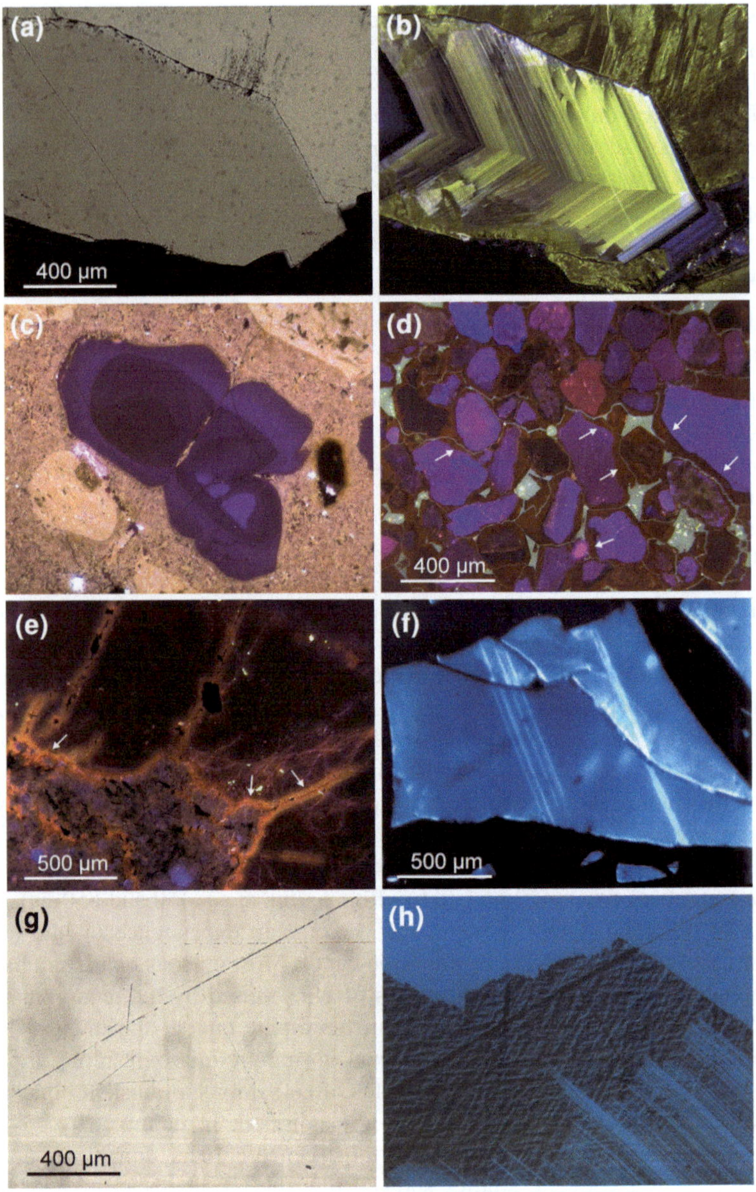


brownish (e.g., Ramseyer et al. 1988; Ramseyer and Mullis 1992; Götze et al. 2001; Götze 2009a). Radiation halos in quartz due to lattice damage by alpha-particles are also characterized by a strong 650 nm CL emission band (e.g., Owen 1988; Meunier et al. 1990; Ramseyer et al. 1988; Götze et al. 2001; Komuro et al. 2002; Botis et al. 2005; Krickl et al. 2008).

◄ **Fig. 1.3** Micrographs in polarized light and cathodoluminescence (CL) of different quartz samples demonstrating the potential of CL studies in natural and synthetic quartz; **a, b** Polarized light—CL pair of a hydrothermal quartz crystal from Chemnitz, Germany showing distinct growth zoning, which is not visible in transmitted light; **c** CL micrograph of quartz in a rhyolite from Euba, Germany; the CL image reveals growth zoning and syneusis of two quartz phenocrysts, which show strong features of resorption; **d** CL micrograph of a sandstone sample from Dohna, Germany; different CL colours of the detrital quartz grains point to varying source rocks; authigenic quartz overgrowths are clearly visible under CL (see *arrows*); **e** CL micrograph of a U-bearing quartz conglomerate from Witwatersrand, RSA; radioactive fluids have caused radiation damage along the grain boundaries and corroded the quartz grains (see *arrows*); **f** CL micrograph of silica glass showing heterogeneities (bright stripes in CL) caused by Al impurities; **g, h** Polarized light—CL pair of a hydrothermal synthetic quartz crystal; the CL image reveals complex internal structures not visible in transmitted light

Applications of the characteristic luminescence properties of quartz in geosciences are numerous. One of the first applications was the evaluation of the provenance of detrital quartz grains in sands and sandstones (Sippel 1968; Zinkernagel 1978; Richter et al. 2003). Furthermore, processes of compaction, brittle deformation, quartz cementation, and porosity evolution in the diagenetic history of reservoir sandstones can also be evaluated using CL (e.g., Housknecht 1991; Millikan and Laubach 2000). CL in combination with trace elements was used to reconstruct the geological evolution of granitic systems, ore-forming processes or the metamorphic history of rocks (e.g., Müller et al. 2000, 2002, 2003a; Graupner et al. 2000; Rusk et al. 2006, 2008; Van den Kerkhof et al. 2004). Furthermore, effects of shock damage in quartz due to impact events could be revealed by CL in terrestrial and extraterrestrial samples (e.g., Sippel 1971; Ramseyer et al. 1992; Götze 2009b).

CL studies can also be helpful in technical applications such as for the evaluation of the quality of $SiO_2$ raw materials and technical $SiO_2$ products (Götze 2000). Impurities and heterogeneities in high-purity quartz and $SiO_2$ materials are often only detectable by CL (see Fig. 1.3 f–h). Moreover, $SiO_2$ plays an important role in many technologically important applications such as silicon semiconductor device technology, optics, or $SiO_2$ insulator layers in electronic devices. The optical, electrical and mechanical properties of these high-tech materials are dependent on the presence and/or generation of defects (imperfections and impurities), which can be detected and characterized by CL spectroscopy (e.g., Barfels 2001; Stevens-Kalceff 2009). For instance, the generation of defects and traps and the related electronic transitions can significantly influence the usability of those materials. Characterization of the defect microstructure of silicon dioxide ($SiO_2$) allows the performance of these applications to be optimized (Fig. 1.4).

#### 1.2.2.3 Trace Elements

Quartz is one of the purest minerals in the Earth crust. This is due to the fact that only a limited number of ions can substitute for $Si^{4+}$ in the crystal lattice or can be

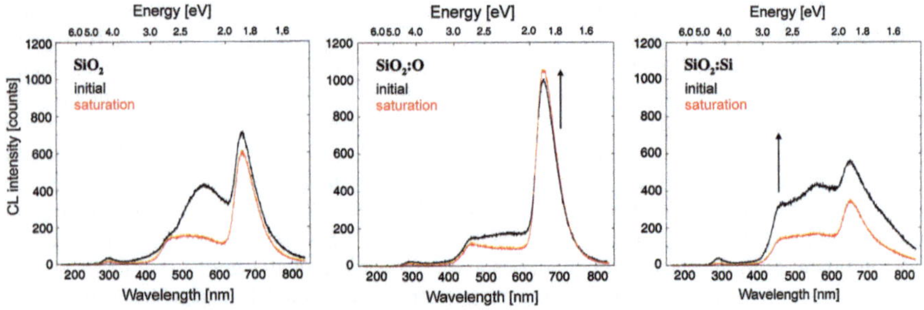


**Fig. 1.4** Cathodoluminescence spectra of undoped $SiO_2$ layers ($SiO_2$) and those implanted by oxygen ($SiO_2$:O) and silicon ($SiO_2$:Si), respectively (*black lines* = initial spectra, *red lines* = spectra after electron irradiation); the CL spectra show a significant increase of the red emission band (650 nm) after doping with oxygen and an increased blue emission (450 nm) after doping with silicon (data from Barfels 2001)

incorporated in interstitial positions. Therefore, most chemical elements are present in quartz in concentrations below 1 ppm (compare Müller and Wanvik, this volume, Fig. 4.4, p. 76).

The structural incorporation in a regular $Si^{4+}$ lattice position was proved for $Al^{3+}$, $Ga^{3+}$, $Fe^{3+}$, $Ge^{4+}$, $Ti^{4+}$ and $P^{5+}$ (e.g. Weil, 1984, 1993), in which Al is probably the most frequent ion (up to a few 1000 ppm) because of its common occurrence in the Earth crust and the similar ionic radii of $Si^{4+}$ and $Al^{3+}$. Titanium and iron may be both incorporated as substitutional ion for Si and bound on mineral microinclusions such as rutile or iron oxides. Other cations such as $H^+$, $Li^+$, $Na^+$, $K^+$, $Cu^+$ or $Ag^+$ can be incorporated in interstitial positions as charge compensating ions, or as in the case of Na and K also hosted by fluid inclusions (e.g., Bambauer 1961; Perny et al. 1992; Plötze 1995; Müller et al. 2003b; Götze et al. 2004; Miyoshi et al. 2005; Jourdan 2008; Müller and Koch-Müller 2009; Lehmann et al. 2009). For most elements in quartz (e.g., Ca, Mg, Ba, Sr, Rb, REE, Mn, U, Th) however, their capture by microinclusions of fluids or minerals is most important (Gerler 1990; Blankenburg et al. 1994; Götze et al. 2004; Götze 2009a and references therein).

In geosciences, trace elements are considered important petrogenetic indicators for interpreting the conditions of mineral formation, to reveal the provenance of quartz, or to reconstruct the genesis of ore deposits and the origin of metal-bearing fluids (e.g., Bambauer 1961; Dennen 1964, 1966, 1967; Walenczak 1969; Lyakhovich 1972; Suttner and Leininger 1972; Stenina et al. 1984; Hallbauer 1992; Heynke et al. 1992; Götze and Lewis 1994; Götze and Plötze 1997; Monecke et al. 1999, 2000, 2002; Götze and Zimmerle 2000; Larsen et al. 2000, 2004, 2009; Müller et al. 2002, 2003a,b, 2008; Götze et al. 2004; Kostova et al. 2004; Götze 2009a).

On the other hand, the knowledge about trace element contents in quartz and the different mechanisms of incorporation is important for the strategy of processing of quartz and $SiO_2$ raw materials and their application in the industry (e.g.,

**Table 1.3** Chemical composition of high-purity quartz raw materials for different applications (data from Blankenburg et al. 1994)

| | Fibre optics | Special optical glass |
|---|---|---|
| $SiO_2$ | >99.9 wt.-% | >99.8 wt.-% |
| $Fe_2O_3^{*)}$ | <2.0 ppm | <20 ppm |
| $TiO_2$ | <1.5 ppm | <25 ppm |
| $Al_2O_3$ | | <500 ppm |
| $Cr_2O_3$ | <0.01 ppm | <0.1 ppm |
| CoO | <0.05 ppm | <0.05 ppm |
| CuO | <0.05 ppm | <0.1 ppm |
| MnO | <0.1 ppm | <1.0 ppm |
| NiO | <0.1 ppm | <0.15 ppm |
| $V_2O_5$ | <0.5 ppm | <15 ppm |

*) $Fe_2O_3$ contents for crystal glass are <150 ppm, window glass <0.1 mass-%, bottle glass 0.5–4 mass-%

Blankenburg et al. 1994; Götze 1997; Müller et al. 2005). Correlations of impurity concentrations and the temperature of formation of the host quartz indicate that lowest trace-element contents can be expected for the temperature range between ca. 480 and 530°C (Wünsch 1987; Blankenburg et al. 1994). In conclusion, pegmatite quartz, metamorphogenically mobilised quartz veins and lenses as well as some hydrothermal quartz veins have the greatest potential for high-purity quartz raw materials.

The concentration limits of certain trace elements in $SiO_2$ raw materials for high-tech applications are very low (Tables 1.3, 1.5) and often require synthetic instead of natural $SiO_2$ material. For instance, the production of lamp tubing and optics requires high-purity materials with Al concentrations below 20 ppm (compare Table 1.3). Semiconductor base materials and crucibles use material with Al < 10 ppm, in micro-electronic devices U and Th concentrations should be below 2 ppb, and the production of solar silicon needs raw materials with very low concentrations of P and B (Table 1.5).

#### 1.2.2.4 Fluid and Mineral Inclusions in Quartz

Residues of mineralizing fluids and/or paragenetic minerals can be included in quartz crystals during growth. Furthermore, exsolution processes after crystallization may result in the formation of microinclusions such as in the case of rutile. The inclusions provide information about formation paragenesis, temperature of formation, chemistry of mineralizing fluids, etc. This information is frequently used for the reconstruction of mineral forming processes. Numerous studies have shown that quartz of different origin may be distinguished using distribution and physiography of inclusions, temperature of homogenization, and chemical composition of fluid inclusions (e.g., Ermakov 1950; Roedder 1984; Shepherd et al.

1985; Leeder et al. 1987; Van den Kerkhof and Hein 2001). Recently, a couple of modern analytical methods such as microchemical analysis, capillary electrophoresis, infrared spectroscopy, Raman spectroscopy, isotope measurements, INAA or LA-ICP-MS is being used to provide data concerning the chemical composition of fluid inclusions (e.g., Klemm 1986; Gerler and Schnier 1989; Götzinger 1990; Ghazi et al. 1993; Hanson et al. 1996; Hallbauer 1997; Channer et al. 1999; Flem et al. 2002; Müller et al. 2003b; Götze et al. 2004).

The information about type and number of inclusions provides important data about impurities in $SiO_2$ raw materials. Mineral and fluid inclusions can significantly influence the trace-element composition of quartz. On the other hand, the type and amount of fluid inclusions can influence the melting behaviour of $SiO_2$ raw materials (e.g. Gemeinert et al. 1992). Traces of refractory minerals (e.g. zircon, aluminium silicates) have also to be considered if quartz raw materials are used in melting processes (e.g. glass industry).

## 1.3 Origin and Nomenclature of $SiO_2$ Rocks

Despite the simple chemical formula $SiO_2$ of quartz, the system of silica and $SiO_2$ rocks is very complex. This is on one hand due to the various silica modifications and quartz/silica varieties and on the other hand because of various $SiO_2$ rocks with partially unclear nomenclature. The basis for a useful classification is probably given by the petrological classification into magmatic, metamorphic and sedimentary rocks.

### *1.3.1 Magmatic and Metamorphic $SiO_2$ Rocks*

Among the group of magmatic and metamorphic rocks, only a few rock types exist, which may almost exclusively consist of $SiO_2$: pegmatite bodies, hydrothermal veins and metamorphic quartzite. In addition, a special type of granite—so called alaskite—is used as raw material for the extraction of high-purity quartz. ***Alaskite*** is a leucocratic variety of granite composed mainly of quartz and alkali feldspar, with less than 5% of dark-coloured minerals.

Postmagmatic processes may accumulate large amounts of high-purity quartz in pegmatites and hydrothermal deposits (Rösler 1981). ***Quartz pegmatites*** originate from residual melts of plutonic magmas, which are enriched in volatile components. Large parts of such pegmatite bodies can consist of high-purity quartz, often in paragenesis with feldspar, mica and/or some rare minerals. ***Quartz veins*** represent mineralization from hydrothermal fluids along wrench-fault systems.

Metamorphic processes can also mobilise and precipitate high-purity quartz, for instance in form of veins and lenses in the shadow of tectonic folds (e.g., Wünsch 1987; Heynke et al. 1992). Such ***metamorphogenic mobilisates*** lost most impurities during the mobilisation and crystallization processes resulting in a high-purity $SiO_2$ material. In contrast, ***quarzite*** is a bedded massive metamorphic rock with at least 90% quartz, formed from pre-existing quartz-rich sandstones by compaction and metamorphosis. A less focused nomenclature for quartzite is used in the industry. Here, all very hard, resistant rocks with >96% $SiO_2$ are called quartzite, independent on the formation history.

## *1.3.2 Sedimentary $SiO_2$ Rocks*

Among the sedimentary $SiO_2$ rocks, the groups of siliceous rocks and siliciclastic rocks should be distinguished based on genetic aspects. ***Siliceous rocks*** are rocks with >50% authigenic silica minerals such as opal-A, opal-CT, chalcedony, quartzine and/or (micro-) quartz (Füchtbauer 1988). In contrast, the group of ***siliciclastic rocks*** includes rocks with preferentially detrital quartz from previously existing rocks, accumulated and compacted during weathering, transport, sedimentation and diagenesis (Füchtbauer 1988).

### 1.3.2.1 Siliceous Rocks

Siliceous rocks consisting almost completely of opal-CT are called ***porcellanite***, whereas those with (micro-)quartz and chalcedony belong to the group of chert. According to Knauth (1994), the term ***chert*** (or ***hornstone***) is used for rocks with interlocking grains of complexly twinned, hydrous, granular microcrystalline quartz that has replaced pre-existing sediments such as opal, carbonate, or evaporate minerals (Fig. 1.5). The formation of chert can be related to:

(a) diagenetic transformation sequences from siliceous oozes ($SiO_2$ from diatoms, radiolarian or sponges and also volcanic glass): opal-A→ opal-CT→ microquartz or
(b) direct diagenetic precipitation of microquartz (replacement of carbonate). According to the siliceous precursors, radiolarite, diatomite or spiculite are formed. The terms ribbon chert, novaculite and lydite are synonymously used for radiolarite (compare Fig. 1.6).

Among the massive cherts, Precambrian cherts are sometimes named due to their origin as stromatolithic chert, which often occurs together with iron oxides in deposits of banded iron ore formations (Knauth 1994). It is recently assumed that microorganisms play an important role in the formation of these siliceous rocks (Konhauser 2007). ***Nodular chert*** is formed during the diagenesis of carbonaceous sediments by the dissolution and replacement of carbonate in shallow marine sediments. The term ***flint*** (French: silex; German: Feuerstein) refers to nodular chert in carbonate rocks, which are of preferentially Cretaceous and Jurassic age (Füchtbauer 1988).

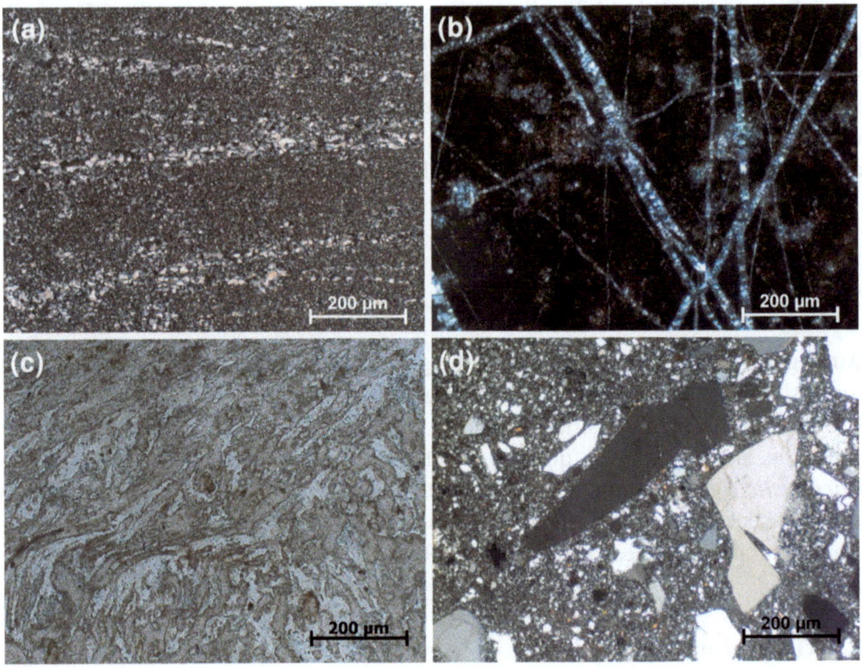


**Fig. 1.5** Micrographs in polarized light (except **c** = transmitted light) of different siliceous rocks showing typical microstructures. **a** Massive chert from Texas, USA with fine-grained granular quartz and chalcedony; **b** Lydite (Kieselschiefer) from Lommatzsch, Germany with extremely fine-grained silica matrix and cross-cutting microcrystalline quartz veins of different generations; **c** Siliceous sinter from Yellowstone, USA showing partially a layered microstructure; **d** Tertiary quartzite (silcrete) from Glossen, Germany with fine-grained siliceous matrix and large detrital quartz grains

Another type of siliceous rocks is ***siliceous sinter***. This is a porous, layered and fine-grained siliceous rock, which originates from the evaporation of silica in hot springs (Fig. 1.5c). ***Tripoli*** (or ***polishing slate***) is a fine-grained, highly porous siliceous rock with inorganic silica source. These fine-grained limnic sediments are preferentially formed during Tertiary.

A special group of siliceous rocks is represented by ***silcretes***. This group of terrestrial siliceous rocks is especially reported from Australia (silcrete) and Europe, where the rocks are called ***Tertiary quartzite*** (because of the preferred occurrence in Tertiary sediments) or ***cement quartzite*** (in contrast to the metamorphic quartzite). These rocks originate from the silicification of pre-existing rocks (mostly sediments) by silica from weathering solutions and may cover large areas (Walther 1993). Probably the term ***limnoquartzite*** (or opalite) also describes this type of rocks considering that a significant amount of silica may originate from volcanic and/or hydrothermal activities.

Because of the intergrowth of both detrital and newly precipitated quartz and silica material, these rocks represent more or less the transition between siliceous

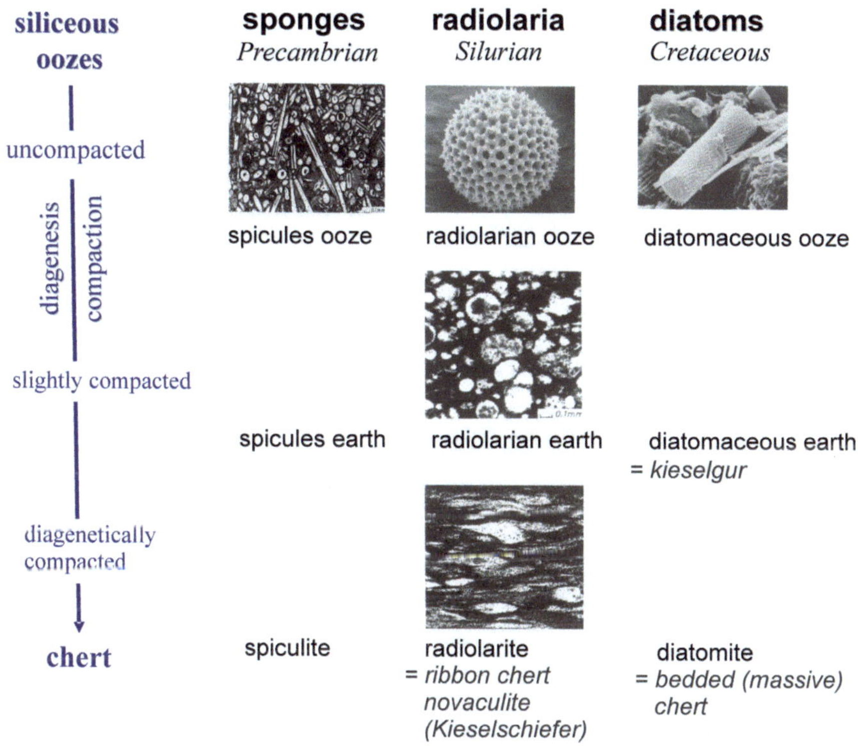


**Fig. 1.6** Formation of different types of chert by the diagenesis of opal-A precursors (siliceous oozes) during burial and solution-reprecipitation steps via opal-A →opal-CT →microquartz; the geological ages relate to the first occurrence; some of the images are modified from Füchtbauer (1988)

and siliciclastic rocks. The texture and appearance of the silcretes drastically vary due to the amount of detrital material and the kind of cementing material (Figs. 1.5d, 1.7).

#### 1.3.2.2 Siliciclastic Rocks

Siliciclastic rocks are accumulations of physically and chemically resistant quartz, feldspar and rock fragments. The general mineral composition and textural characteristics depend on the properties of the primary host rocks and the conditions of weathering, transport, deposition and diagenesis during rock formation (Pettijohn et al. 1987). Therefore, the mineralogical and chemical characterization and evaluation of siliciclastic rocks as potential silica raw materials is very complex (e.g., Götze and Blankenburg 1990; Götze 1997; Götze and Siedel 2007).

Rapid erosion as well as cold and dry climate causes mechanical destruction of the host rocks and minor chemical alteration with slow decomposition of unstable

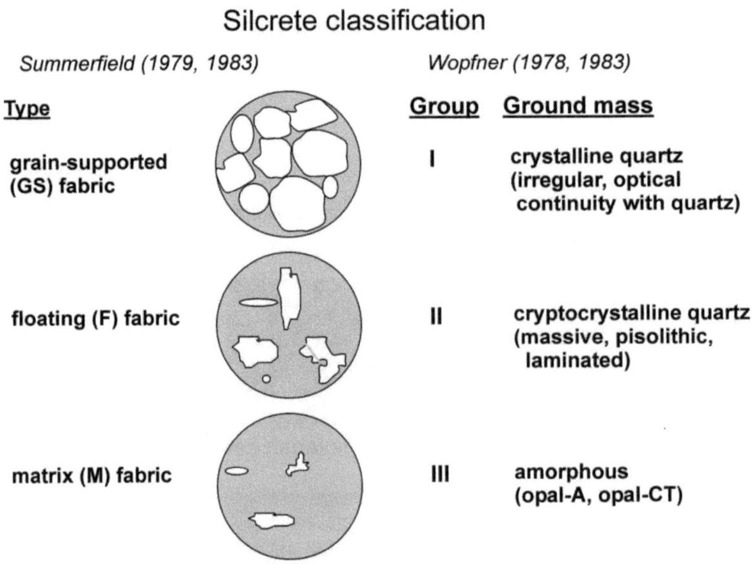


**Fig. 1.7** Classification criteria of silcretes based on textural features (*left*—according to Summerfield 1979, 1983) and the type of cementing material (*right*—according to Wopfner 1978, 1983)

minerals. This results in complex mineral composition and reflects at least partially the primary composition of the host rock(s). In contrast, relatively slow erosion together with warm and wet climatic conditions cause decomposition of unstable minerals and result in a sedimentary rock with high amounts of detrital quartz. Sandstones (Fig. 1.8) have in average 65 vol% quartz (Blatt et al. 1980), but it may be ≥99% in some mature sands (Götze 1997).

In addition, the granumlometric properties of the sedimentary rocks, e.g. grain-size distribution, roundness and surface properties of quartz grains are especially influenced by the distance and intensity of the transport conditions and provide information concerning the geological environment during rock formation. During diagenesis, compaction/cementation as well as alteration and neoformation of minerals can significantly change the properties of the siliciclastic rocks. Diagenetic and metamorphic processes can cause a compaction of sands/sandstones and a transformation into quartzite (Fig. 1.8d).

## 1.4 Formation and Industrial Use of $SiO_2$ Raw Materials

Natural $SiO_2$ raw materials represent a complex group of industrial rocks and minerals, which are used as single crystals and polycrystalline material or compact $SiO_2$ rocks in the industry. The manifold applications include perfect crystals of piezo- and optical quartz (recently often synthetic crystals), high-purity quartz

**Fig. 1.8** Micrographs in polarized light of different siliciclastic rocks showing typical mineral composition and microstructures, respectively; **a** Arcosic sandstone from Altendorf, Germany consisting of mainly quartz, feldspar and some sheet silicates; **b** Glauconitic sandstone from Goslar, Germany with rounded glauconite grains and infiltrated illite on grain surfaces; **c** Mature sandstone from Lohmen, Germany consisting almost exclusively of detrital quartz and diagenetic silica cement; **d** Metamorphic quartzite (Dubrau quartzite, Germany) formed during high pressure (and temperature) from a quartz rich sediment; note the typical sutured grain contacts

sands and gravels in the foundry and glass industry as well as for the production of semiconductor silicon and silicon alloys, or quartzites for refractory materials (Table 1.4). Knowledge of the interrelation between genesis—specific properties—parameters for technical application of the raw material is necessary for a successful use in many industrial applications. Therefore, for the evaluation of the usability of $SiO_2$ raw materials for a specific application, the limiting properties, e.g. chemical purity, are essential (e.g., Blankenburg et al. 1994; Götze 1997; Müller et al. 2005).

## *1.4.1 Magmatic and Postmagmatic Quartz*

Although more than 90% of quartz and other $SiO_2$ minerals of the lithosphere are present in magmatic rocks (Rösler 1981), only about 1% of industrially used $SiO_2$ raw materials derive from this source. The majority of quartz in $SiO_2$-rich igneous and volcanic rocks (granite, rhyolite) is intergrown with other rock-forming

**Table 1.4** Interrelation between genesis and specific properties of different types of $SiO_2$ raw materials and their preferred application in the industry

| Quartz type | Properties | Preferred application |
|---|---|---|
| ***Magmatic/postmagmatic*** | | |
| Quartz of alaskite ("Iota quartz") | Chemical purity | High-purity $SiO_2$ material, optics, lamp tubing, semiconductor and solar silicon |
| Pegmatite and Hydrothermal quartz | Chemical purity, perfect crystal order | Optical and piezo quartz, quartz synthesis ("lascas"), semiconductor and solar silicon, silicon alloys, glass |
| ***Metamorphic*** | | |
| Quartzite | $SiO_2$ up to >98%, lumpy | Refractory materials, silicon and Si alloys (e.g. FeSi) |
| Metamorphogenic quartz mobilisates | Chemical purity | Quartz synthesis ("lascas") |
| ***Sedimentary*** | | |
| Quartz sands | Chemical purity, granulometric properties | Glass and foundry industry, cristobalite, quartz powder, silica glass, SiC, |
| Quartz gravel | Chemical purity, grain size | Silicon and Si alloys (e.g. FeSi), building industry |
| Sedimentary quartzite | Chemical purity, cryptocrystalline silica | Refractory materials (silica stones) |

**Table 1.5** Chemical composition (in ppm) of ultra-high purity Iota quartz standards (IOTA 2005) and high purity quartz HPQ (Norwegian Crystallites AS 2006)

| | Iota Std | Iota CG | Iota 4 | Iota 6 | HPQ*) |
|---|---|---|---|---|---|
| Al | 16.2 | 14.7 | 8.6 | 8.0 | <30 |
| Ti | 1.3 | 1.1 | 1.4 | 1.4 | <10 |
| Fe | 0.2 | 0.2 | 0.3 | 0.2 | <3 |
| Na | 0.9 | 1.0 | 0.3 | 0.1 | <8 |
| Zr | 1.3 | 1.3 | <0.01 | <0.01 | – |
| P | 0.1 | 0.1 | 0.05 | 0.05 | <2 |
| B | 0.08 | 0.08 | 0.04 | 0.04 | <1 |

*) Σ Al, Ti, Fe, Ca, K, Na, Li, P, B < 50 ppm

silicates. Therefore, quartz from these rocks does not play an important role as raw material. The only exception is the extraction of high-purity silica material by chemical leaching from alaskites (so called Iota quartz—Jung 1992).

In contrast, pegmatite bodies and hydrothermal veins may provide large amounts of high-purity quartz (Table 1.5). Such deposits can reach dimensions of several tenths or hundreds of meters and with extremely low concentrations of impurities (Blankenburg et al. 1994). This material is preferentially used as raw material for the hydrothermal quartz synthesis (Table 1.4). High-purity natural pegmatite and hydrothermal quartz crystals have formerly been used directly as piezo quartz or optical quartz.

### *1.4.2 Metamorphic Quartz*

Quartz from metamorphic rocks represents only about 3% of the whole quartz in the lithosphere (Rösler 1981). In addition, most of these quartz types are relatively fine-grained, intimately intergrown with other rock-forming minerals and of low crystallographic perfection. In conclusion, most quartz from metamorphic rocks is not usable as high-quality $SiO_2$ raw material.

However, metamorphic quartzites (and partially siliceous slates) of high chemical purity (>98 mass% $SiO_2$) can be used as raw materials for the refractory industry (production of silica stone—Table 1.4) and sometimes also as high-purity quartz for high-technology industries (e.g. Müller et al. 2007). Moreover, metamorphogenic quartz mobilisates often represent a high-purity $SiO_2$ material that can be used e.g. as raw material for single-crystal growth. The type of "alpine quartz veins", which often contains clear rock crystals, also belongs to this type of metamorphic quartz. However, this quartz type often contains elevated concentrations of trace elements (e.g. Bambauer 1961).

### *1.4.3 Sedimentary $SiO_2$ Rocks*

Sedimentary $SiO_2$ rocks are the main suppliers of $SiO_2$ raw materials in the industry. In Germany, for instance, more than 80% of industrially used quartz raw materials are high-purity quartz sands for the foundry and glass industry (Blankenburg et al. 1994). High-purity quartz sands represent an extraordinary geological material, which is characterized by specific properties such as extreme chemical purity and grain-size distribution. Such deposits have formed under specific geological conditions, e.g., intensive weathering, beach deposits and reworked quartz rich sediments, which prevailed during the Cretaceous and Tertiary of Central Europe (Götze 1997).

One of the most important industrial applications of high-purity quartz sands is the use as raw material for the production of silicate glasses, where up to 80 mass% of quartz are being used. Extreme chemical purity (compare Table 1.3) and a consistent grain-size distribution are necessary for glass sands. Quartz sands are also applied in the foundry industry for the preparation of casting forms with certain binding agents such as clay and bentonite, water glass and/or organic materials. Because of the dependence of the quality of the forms on the perfect interaction between the quartz grains and the binding material, optimal granulometric properties (grain size, grain shape, surface properties) of the raw materials are required. In addition, high-purity quartz sands are used for the production of silica glass, silicon carbide (SiC), cristobalite and quartz powder (Table 1.4).

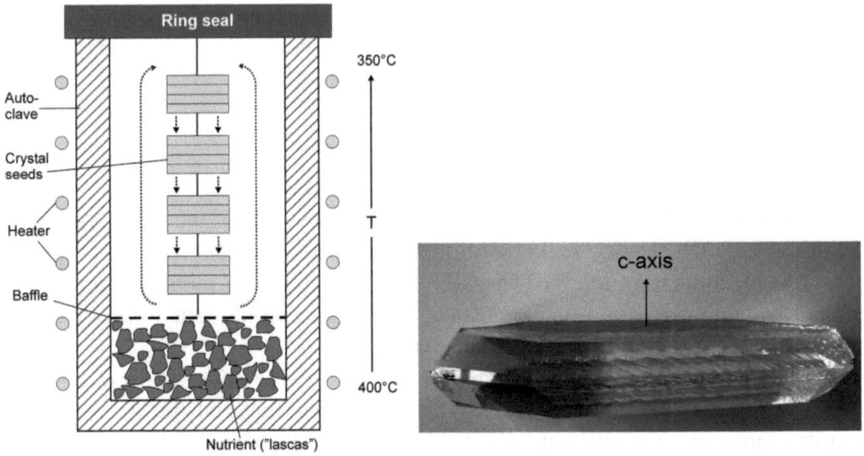


**Fig. 1.9** Scheme of an autoclave for hydrothermal growth of quartz single crystals (*left*) and synthetic quartz crystal (*right*—width ca. 15 cm) showing typical morphology. Note the development of large basal faces, which are uncommon in natural quartz crystals

Another sedimentary $SiO_2$ material for industrial applications is represented by sedimentary quartzites (silcretes/Tertiary quartzites). These sedimentary quartzites are characterized by high chemical purity and elevated amounts of cryptocrystalline and/or amorphous silica cement (see Sect. 1.3.2). This material is preferentially used in the refractory industry as raw material for the production of silica stones. The fabrication of silica stones involves a high-temperature process at ca. 1500°C, where primary quartz transforms into cristobalite and tridymite. The fine-grained silica cement enhances a more or less complete transformation of quartz, which is necessary for high thermal and mechanical stability of silica stones in high-temperature applications (prevention of "after-expansion").

## *1.4.4 Synthetic $SiO_2$ Raw Materials*

Modern technologies require extreme physical and chemical properties, which cannot be achieved by natural raw materials. The need for perfect crystalline and extremely pure quartz crystals (e.g. Al $<$ 1 ppm) resulted in the development of hydrothermal synthesis of quartz single crystals in the 1950s (e.g., Nacken 1950; Mosebach 1955). Although quartz frequently occurs in nature, the quality of most natural quartz is too low to satisfy the requirements for specific optical and/or electrical applications. Outstanding properties such as the refraction index, optical dispersion and activity, piezoelectricity or the transparency from 150 to 3000 nm are being used in several high-tech applications. Some 1,000 t of synthetic quartz crystals (Fig. 1.9) are produced annually in autoclaves under hydrothermal conditions (350–400°C, 100–120 MPa). High-purity pegmatite, hydrothermal and metamorphic quartz is preferentially used as raw material for crystal growth (so called "lascas").

The occurrence of natural silica glass is relatively rare and is restricted to places, where vitreous silica is produced due to heat and/or pressure by lightning of quartz sands (fulgurites-lechatelierite) or meteoritic impacts (impact glass). In contrast, the use of technical silica glass is widespread because of the outstanding physical and chemical properties (Beall 1994). In general, quartz sand can be used as raw material for the production of silica glass, but in the case of ultra-pure material the use of synthetic silica raw material is preferred. Silica glass with trace element contents below 1 ppm can be produced with different techniques using tetrachlorsilane ($SiCl_4$) or organic Si compounds as starting material.

## 1.5 Conclusions

$SiO_2$ minerals and rocks are important constituents of the Earth's crust and they play an important role as usable materials since the beginning of human being. Owing to their abundance and properties, $SiO_2$ minerals and rocks have been used in different applications such as tools, weaponries, jewelleries or building materials. Recently, quartz and other silica minerals and rocks cover a broad field of geological and technical applications including high-tech materials. The identification and characterization of these $SiO_2$ materials require both a clear nomenclature and a valuable analytical approach.

The classification of silica minerals and rocks is based on the mineralogical and petrographical nomenclatures. Accordingly, we have to differentiate between $SiO_2$ minerals (including different varieties) and $SiO_2$ rocks. Quartz is the most important silica mineral in respect to its appearance in the lithosphere and the industrial use. Therefore, investigations concerning characteristic properties in dependence on the specific conditions of formation can help to evaluate the potential of raw material for industrial applications. Among the $SiO_2$ rocks, in particular sedimentary rocks have to be considered as potential raw material. The group of sedimentary $SiO_2$ rocks can be subdivided into the two sub-groups of siliceous rocks and siliciclastic rocks. The properties of these rocks (chemistry, mineral composition, texture, etc.) mainly depend on the geological history.

In recent years, the development and application of advanced analytical methods have provided a large amount of new data concerning the structure and properties of natural and synthetic $SiO_2$ material, which can help to reconstruct geological processes and to determine the viability of specific industrial applications.

**Acknowledgments** I gratefully acknowledge the comments of K. Ramseyer (Bern) and an anonymous reviewer, which helped to improve the quality of the paper.

# References


Baker JM, Robinson PT (1983) EPR of a new defect in natural quartz: possibly $O_2^-$. Solid State Commun 48:551–554

Bambauer HU (1961) Spurenelemente und $\gamma$-Farbzentren in Quarzen aus Zerr-klüf-ten der Schwei-zer Alpen. Schweizerische Mineralogische Petrographische Mitteilungen 41:335–367

Barfels T (2001) Kathodolumineszenz amorpher und kristalliner Modifikationen von $SiO_2$ und $GeO_2$. Ph.D. thesis, University Rostock, p 168

Beall GH (1994) Industrial applications of silica. Rev Mineral 29:469–506

Blankenburg H-J, Götze J, Schulz H (1994) Quarzrohstoffe. Deutscher Verlag für Grundstoffindustrie, Leipzig-Stuttgart, p 296

Blatt H, Middleton GV, Murray RC (1980) Origin of sedimentary rocks, 2nd edn. Prentice-Hall, Inc., Englewood Cliffs, p 782

Botis S, Nokhrin SM, Pan Y, Xu Y, Bonli T (2005) Natural radiation-induced damage in quartz. I. Correlations between cathodoluminescence colors and paramagnetic defects. Can Mineral 43:1565–1580

Channer DM DeR, Bray CJ, Spooner ETC (1999) Integrated cation-anion/volatile fluid inclusion analysis by gas and ion chromatography; methodology and examples. Chem Geol 154:59–82

Cohen AJ (1956) Color centers in alpha-quartz. Part I. Smoky quartz. J Chem Phys 25:908–914

Demars C, Pagel M, Deloule E, Blanc P (1996) Cathodoluminescence of quartz from sandstones: interpretation of the UV range by determination of trace element distributions and fluid-inclusion P-T-X properties in authogenic quartz. Am Mineral 81:891–901

Dennen WH (1964) Impurities in quartz. Geol Soc Am Bull 75:241–246

Dennen WH (1966) Stochiometric substitution in natural quartz. Geochim Cosmochim Acta 30:1235–1241

Dennen WH (1967) Trace elements in quartz as indicators of provenance. Geol Soc Am Bull 78:125–130

Ermakov NP (1950) Research on the nature of mineral-forming solutions (in Russ.). University of Kharkov Press, Kharkov, p 460

Flem B, Larsen RB, Gromstvedt A, Mansfeld J (2002) In situ analysis of trace elements in quartz by using laser ablation inductively coupled plasma mass spectrometry. Chem Geol 182:237–247

Friebele EJ, Griscom DL, Stapelbroek M, Weeks RA (1979) Fundamental defect centers in glass: the peroxy radical in irradiated, high-purity, fused silica. Phys Rev Lett 42:1346–1348

Füchtbauer H (1988) Sediments and sedimentary rocks. Schweizerbart, Stuttgart

Gemeinert M, Gaber M, Hager I, Willfahrt M, Bortschuloun D (1992) On correlation of gas-liquid-inclusion's properties and melting behaviour of different genetic quartzes for production of transparent fused silica. Neues Jahrbuch Mineralogie, Abhandlungen 165:19–27

Gerler J (1990) Geochemische untersuchungen an hydrothermalen, metamorphen, graniti-schen und pegmatitischen Quarzen und deren Flüssigkeitseinschlüssen. Ph.D. thesis, University Göttingen, p 169

Gerler J, Schnier C (1989) Neutron activation analysis of liquid inclusions exemplified by a quartz sample from the Ramsbeck Mine, F.R.G. Nucl Geophys 3:41–48

Ghazi AM, Vanko DA, Roedder E, Seeley RC (1993) Determination of rare earth elements in fluid inclusions by inductively coupled plasma-mass spectrometry (ICP-MS). Geochim Cosmochim Acta 57:4513–4516

Götze J (1997) Mineralogy and geochemistry of German high-purity quartz sands. In: Papunen H (ed) Mineral deposits: research and exploration. Balkema, Rotterdam, pp 721–724

Götze J (2000) Cathodoluminescence microscopy and spectroscopy in applied mineralogy. Freiberger Forschungsheft, C 485 Geowissenschaften, p 128

Götze J (2009a) Chemistry, textures and physical properties of quartz—geological interpretation and technical application. Mineral Mag 73:645–671

Götze J (2009b) Cathodoluminescence microscopy and spectroscopy of lunar rocks and minerals. In: Gucsik A (ed) Cathodoluminescence and its application in the planetary sciences. Springer, Berlin, pp 87–110

Götze J, Blankenburg H -J (1990) Zur Methodik der komplexen mineralogisch–geochemischen Charakterisierung von Glassanden. Sprechsaal International Ceramics & Glass Magazine 123, 8:795–803, 9:928–940, 12:1184–1189

Götze J, Lewis R (1994) Distribution of REE and trace elements in size and mineral fractions of high purity quartz sands. Chem Geol 114:43–57

Götze J, Plötze M (1997) Investigation of trace-element distribution in detrital quartz by electron paramagnetic resonance (EPR). Eur J Mineral 9:529–537

Götze J, Zimmerle W (2000) Quartz and silica as guide to provenance in sediments and sedimentary rocks. Contributions to Sedimentary Petrology, vol 21. Schweizerbart'sche Verlagsbuchhandlung, Nägele & Obermiller, Stuttgart, 91 S

Götze J, Siedel H (2007) A complex investigation of building sandstones from Saxony (Germany). Mater Charact 58:1082–1094

Götze J, Plötze M, Fuchs H, Habermann D (1999) Defect structure and luminescence behaviour of agate—results of electron paramagnetic resonance (EPR) and cathodoluminescence (CL) studies. Mineral Mag 63:149–163

Götze J, Plötze M, Habermann D (2001) Cathodoluminescence (CL) of quartz: origin, spectral characteristics and practical applications. Mineral Petrol 71:225–250

Götze J, Plötze M, Graupner T, Hallbauer DK, Bray C (2004) Trace element incorporation into quartz: a combined study by ICP-MS, electron spin resonance, cathodoluminescence, capillary ion analysis and gas chromatography. Geochim Cosmochim Acta 68:3741–3759

Götze J, Plötze M, Trautmann T (2005) Structure and luminescence characteristics of quartz from pegmatites. Am Mineral 90:13–21

Götzinger MA (1990) Determination of aqueous salt solutions in fluid inclusions by infrared investigations. Neues Jahrbuch Mineralogie, Monatshefte H.1:1–12

Graetsch H (1994) Structural characteristics of opaline and microcrystalline silica minerals. Rev Mineral 29:209–232

Graupner T, Götze J, Kempe U, Wolf D (2000) Cathodoluminescence imaging as a tool for characterization of quartz and trapped fluid inclusions in multistage deformed mesothermal Au-quartz vein deposits: a case study from the giant Muruntau Au-ore deposit (Uzbekistan). Mineral Mag 64:1007–1016

Griffiths JHE, Owen J, Ward IM (1954) Paramagnetic resonance in neutron-irradiated diamond and smoky quartz. Nature 173:439–442

Griscom DL (1985) Defect structure of glasses. J Non-Cryst Solids 73:51–77

Hallbauer DK (1992) The use of selected trace elements in vein quartz and quartz pebbles in identifying processes of formation and source rocks. Geologcal Society of South Africa 24th Congress, Bloemfontein, Abstracts, 157–159

Hallbauer DK (1997) The application of capillary ion analysis to the geochemistry of natural aqueous fluids and in particular to the analysis of fluid inclusions in minerals. In: Proceedings of the 30th International Geological Congress, vol 9, pp 409–424

Hanson B, Delano JW, Lindstrom DJ (1996) High-precision analysis of hydrous rhyolitic glass inclusions in quartz phenocrysts using the electron microprobe and INAA. Am Mineral 81:1249–1262

Heany PJ, Prewitt CT, Gibbs GV (1994) Silica—physical behaviour, geochemistry and materials application. Reviews in Mineralogy, vol 29. Mineralogical Society of America, Washington, p 606

Heynke U, Leeder O, Schulz H (1992) On distinguishing quartz of hydro-thermal or metamorphogenic origin in different monomineralic veins in the eastern part of Germany. Mineral Petrol 46:315–329

Houseknecht DW (1991) Use of cathodoluminescence petrography for understanding compaction, quartz cementation, and porosity in sandstones. In: Baker CE, Kopp OC (eds) Luminescence microscopy: quantitative and qualitative aspects. SEPM, Dallas, pp 59–66

Ioannou SE, Götze J, Weiershäuser L, Zubowski SM, Spooner ETC (2003) Cathodoluminescence characteristics of Archean VMS-related quartz: Noranda, Ben Nevis, and Matagami districs, Abitibi Subprovince, Canada. G3 Online Publication, 5(2), doi:10.1029/2003GC000613
IOTA® (2005) IOTA® high-purity quartz. http://www.iotaquartz.com/welcome.html. Accessed 20 May 2005
Jourdan A-L (2008) Elemental and isotopic zoning in natural alpine quartz. Ph.D. thesis, University Lausanne
Jung L (1992) High purity natural quartz. Part I: high purity natural quartz for industrial use. Part II: High purity natural quartz markets for suppliers and users. Quartz Technology. Liberty Corner, New Jersey, p 657
Klemm W (1986) Beiträge zur analytischen Geochemie von Gas-Flüssigkeits-Einschlüssen in hydrothermalen Mineralen. Habilitation thesis, TU Bergakademie Freiberg
Knauth LP (1994) Petrogenesis of chert. In: Heaney PJ, Prewitt CT, Gibbs GV (eds) Silica—physical behaviour, geochemistry and materials applications. Reviews in Mineralogy, vol 29. Mineralogical Society of America, Washington, pp 233–258
Komuro K, Horikawa Y, Toyoda S (2002) Development of radiation-damage halos in low-quartz: cathodoluminescence measurement after He+ ion implantation. Mineral Petrol 76:261–266
Konhauser K (2007) Introduction to geomicrobiology. Blackwell Publishing, Malden, p 425
Kostov RI, Bershov LV (1987) Systematics of paramagnetic electron-hole centres in natural quartz (in Russian). Izvestiya Akademii nauk USSR, Seria Geologia 7:80–87
Kostova B, Pettke T, Driesner T, Petrov P, Heinrich CA (2004) LAICP-MS study of fluid inclusions in quartz from the Yuzhna Petrovitsa deposit, Madan ore field, Bulgaria. Swiss Bull Mineral Petrol 84:25–36
Krickl R, Nasdala L, Götze J, Grambole D (2008) Alteration of $SiO_2$ caused by natural and artificial alpha-irradiation. Eur J Mineral 20:517–522
Larsen RB, Polvé M, Juve G (2000) Granite pegmatite quartz from Evje-Iveland: trace element chemistry and implications for high-purity quartz formation. Norges Geologiske Undersøkelse Bulletin 436:57–65
Larsen RB, Henderson I, Ihlen PM, Jacamon F (2004) Distribution and petrogenetic behaviour of trace elements in granitic pegmatite quartz from South Norway. Contrib Mineral Petrol 147:615–628
Larsen RB, Jacomon F, Kronz A (2009) Trace element chemistry and textures of quartz during the magmatic hydrothermal transition of Oslo Rift granites. Mineral Mag 73:691–705
Leeder O, Thomas R, Klemm W (1987) Einschlüsse in Mineralen. VEB Deutscher Grundstoffverlag, Leipzig
Lehmann K, Berger A, Götte T, Ramseyer K, Wiedenbeck M (2009) Growth related zonations in authigenic and hydrothermal quartz characterized by SIMS-, EPMA-, SEM-CL- and SEM-CC-imaging. Mineral Mag 73:633–644
Lyakhovich VV (1972) Trace elements in rock-forming minerals of granitoides (in Russian). Nedra, Moscow, p 200
Mackey JH (1963) EPR study of impurity-related color centers in germanium-doped quartz. J Chem Phys 39:74–83
Meunier JD, Sellier E, Pagel M (1990) Radiation-damage rims in quartz from uranium-bearing sandstones. J Sediment Petrol 60:53–58
Milliken KL, Laubach SE (2000) Brittle deformation in sandstone diagenesis as revealed by scanned cathodoluminescence imaging with application to characterization of fractured reservoirs. In: Pagel M, Barbin V, Blanc P, Ohnenstetter D (eds) Cathodoluminescence in geosciences. Springer, Berlin, pp 225–243
Mineeva RM, Bershov LV, Petrov I (1991) EPR of surface-bound $Fe^{3+}$ ions in polycrystalline quartz (in Russian). Dokladi Akademii Nauk SSSR 321:368–372
Miyoshi N, Yamaguchi Y, Makino K (2005) Successive zoning of Al and H in hydrothermal vein quartz. Am Mineral 90:310–315

Monecke T, Kempe U, Petersen S, Götze J, Herzig P, Wolf D (1999) Trace element characteristics of quartz from the TAG hydrothermal mound (Mid-Atlantic Ridge at 26E08′N). In: Mineral deposits: processes to processing. Balkema, Rotterdam, pp 551–554
Monecke T, Bombach G, Klemm W, Kempe U, Götze J, Wolf D (2000) Determination of trace elements in quartz standard UNS-SpS and in natural quartz by ICP-MS. Geostand Newsl 24(1):73–81
Monecke T, Kempe U, Götze J (2002) Genetic significance of the trace element content in metamorphic and hydrothermal quartz: a reconnaissance study. Earth Planet Sci Lett 202:709–724
Mosebach R (1955) Neue Ergebnisse auf dem Gebiet der hydrothermalen Forschung. Chemiker Zeitung 79:583–599
Müller A, Koch-Müller M (2009) Hydrogen speciation and trace element contents of igneous, hydrothermal and metamorphic quartz from Norway. Mineral Mag 73:569–584
Müller A, Seltmann R, Behr HJ (2000) Application of cathodoluminescence to magmatic quartz in tin granite—case study from the Schellerhau Granite Complex, Eastern Ertgebirge, Germany. Mineralium Deposita 35:169–185
Müller A, Kronz A, Breiter K (2002) Trace elements and growth patterns in quartz: a fingerprint of the evolution of the subvolcanic Podlesi Granite System (Krušne Hory, Czech Republic). Bull Czech Geol Surv 77(2):135–145
Müller A, Rene M, Behr H-J, Kronz A (2003a) Trace elements and cathodoluminescence of igneous quartz in topaz granites from the Hub Stock (Slavkovský Les Mts., Czech Republic). Mineral Petrol 79:167–191
Müller A, Wiedenbeck M, Van den Kerkhof AM, Kronz A, Simon K (2003b) Trace elements in quartz—a combined electron microprobe, secondary ion mass spectrometry, laser-ablation ICP-MS, and cathodoluminescence study. Eur J Mineral 15:747–763
Müller A, Ihlen PM, Kronz A (2005) Potential resources of quartz and feldspar raw material in Sørland IV: relationships between quartz, feldspar and mica chemistry and pegmatite type. Norwegian Geological Survey Report 2005.075, Trondheim, p 94
Müller A, Ihlen PM, Wanvik JE, Flem B (2007) High-purity quartz mineralisation in kyanite quartzites, Norway. Mineralium Deposita 42:523–535
Müller A, Ihlen PM, Kronz A (2008) Quartz chemistry in polygeneration Sveconorwegian pegmatites, Froland, Norway. Eur J Mineral 20:447–464
Nacken R (1950) Hydrothermale Mineralsynthese zur Züchtung von Quarzkristallen. Chemiker Zeitung 74:745–749
Nishikawa H, Watanabe E, Ito D, Ohki Y (1994) Decay kinetics of the 4.4 eV photoluminescence associated with the two states of oxygen-deficient-type defect in amorphous $SiO_2$. Phys Rev Lett 72:2101–2104
Norwegian Crystallites AS (2006) Norwegian crystallites AS-products crystal quartz analyses. http://www.norcryst.no. Accessed 20 Sept 2006
O'Brien MCM (1955) The structure of the colour centres in smoky quartz. In: Proceedings of the royal society, A 231:404–414
Owen MR (1988) Radiation-damage halos in quartz. Geology 16:529–532
Pan Y, Nilges MJ, Mashkovtsev RI (2009) Radiation-induced defects in quartz: a multifrequency EPR study and DFT modelling of new peroxy radicals. Mineral Mag 73:519–536
Perny B, Eberhardt P, Ramseyer K, Mullis J (1992) Microdistribution of aluminium, lithium and sodium in quartz: possible causes and correlation with short-lived cathodoluminescence. Am Mineral 77:534–544
Pettijohn FJ, Potter PE, Siever R (1987) Sand and Sandstone. Springer, New York, p 553
Plötze M (1995) EPR investigations of quartz, scheelite and fluorite from high-thermal trace-metal mineralization (in German). Ph.D. thesis. TU Bergakademie Freiberg, p 141
Plötze M, Wolf D (1996) EPR- and TL spektren von quarz: bestrahlungsabhängigkeit der $[TiO_4^-/Li^+]^0$-Zentren. European Journal of Mineralogy 8, Bh. 1:217
Rakov LT, Milovidova ND, Kuvshinova KA, Moiseev BM (1985) EPR investigations of germanium centers in natural polycrystalline quartz (in Russian). Geokhimiya 9:1339–1344

Ramseyer K, Mullis J (1990) Factors influencing short-lived blue cathodoluminescence of alpha-quartz. Am Mineral 75:791–800
Ramseyer K, Baumann J, Matter A, Mullis J (1988) Cathodoluminescence colours of alpha-quartz. Mineral Mag 52:669–677
Ramseyer K, AlDahan AA, Collini B, Landström O (1992) Petrological modifications in granitic rocks from the Siljan impact structure: evidence from cathodoluminescence. Tectonophysics 216:195–204
Richter DK, Götte Th, Götze J, Neuser RD (2003) Progress in application of cathodoluminescence (CL) in sedimentary geology. Mineral Petrol 79:127–166
Rinneberg H, Weil JA (1972) EPR studies of $Ti^{3+}$-$H^{+}$ centers in X-irradiated alpha-quartz. J Chem Phys 56:2019–2028
Rink WJ, Rendell H, Marseglia EA, Luff BJ, Townsend PD (1993) Thermoluminescence spectra of igneous quartz and hydrothermal vein quartz. Phys Chem Mineral 20:353–361
Roedder E (1984) Fluid inclusions. Rev Mineral 12:1–645
Rösler HJ (1981) Lehrbuch der Mineralogie. 2. Aufl., VEB Deutscher Verlag für Grundstoffindustrie, Leipzig, p 833
Rusk BG, Reed MH, Dilles JH, Kent AJR (2006) Intensity of quartz cathodoluminescence and trace-element content in quartz from the porphyry copper deposit at Butte, Montana. Am Mineral 91:1300–1312
Rusk B, Lowers HA, Reed MH (2008) Trace elements in hydrothermal quartz: relationships to cathodoluminescent textures and insights into vein formation. Geology 36:547–550
Rykart R (1995) Quarz-Monographie. 2. Aufl., Ott Verlag Thun, p 462
Shepherd, Rankin TJ, Alderton AH, Alderton DHM (1985) A practical guide to fluid inclusion studies. Blackie & Sons, Glasgow
Sippel RF (1968) Sandstone petrology evidence from luminescence petrography. J Sed Petrol vol 38, pp 530–554
Sippel RF (1971) Luminescence petrography of the Apollo 12 rocks and comperative features in terrestrial rocks and meteorites. In: Proceedings of the second lunar science conference, vol 1, pp 247–263
Stegger P, Lehmann G (1989) The structures of three centers of trivalent iron in alpha-quartz. Phys Chem Mineral 16:401–407
Stenina NG, Bazarov LS, Shcherbakova MY, Mashkovtsev RI (1984) Structural state and diffusion of impurities in natural quartz of different genesis. Phys Chem Mineral 10:180–186
Stevens-Kalceff MA (2009) Cathodoluminescence microcharacterization of point defects in α-quartz. Mineral Mag 73:585–606
Stevens-Kalceff MA, Phillips MR, Moon AR, Kalceff W (2000) Cathodoluminescence microcharacterisation of silicon dioxide polymorphs. In: Pagel M, Barbin V, Blanc P, Ohnenstetter D (eds) Cathodoluminescence in geosciences. Springer, Berlin, pp 193–224
Strunz H, Tennyson C (1982) Mineralogical Tables. Akademische Verlagsgesellschaft Geest & Portig, Leipzig, p 621
Summerfield MA (1979) Origin and palaeoenvironmental interpretation of sarsens. Nature 281:137–139
Summerfield MA (1983) Silcrete. In: Goudie AS, Pye K (eds) Chemical sediments and geomorphology. Academic Press, San Diego, pp 59–91
Suttner L, Leininger RK (1972) Comparison of the trace element content of plutonic, volcanic and metamorphic quartz from Southwestern Montana. Geol Soc Am Bull 83:1855–1862
Van den Kerkhof AM, Hein UF (2001) Fluid inclusion petrography. Lithos 55:27–47
Van den Kerkhof AM, Kronz A, Simon K, Scherer T (2004) Fluid-controlled quartz recovery in granulite as revealed by cathodoluminescnec and trace element analysis (Bamble sector, Norway). Contrib Mineral Petrol 146:637–652
Walenczak Z (1969) Geochemistry of minor elements dispersed in quartz. Archi-wum Mi-nera-lo-giczne 28:189–335
Walther H (1993) Silcretes in Germany and Australia. Ph.D. thesis, TU Bergakademie Freiberg, p 126

Weil JA (1984) A review of electron spin spectroscopy and its application to the study of paramagnetic defects in crystalline quartz. Phys Chem Mineral 10:149–165
Weil JA (1993) A review of the EPR spectroscopy of the point defects in α-quartz: the decade 1982–1992. In: Helms CR, Deal BE (eds) Physics and Chemistry of $SiO_2$ and the Si-SiO interface 2. Plenum Press, New York, pp 131–144
Wopfner H (1978) Silcretes of northern South Australia and adjacent regions. In: Langford-Smith T (ed) Silcrete in Australia, Armidale, pp 93–141
Wopfner H (1983) Environment of silcrete formation: a comparison of examples from Australia and the Cologne Embayment, West Germany. In: Wilson RCL (ed) Residual deposits: surface related weathering processes and materials. Geological Society of London Special Publications 11, pp 151–158
Wright PM, Weil JA, Buch T, Anderson JH (1963) Titanium colour centers in rose quartz. Nature 197:246–248
Wünsch K (1987) Mineralogische, geochemische und strukturelle Untersuchungen an metamorphogenen Quarzmobilisaten. Ph.D. thesis, TU Bergakademie Freiberg, Germany
Zinkernagel U (1978) Cathodoluminescence of quartz and its application to sandstone petrology. Contributions to Sedimentology, 8. Schweizerbart'sche Verlagsbuchhandlung, Nägele and Obermiller, Stuttgart, p 69

# Chapter 2
# Assessment of High Purity Quartz Resources

**Reiner Haus, Sebastian Prinz and Christoph Priess**

**Abstract** Very high purity quartz for advanced high-tech applications is currently sourced from just a few locations around the world. Such is the expected growth in demand that more sources are required to be found. For successful high purity raw quartz resource identification detailed analysis and appropriate process technology selection is essential. This article reviews general aspects of high purity quartz deposits, exploration requirements, quality evaluation of raw quartz, and provides basic insight into the different specifications and market developments of the high-tech end-user industries reliant on very high purity refined quartz products.

## 2.1 Introduction

Quartz is one of the most abundant minerals. It occurs in many different settings throughout the geological record (Götze 2009). As the prime source of silica the mineral has wide large volume application in the manufacture of glass, ceramics, refractory materials and other traditional uses (Blankenburg et al. 1994). However, only very few deposits are suitable in volume, quality and amenability to tailored refining methods for speciality high purity applications. As such high purity quartz has become one of today's key strategic minerals with applications in high-tech industries that include semiconductors, high temperature lamp tubing, telecommunications and optics, microelectronics, and solar silicon applications (Blankenburg et al. 1994; Haus 2005, 2010; Moore 2005; Dal Martello et al. 2011a, b).

R. Haus (✉) · S. Prinz · C. Priess
Dorfner Analysenzentrum und Anlagenplanungsgesellschaft mbH (ANZAPLAN), Scharhof 1, 92242 Hirschau, Germany
e-mail: reiner.haus@dorfner.com

J. Götze and R. Möckel (eds.), *Quartz: Deposits, Mineralogy and Analytics*, Springer Geology, DOI: 10.1007/978-3-642-22161-3_2,

Whilst most processing plants for standard quartz applications deploy off-the-shelf technology solutions, quartz for high-tech applications requires tailored processing techniques and specially designed equipment to achieve essential high purity specifications. Beneficiation of raw quartz into refined high-purity products involves several refinement steps which need to be adapted to effectively minimise the specific impurities of the individual raw quartz feed to comply with stringent end-use specifications (Haus 2005). As a result, high purity quartz with total impurity levels less than 20 ppm may be achieved so creating a highly valuable raw material which commands up to 5 EUR/kg.

## 2.2 Present Quartz Supplying Countries

Globally, the export of lump quartz has been diminishing for a variety of reasons. Both suppliers and consumers increasingly require quartz granules with the highest quality specifications. This is beneficial for suppliers who add more value to the quartz product by improving refining techniques. For consumers it saves processing cost and enables them to concentrate on their core business instead of investing in purifying steps to improve the raw material.

Today, US-based Unimin Corp./Sibelco still dominates the global high purity quartz market from operations in North Carolina, USA. One of the few alternative suppliers, Norsk Mineral's Norwegian Crystallites has been producing high purity quartz from its Drag plant in western Norway and several underground and open pit mines since mid-1996 when the company changed ownership. In 2011 Imerys SA has combined its US-based Spruce Pine companies KT Feldspar and The Feldspar Corp. (TFC) with Norwegian Crystallites now owned and operated by The Quartz Corp, a joint venture between Imerys and Norsk Mineral.

Potential new entrants into the high purity quartz world market are still under development. Moscow-based JSC Polar Quartz (Moore 2005) has raw material supplies based on quartz deposits on the eastern slopes of the sub-polar Ural Mountains and Kyshtym Mining's (KGOK) crystal quartz deposit is situated on the eastern slopes of the South Ural Mountains. In the Soviet era Kyshtym's plant supplied 60% of domestic high purity quartz demand used to make clear glass for microelectronics and optical applications. Little is known about Chinese Donghai Pacific Quartz although it is understood to serve domestic markets.

In the early 1970s Brazil was the world's main supplier of high purity quartz based on lascas; a term used to describe manually beneficiated rock crystal. Up to 1974, when the Brazilian government imposed an embargo on exports of lump quartz, export levels rose to in excess of 10,000 tpa. Brazil appears to have irretrievably lost its position in the quartz consuming world. Additionally, its infrastructure does little to attract investment of the foreign capital necessary to develop the quartz industry over the medium term. It is also known that the Brazilian high quality quartz materials show considerable variations in quality compared with raw quartz from deposits in other parts of the world. This inhibits the resurrection and further development of Brazilian quartz for high quality products. Potential deposits in Madagascar and Angola also suffer from poor infrastructure and lack of interest on the part of their governments.

In Asia, Japan's influence is large especially in Southeast India and in Sri Lanka. Japan once imported quartz lumps from these regions. However, government-led efforts on the part of the quartz supplier countries were launched to stop the export of unrefined quartz lumps and to support the development of quartz processing within the country. Today these countries specialise in the production of high purity filler materials for epoxy moulded compounds (EMC) low in uranium and thorium used in the manufacture of computer chips.

Given its strategic relevance to the semiconductor and photovoltaic industries many more high purity quartz deposits are under development. However, the exploration and exploitation of new suitable quartz deposits is hampered by quality assurance regulations, which are globally applied. Whilst traditional sources of quartz have been questioned in quality terms, it is in their interest to demonstrate that their material is tested and meets the standards of silica glass production. Suppliers from new quartz deposits have yet to achieve this status. This is achieved normally in several consecutive cycles of tests with increasing quantities of test material. These tests must be carried out for many of the product groups in the various application areas.

These pre-business services are expensive but made necessary by increasingly strict quality regulations. They require certain personnel expertise from the potential new suppliers (e.g. for the provision of relevant raw material) and highly specialised cooperation partners well recognised within the industry.

## 2.3 Typical High Purity Quartz Applications

As mentioned in the introduction the applications of high purity quartz in the high tech industries are manifold. Main applications are in the semiconductor, high temperature lamp tubing, telecommunication and optics, microelectronics, and solar silicon industries.

The semiconductor industry places the most stringent requirements on quartz purity. From single crystal silicon growth in quartz crucibles via the Czochralski process to the handling and processing of wafers in clean rooms, high purity quartz ware is employed. Fused quartz is the basic material for quartz ware used in the semiconductor industry since it combines excellent high temperature properties (i.e. thermal shock resistance and thermal stability) and high purity in a unique way. It withstands the high temperature gradients and high rates of heat transfer in rapid thermal processing which are commonly applied to wafers in order to modify their properties. The high purity of the quartz prevents contamination of wafers during the different processing steps.

The application of high purity quartz as basic material in high temperature lamp tubing takes advantage of its high transmission characteristic and its exceptional

**Fig. 2.1** High temperature Xenon lamp made from high purity quartz

**Fig. 2.2** High purity quartz is applied as filler material in epoxy moulding compounds (EMC) used as cover in the microelectronics chip production

thermal shock resistance and thermal stability. It is used in the high performance, high temperature lamp manufacturing sectors for UV lamps, mercury, xenon and halogen bulbs, and high intensity discharge lamps (Fig. 2.1).

Silica glass is widely used as basic material for optical fibers and additional optoelectronic devices in the telecommunications industry. It is used in the optical industry in microlithographic applications, excimer laser optics, beamers and other specialised applications. In the microelectronics industry, a major application is as filler material in epoxy moulding compounds (EMC) for electronic components (Fig. 2.2).

Silicon is the most common material for the production of solar cells in the photovoltaic industry either in mono- or polycrystalline form.

Specific requirements as to tolerable limiting values differ from industry to industry. In the lamp tubing and optics industries aluminium content in the refined quartz concentrate should not exceed 20 ppm, other metals should be less than 1 ppm, and total impurities less than 30 ppm. For semiconductor base materials and crucibles aluminium content should be even lower, specified to less than 10 ppm, other metals less than 0.1 ppm, and total impurities not to exceed 15 ppm.

Feedstock for solar silicon used in the photovoltaic industry should generally have boron and phosphorus contents in the sub-ppm range since these elements are most difficult to remove and negatively affect the performance of the solar cells (Geerligs et al. 2002).

For microelectronics applications, e.g. in epoxy moulding compounds, uranium and thorium, both responsible for soft errors by alpha radiation, should be less than 2 ppb, and in low alpha applications, even less than 0.5 ppb.

## 2.4 Demand Situation

The global outlook for high purity quartz using industries is still positive with growth rates between 3 and >5%, specifically in certain lighting, semiconductor and photovoltaic applications.

### *2.4.1 Lighting Industries*

In the USA, the EU and many more countries the sale of general service incandescent lamps is banned. Compact fluorescent lamps (CFLs) are expected to replace most of them owing to government advice. Demand for CFLs will surge more than 20% annually through 2013 as consumers transfer to the new technology. However, halogen lamps are also forecast to grow especially fast through 2013, as, they will benefit from the ban too. Although less efficient and shorter-lived than CFLs, halogen lamps outperform CFLs in terms of colour rendering and they do not contain mercury. In addition, new halogen-based product inventions are already close to the energy efficiency of CFLs and fit well with general lamp connectors. Demand for halogen lamps will also benefit from recovering motor vehicle production levels, as they find widespread use in headlamps.

Metal halide lamps (Xenon) will lead the high intensity discharge (HID) product segment. As mercury vapour lamps are phased-out of the market, metal halide lamps will gain market share in outdoor lighting applications. Furthermore, metal halide lamps are increasingly finding use in high-end (premium cars) motor vehicle headlamps. Expectations of European lamp manufacturers are for growth rates in the 3–5% range worldwide over the next 3–5 years. Rapid development in the BRIC countries also supports these expectations.

Since both, halogen- and xenon-based lamps are made of high purity quartz glass to withstand the high temperatures and to offer high transmission characteristics as well as exceptional shock resistance (Chap. 3), the demand in high purity quartz will benefit from the increasing demand in the lamp tubing and automotive industries.

## 2.4.2 Semiconductor Industry

The demand for high purity quartz products in the semiconductor industry is closely related to new fabrication plants (fabs) to be built, consuming most of the semiconductor base materials, as well as to the production of monocrystalline silicon (crucibles).

SEMI's World Fab Forecast (www.semi.org) recently predicted steady growth in capacity, about 9% for 2011 and 7% for 2012. In 2010, about seven facilities were expected to begin construction; of these, five are considered high-volume fabs.

The outlook of 5.5% revenue growth for total fab materials is in-line with current expectations of modest single-digit growth for the semiconductor industry in 2011. Stronger demand for electronics, especially mobile products, could result in even higher than expected growth for semiconductor products in 2011; therefore increasing the amount of wafers processed and materials consumed.

The record levels for semiconductor sales, in both revenues and units, generated tremendous growth in the semiconductor materials market. Total silicon wafer shipments, from wafer suppliers to the fabs, grew 40% in 2010. Growth was strong for all wafer diameters. With the broad-based recovery in the semiconductor industry, both 150 and 200 mm shipments increased by rates comparable to those for 300 mm wafer shipments last year. Overall, 6% wafer shipment growth is currently estimated for 2011 with 300 mm shipments likely growing in the 11–13% range.

## 2.4.3 Photovoltaic Industry

In the solar industry all signs point to further growth in the coming years.

Production of c(crystalline)-Si in the 2007–2011 period saw CAGR of 45% while Si-production for 2010–2014 is forecast to slow to CAGR in the 20–30% range reaching 300,000–400,000 tpa in 2014 (2010: 170,000 tpa). In 2010 solar cell production was up 118%, compared with 2009, when more solar cells were produced than the combined total in all prior years.

Grid parity has now been achieved in a number of countries. Based on the demand in high efficiency cells, mono c-Si is expected to keep around 30% share at least for the next 3 years.

However, new multi-pulling production techniques (recharging) and larger diameters of the monocrystals will demand more voluminous crucibles, i.e. less crucible silica glass per kg of monocrystal produced (volume to area ratio). Hitherto, mainly small furnaces with crucible diameters between 16″ and 22″ were used for PV applications. More recently, however, the same sizes for crucibles as are used in the semiconductor industry are being requested. Therefore, the increase in high-purity quartz granule consumption will not equal the upcoming growth in c-Si monocrystal production, but it is still estimated at above 5% p.a.

Manufacturers of solar silicon are looking to remain competitive by pushing production costs down. The result might be a process of consolidation: in 2012, the four largest producers alone—OCI, Hemlock, GCL and Wacker—could probably cover worldwide demand. Prices, which are currently still dominated by demand, should then fall and relate more to costs.

This could attract new technologies which are based on high purity materials and less investment and operating costs to enter the solar silicon market. On the other hand it will raise the demand in high quality quartz, specifically low in boron and phosphorus.

### 2.4.4 *Optical Fibers*

Requirement for continually improving broadband connections has increased demand for synthetic silica glass used in fibre optics for optical data transmission. In fibre optics almost all waveguides are made of synthetic quartz/silica, handles and leader cornings, however, are manufactured of natural quartz.

## 2.5 Resource Estimation

Expertise in geology, chemistry, and mineral processing is required to develop the full potential of a high purity quartz deposit. Related to the geological conditions at the time of formation, quartz often contains fluid and/or mineral inclusions (Roedder 1984; Rykardt 1995; Hyrsl and Niedermayr 2003; Heaney et al. 1994) that need to be removed to meet stringent requirements for high purity applications. In addition detailed information about structural impurities in quartz (Flem et al. 2002; Götze et al. 2004) are of interest since they may define the economic limits in purifying the raw quartz. In order to achieve maximum value, process technologies need to be adapted to the specific characteristics of each quartz deposit. Remnant fluids, from which quartz crystallised, occur trapped within individual crystals as do different types of mineral inclusions such as iron oxides, phosphates, silicates and heavy minerals. These inclusions need to be identified by analytical methods in order to determine what beneficiation processes will be used to remove them, as they are undesirable impurities that detract from the quality of potential quartz products.

Whilst chemistry is important in determining quartz quality it is equally important to identify quantity in order to determine how long, at a given production rate, a potential mine may produce economically viable product for at least, say, 20 years as a realistic resource/venture.

**Fig. 2.3** Typical vein quartz deposit (Africa), width of the picture approx. 15 m

Hydrothermal vein-like quartz deposits (Fig. 2.3) usually comprise zones reflecting different periods of mineralisation. Zones vary enormously in thickness, with individual quartz veins ranging from less than 1 up to 500 kt. Massive quartz deposits in similarly zoned pegmatites, essentially very coarsely crystalline hydrothermal host rock fissure fillings, are basically of larger volumes (up to 5 Mt) while metamorphic quartzite bodies are usually of even larger dimensions but of lower quality. In contrast, silica sand deposits are typically loosely consolidated and are of little prospect for economically viable high purity applications.

International standards, e.g. Australia's JORC or Canada's NI 43-101 codes, prescribe procedures to progressively identify, in terms of quality and quantity, the calibre of a given mineral resource from initial inferred status through increasing levels of detailed exploration and expenditure to proven status primarily to attract and justify continuing investment in resource development before the start of extractive operations.

The main activity of geological assessment is field exploration including the mineralogical and chemical trace element analysis of samples collected from outcrops and/or drill cores (Fig. 2.4) showing the variation of quartz quality within the deposit. Drilling based on geological mapping and appropriate remote sensing techniques, determine the three dimensional extent of the mineralised body.

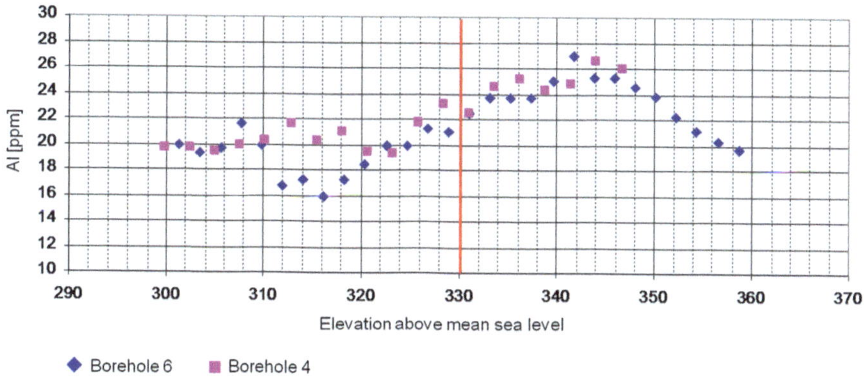


**Fig. 2.4** Aluminium distribution in the vertical drill core of a massive pegmatite quartz deposit, elevation in m

In order to reduce exploration costs information from initial drilling is combined with geophysical data derived from methods such as seismic, gravimetric, geoelectric or geomagnetic field surveys.

Geophysical resistivity profiles apparent lateral variations in resistivity of rocks using a specific electrode array over constant distances in specific locations. Vertical Electrical Sounding (VES) provides information on vertical variations in resistivity within a geological formation. Apparent resistivity values are correlated with the geological formations present to provide an interpretation of the extent of a deposit (Fig. 2.5). This assists efficient drill hole siting, the results of which ideally confirm and complement, in greater detail, resistivity data.

Once a quartz deposit has been identified the precise quality of its mineral content and potential for quality improvement are key factors in determining its economic value. Representative samples are taken for detailed investigation to evaluate the potential of the raw material to be processed into a high value refined product. Determinative mineralogical techniques characterise any fluid and/or mineral inclusions that need to be removed by applying tailor-made processes. The combination of mineralogical characterisation across the whole deposit by means of representative sample analysis with the identification of appropriate specific processes to remove impurities is crucial to the thorough evaluation of any raw quartz deposit for high purity and high value applications.

## 2.6 Analytics

Naturally occuring high purity quartz always has inclusions which are present either in the form of finely dispersed solids (mineral phases such as e.g. muscovite, rutile, calcite) or fluid (liquid and gaseous) inclusions which can provide valuable

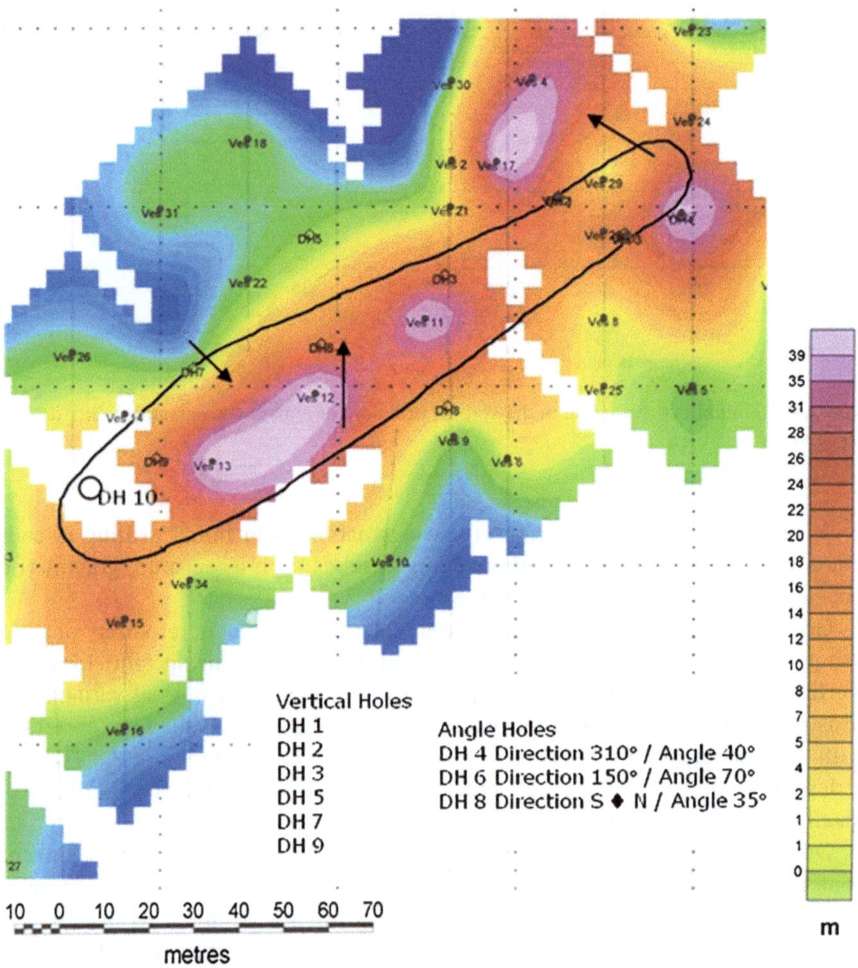


**Fig. 2.5** Interpretation of lateral resistivity profiling and Vertical Electrical Sounding (VES) data presenting the thickness and subsurface distribution of a quartz body

insight into the conditions prevalent at the time of formation of the quartz. By the analysis of trace elements and impurities the conditions of formation of the quartz can be unraveled. Typically the bulk sample will be chemically characterized first. Then the size and distribution as well as chemical composition of the mineral and fluid inclusions are investigated in detail. The different steps in the analysis of raw quartz samples are discussed in the following.

### 2.6.1 Bulk Chemical Analysis

Characterisation begins with bulk chemical analysis of representative samples of raw quartz. The results serve as a reference point for all processing tests in the later stages of the quartz assessment procedure. Special care needs to be taken at all stages of the analytical process:

- Contamination free comminution
- Acid digestion
- Trace element analysis by ICP-OES/MS or LA-ICP-MS

The technique most appropriate for detection of major chemical impurities is X-ray fluorescence (XRF), minor impurities down to the sub-ppm level are commonly detected by inductively-coupled-plasma optical emission spectrometry/mass-spectrometry (ICP-OES/MS) or laser-ablation ICP-MS. Laser ablation ICP-MS has the advantage, that the sample does not have to be digested, but has the disadvantage that a very small volume is probed, which is not representative of the bulk sample and may give misleading results.

Besides ICP-OES/MS there are complementary methods for the detection of trace elements in quartz such as electron spin resonance, cathodoluminescence, capillary ion analysis and gas chromatography (Götze et al. 2004). During initial characterization, typically 16 trace elements, defined to be most critical depending on the specific target application area, are usually determined:

- Alkali metals, calcium and heavy metals—critical in lamp tubing and semiconductor applications
- Uranium and thorium—critical for microelectronics applications
- Boron and phosphorus—critical in solar silicon applications

Because of the low concentrations of trace elements multiple measurements and external standards such as specified and certified industry products (e.g. Iota Quartz) are required to ensure a very high level of confidence.

### 2.6.2 Characterization of Mineral Inclusions

Although high purity is determined by the bulk composition, more information is needed to set up the most appropriate beneficiation route. Typical properties of raw quartz that must be characterized are:

- Size of inclusions
- Chemical composition of inclusions
- Spatial distribution of inclusions and localization of isomorphic substitutional elements (e.g. Al, Ti) in the quartz crystal lattice

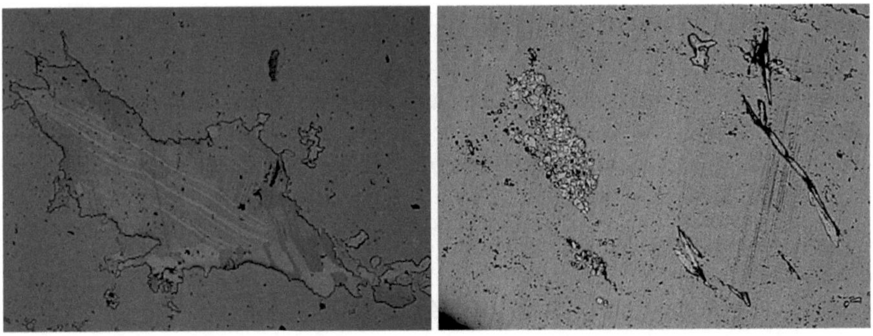

**Fig. 2.6** *Left*: photomicrograph (reflected light) of a large xenomorphic calcite crystal with wavy deformation twins. Note also the presence of smaller irregularly shaped crystals nearby. Width of the picture is 600 μm; *Right*: photomicrograph (reflected light) of quartz with a cluster of apatite crystals on the *left* and tremolite fibres on the *right* side. Picture width is 2.3 mm

Optical microscopic analysis of raw quartz is the starting point. It allows rapid detection and a first overview of the sample texture and structure as well as impurities such as fluid and solid phase inclusions. An example of large xenomorphic calcite crystals with characteristic deformation twinning and apatite and tremolite inclusions determined by optical microscopy within the quartz matrix are shown in Fig. 2.6.

Impurities which are directly accessible to the analytics can most conveniently be detected by Micro X-ray fluorescence. An example of K-feldspar in quartz identified by Micro XRF is shown in Fig. 2.7. EPMA is another technique which is suitable for the precise identification of the chemical composition of minor inclusions such as K-feldspar in quartz shown in Fig. 2.8. Trace element contents ranging in the 100 ppm's can be detected by conventional EPMA. When improved background modeling, new quantitative "blank" correction routines and multiple spectrometers to improve the geometric efficiency are used, detection limits for Ti and Al as low as 2–3 ppm and 6–7 ppm can be achieved (Donovan et al. 2011).

However, critical questions such as the localization of aluminium (isomorphic substitution for silicon in the quartz crystal lattice and/or mineral inclusions) can only be determined by combining optical microscopy with advanced spectroscopic methods (Tlili et al. 1989), e.g. Raman spectroscopy (Fig. 2.9), or electron paramagnetic resonance (EPR) spectroscopy. In Raman spectroscopy the sample is irradiated by laser light. A small part of the scattered light shows a shifted frequency with respect to the primary light resulting from inelastic scattering processes. From the resulting spectrum the oscillation frequency of the molecules can be determined.

To unambiguously identify if substitutional elements are incorporated into the quartz lattice structure electron paramagnetic resonance is used.

EPR is a method which is able to detect the presence of unpaired electrons in a material. The sample is placed in an external uniform magnetic field and irradiated by microwaves which interact with the unpaired electrons. In quartz unpaired electrons are present in the paramagnetic $[AlO_4]^0$ centre or as electron- and hole centres which are caused by radiation (Götze and Plötze 1997; Nuttall and Weil

**Fig. 2.7** Photograph of K-feldspar in quartz (*whitish spots* in the area framed by *black ink dots*) and corresponding analysis by Micro XRF

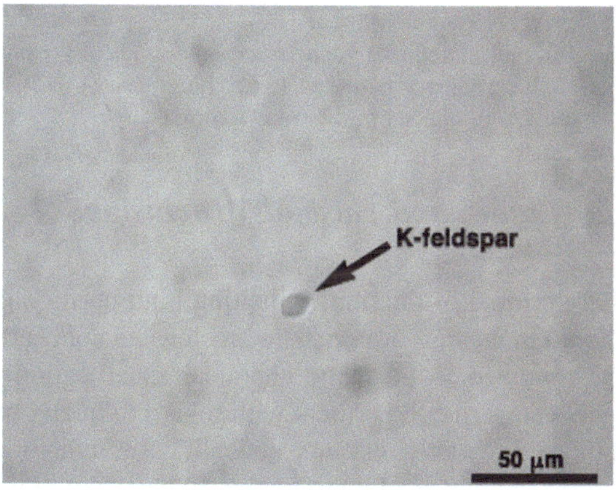


| Mineral | K-feldspar [wt.-%] |
|---|---|
| $SiO_2$ | 64,11 |
| $TiO_2$ | 0,00 |
| $Al_2O_3$ | 18,06 |
| $Cr_2O_3$ | 0,00 |
| $Fe_2O_3$ | 0,00 |
| MnO | 0,00 |
| MgO | 0,00 |
| CaO | 0,00 |
| $Na_2O$ | 0,65 |
| $K_2O$ | 16,37 |
| BaO | 0,05 |
| Total | 99,25 |

**Fig. 2.8** Microphotograph (crossed polarizers) of a small grain of K-feldspar in quartz and corresponding results of electron probe microanalysis

1981; Okada et al. 1971). Figure 2.10 shows an example of EPR spectra of two quartz samples (labeled 1 and 2) from different deposits. In the upper part of the figure the simulated spectra of $[AlO_4]^0$ and $[TiO_4^-/Li^+]^0$ centers are shown. Measured spectra match those of the simulations and clearly confirm the presence of both types of centers in sample 1 (1a, 1b) and the presence of the former in sample 2. The aluminium contents in the samples, as determined by ICP-OES analyses, are 64 ppm and 10 ppm respectively.

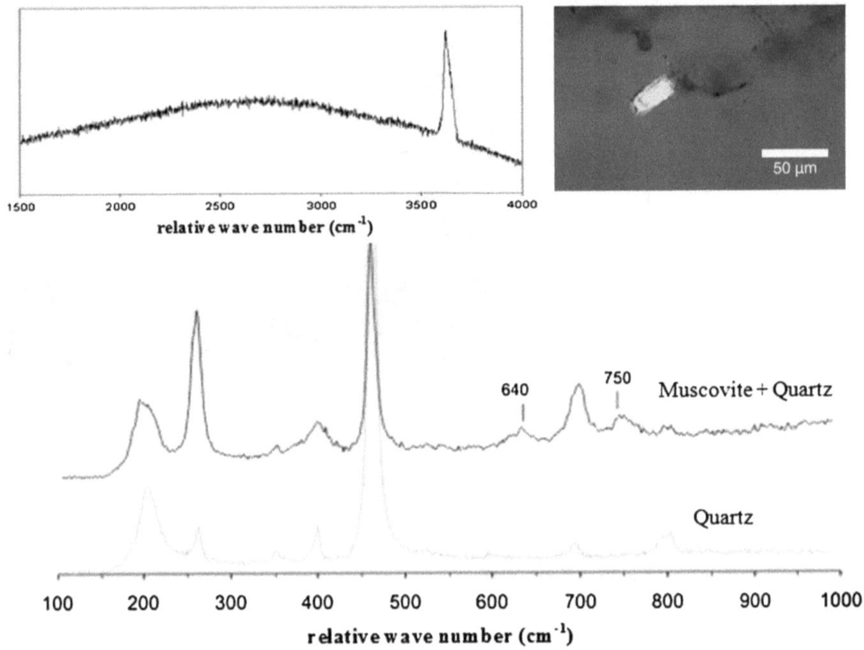


**Fig. 2.9** Laser micro-Raman spectra and photomicrograph (*inset upper right*) of a muscovite inclusion in host quartz (*grey*). Note the presence of the OH stretching peak at ~3630 cm$^{-1}$ (*inset upper left*) that is characteristic for aluminous micas (muscovite)

## 2.6.3 Analysis of Fluid (Liquid and Gaseous) Inclusions

Fluid inclusions are especially critical in relation to melting high purity quartz in glass making. Oftentimes, liquid inclusions are brines with elevated concentrations of alkalis which need to be removed to meet the chemical specification of the glass. In addition, gaseous inclusions may be critical with regard to quartz melting behavior. Quartz, when molten, is highly viscous and all gases which cannot escape or be dissolved will form bubbles leading to quality impairment in the final glass product.

Fluid inclusions are formed during the primary crystallization of quartz from cooling rock forming fluids and also as a result of secondary (re)-crystallisation processes related to tectonic events (Roedder 1984). They can be classified based on descriptive parameters like size, shape, color, refractive index, their origin (primary, secondary) and the number of physical phases (liquid (L), vapor (V), solid (S)) present within the inclusion (Van den Kerkhof and Hein 2001). An example of secondary fluid inclusions with aqueous two-phase liquid–vapor inclusions is shown in Fig. 2.11.

Oftentimes, high purity quartz contains different generations of fluid inclusions, which may contain gas and liquid, and sometimes even very small crystals.

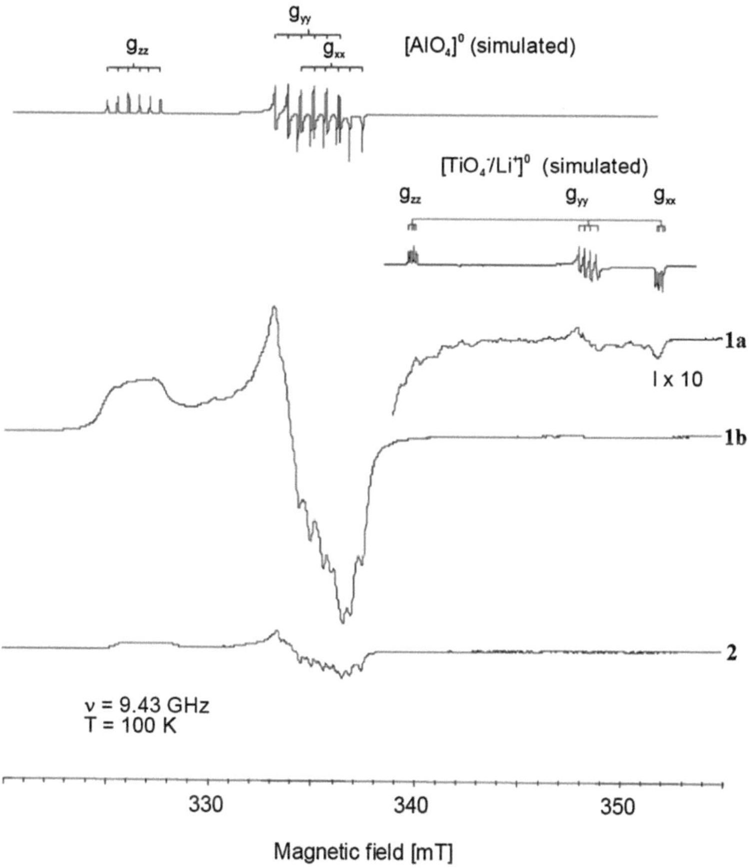


**Fig. 2.10** Determination of substitutional elements Ti and Al in the quartz lattice by EPR; simulated EPR spectra of $[AlO_4]^0$ and $[TiO_4^-/Li^+]^0$ centers are shown in the *top* part; measured spectra of two different quartz samples (labeled 1 and 2) are shown below: sample 1 shows both the typical signature of $[TiO_4^-/Li^+]^0$ and of $[AlO_4]^0$ centers (spectra 1a, 1b), sample 2 indicates only the presence of a minor amount of $[AlO_4]^0$ centers

The microphotograph in Fig. 2.12 shows an example of such an inclusion of an opaque cross-shaped twinned crystal within the fluid inclusion.

Critical for process engineering can be:

- Orientation of the fluid inclusions (along tracks or scattered)
- Chemical composition
- Pressure

The orientation of fluid inclusions is especially important since they may be aligned along tracks that will open during comminution or, if randomly scattered within the quartz crystals, their liberation by mechanical and physical processes will be prohibitively much more difficult. The chemical composition defines the

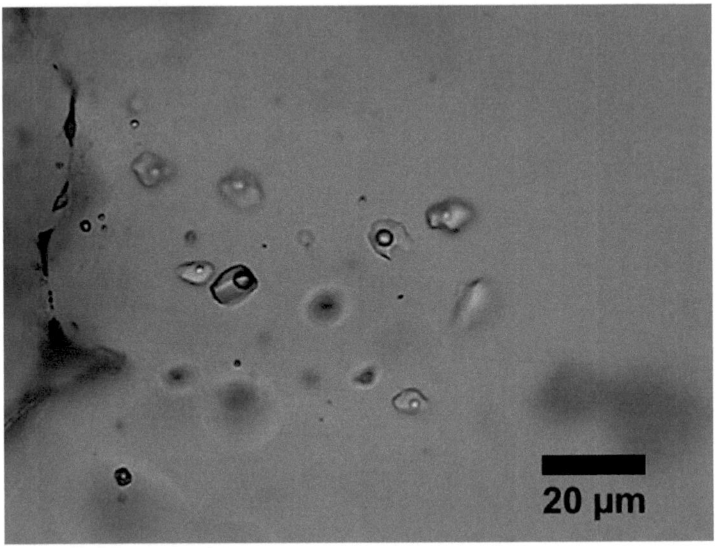


**Fig. 2.11** Secondary aqueous two-phase liquid–vapor inclusions with low degree of fill in quartz. Plane Polarized light

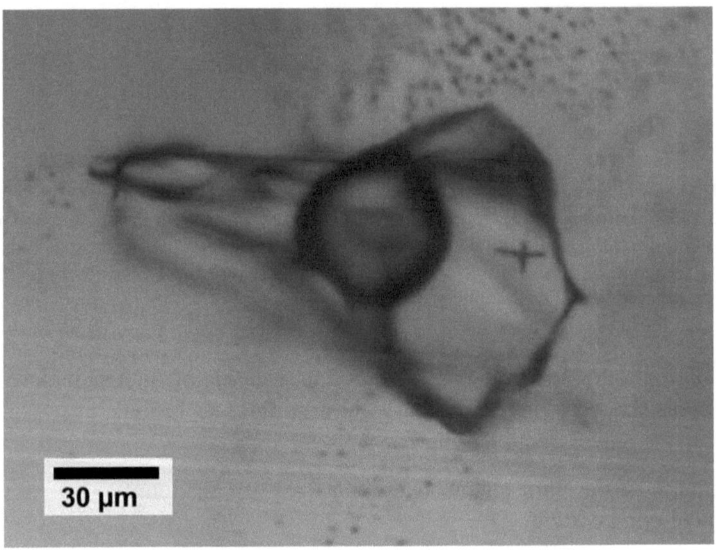


**Fig. 2.12** Microphotograph of a large re-equilibrated two liquid–vapor inclusion with a *cross-shaped black twinned crystal*. Plane polarized light

impurity level when element concentrations enriched in the fluid inclusions are forced in the silica glass during the melting process. Fluid inclusions may form bubbles making the silica glass less valuable. Since silica glass melt is highly viscous, smaller bubbles are not able to rise to the surface of the melt and so survive to unacceptably impair the quality of the silica glass.

In order to define the properties of the fluid inclusions, optical microscopy in combination with micro thermometry and Raman spectroscopy is employed. Specific options for process development, e.g. thermal treatment or opening of fluid inclusions by specific comminution technology, are explored according to the results of these analyses. Some quartz samples consist of clear and milky parts, the latter being richer in fluid inclusions than the former. The size of these parts is determined during the characterisation procedure in order to evaluate options for separating clear from milky quartz, by e.g. optical sorting. All these analyses are a full quality assessment that is a prerequisite for the design of optimum bench, pilot or technical scale processing tests.

## 2.7 High Purity Quartz Processing

Conversion of raw quartz into refined high purity and high value quartz products needs advanced comminution technology. Comminution reduces raw quartz to the required product size and liberates mineral impurities. Further mechanical, physical, chemical and thermal steps are needed to separate or dissolve the impurities to meet final quality requirements. The main stages of the process may be summarised as

- Pre-processing (mechanical)
- Physical processing
- Chemical leaching
- Thermal treatment

### *2.7.1 Pre-processing*

Based on the specific characteristics of the quartz deposit, one or more processing stages are required in order to liberate mineral impurities and fluid inclusions for further physical treatment:

- Initial crushing
- Optical sorting
- Comminution
- Classification to product particle size

Mined quartz lumps need to be washed to remove surface contaminations and crushed to produce granules in the centimetre size range. Crushing and grinding techniques need to consider both, to reduce the wear related contamination of the quartz to a minimum and to selectively liberate the mineral inclusions. Electrodynamic comminution is an enhanced technology to liberate impurities in high purity quartz crystals. It minimizes the amount of undersize particles and contamination. Since crystal boundaries host most of the mineral impurities, downstream processing is more efficient and expected yield increases significantly compared with mechanical comminution techniques. This makes electrodynamic comminution one of the favored methods in high purity quartz processing.

Classification then provides closely defined size fractions beneficial for further processing. Recent developments have proven the effectiveness of optical sorting for the production of high purity quartz.

The technique separates the liberated components of raw materials on the basis of differing colour (or transparency) and shape, and may improve, or even replace, costly selective mining or hand sorting practices. Fully automated sensor based sorting devices can be equipped with color CCD-cameras, X-ray Transmission (XRT) and Near Infrared (NIR) technology for grain detection. After detection, non specified grains are selectively extracted from the bulk flow by a precise pulse of pressurized air from a high performance nozzle system. A specific benefit has been achieved for size fractions well below 40 mm, where manual sorting is not economical. Down to the 3–5 mm range, optical sorting is applied with high efficiency (Fig. 2.13).

Optical sorting may be used to separate differently coloured quartz fractions such as patches of rose quartz, critical in solar applications owing to elevated phosphorus concentrations or smoky quartz with radiation induced discolorations. In addition clear quartz can be separated from milky quartz, being rich in fluid inclusions, and thereby often reduce the alkali content (Table 2.1) and improve the melting characteristics.

Standard size reduction and liberation of differing mineral constituents use jaw and cone crushers in order to reach the product grain sizes. However, these techniques introduce high amounts of contamination, owing to wear, especially into the fine product fractions. For high purity quartz alternative comminution techniques are applied:

- Autogenous grinding
- Electrodynamic fragmentation

In autogenous grinding the high purity quartz is ground on a bed of high purity quartz. By this comminution technology contamination from the wearing internal surfaces of the autogenous mill is minimised.

Electrodynamic fragmentation liberates quartz crystals in the composite rock largely without contamination and with low loss of undersize particles. A high-voltage discharge generates shock waves within the quartz lump causing it to fracture along grain boundaries (Andres et al. 1999; Dal Martello et al. 2011a, b). Since most of the mineral impurities are located along crystal boundaries downstream processing (e.g. chemical treatment) is more efficient. Yields increase significantly compared with mechanical comminution techniques. Moreover, this

**Fig. 2.13** Fraction 3–8 mm, clear quartz (*left*) and milky quartz (*right*) after optical sorting

**Table 2.1** Chemical characteristics of a quartz sample after optical sorting

| | Al (ppm) | Fe (ppm) | Na (ppm) | K (ppm) | Li (ppm) | Ti (ppm) | Zr (ppm) |
|---|---|---|---|---|---|---|---|
| Quartz transparent | 23 | 0.3 | 6.2 | 0.6 | 2.1 | 1.3 | <0.1 |
| Quartz milky | 25 | 0.8 | 10 | 1.6 | 2.4 | 1.4 | <0.1 |

**Table 2.2** Chemical analyses of quartz sample 0.1–0.3 mm after conventional comminution and electrodynamic fragmentation

| | Al (ppm) | Fe (ppm) | Na (ppm) | K (ppm) | Li (ppm) | Ti (ppm) | Zr (ppm) |
|---|---|---|---|---|---|---|---|
| Quartz raw material | 41 | 4.9 | 12 | 15 | 0.5 | 1.3 | <0.1 |
| Conventional comminution | 23 | 464 | 10 | 3.5 | 0.6 | 1.9 | <0.1 |
| Electrodynamic fragmentation | 28 | 1.3 | 13 | 4.7 | 0.6 | 0.5 | <0.1 |

new technology is ideal for selective liberation of gas and liquid inclusion trails within the quartz crystals (Table 2.2).

## 2.7.2 Physical Processing

Impurities liberated during the preparation stage may be separated from quartz owing to their differing responses to physical processes, such as:

- Attrition
- Magnetic separation
- High tension separation
- Flotation

Attrition is applied to clean the surfaces of the quartz particles. Thereby fine particles attached to quartz surfaces, e.g. clay minerals or iron oxide coatings, are either washed away or liberated for their subsequent physical separation. Magnetic separation removes heavy minerals from quartz as they are mostly paramagnetic or ferromagnetic. These minerals are attracted by a magnetic field. Quartz, being diamagnetic, is repelled. Since magnetic susceptibility is strong in the case of ferromagnetic minerals only moderate magnetic field strengths are necessary for their separation, whereas higher field strengths are required to separate paramagnetic minerals.

High tension techniques separate minerals owing to differences in their surface conductance. For this processing step particles are uniformly passed through an electrostatic field. The electrostatic separator consists of a heated chamber where the electrodes are situated. The generated electrostatic field is up to 120 kV. Feed material is activated by heating the sample or by the addition of diluted acids to the feed material prior to heating. Typically, feldspar impurities may be separated from quartz via high tension as a dry alternative to froth flotation.

Froth flotation selectively separates minerals according to differences in their ability to be wetted, enhanced or suppressed by conditioning reagents. Separation takes place in a water-filled medium into which the ore is fed to form a suspension which is agitated to avoid sedimentation processes. A frothing agent is added and air introduced to form rising air bubbles. Hydrophobic/Hydrophobized mineral particles (such as heavy minerals, feldspar or mica) attach to the air bubbles and rise to the surface forming froth whereas hydrophilic (wetted) particles remain below the froth layer in the suspension. The mineral-carrying froth is removed.

Flotation process designs vary in complexity depending primarily on the type of mineral, degree of liberation, and the desired purity of the product.

### 2.7.3 Chemical Treatment

Chemical treatment is an important addition to physical processing methods in order to achieve maximum purity quartz through the removal of surface impurities. Acid washing, leaching and hot chlorination are the three chemical treatment processes.

While acid washing uses less aggressive acids such as hydrochloric or sulphuric acid, leaching uses an advanced hydrofluoric acid process at elevated temperatures, to remove liberated surface impurities most effectively. In addition, those impurities enriched in micro fissures and along dislocations, will be liberated and removed owing to an enhanced dissolution rate of quartz in regions where impurities are concentrated.

In the hot chlorination process, quartz is heated to temperatures of 1,000–1,200°C in a chlorine or hydrogen chloride gas atmosphere. This refining process is suitable to specifically reduce the level of alkali metal impurities

**Table 2.3** Chemical analyses of quartz sample 0.1–0.3 mm after chemical treatment

| | Al (ppm) | Fe (ppm) | Na (ppm) | K (ppm) | Li (ppm) | Ti (ppm) | Zr (ppm) |
|---|---|---|---|---|---|---|---|
| Quartz after magnetic separation | 21 | 0.2 | 3.1 | 1.0 | 2.2 | 1.2 | <0.1 |
| Acid washing | 21 | <0.1 | 2.8 | 0.9 | 2.2 | 1.2 | <0.1 |
| Leaching | 20 | <0.1 | 0.7 | 0.3 | 2.2 | 1.2 | <0.1 |
| Hot chlorination | 21 | <0.1 | 0.2 | <0.1 | 1.6 | 1.1 | <0.1 |

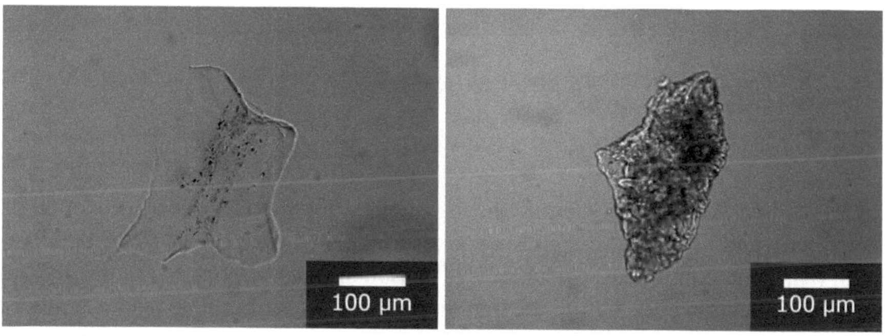


**Fig. 2.14** Comparison of quartz grains of raw quartz, fraction 0.1–0.3 mm (*left*) and quartz after thermal treatment (*right*)

(Table 2.3) which are highly restricted in lamp tubing and semiconductor applications.

### 2.7.4 Thermal Treatment

Thermal treatment (calcination) of high purity quartz improves its melting behaviour owing to the removal of fluid inclusions (Fig. 2.14). Depending on the degree of filling, the pressure and applied calcination technology (atmosphere, temperature etc.) a significant reduction of the bubble content in the silica glass has been observed (Fig. 2.15).

## 2.8 Conclusion

Whilst many industrial minerals deploy off-the-shelf technology solutions most speciality minerals require tailored processes and specially designed equipment that cannot be conceived prior to thorough analytical evaluation. That is why detailed investigations of the specific impurities ubiquitously found in quartz need to be performed before tailor-made processing concepts may be aligned with realistic

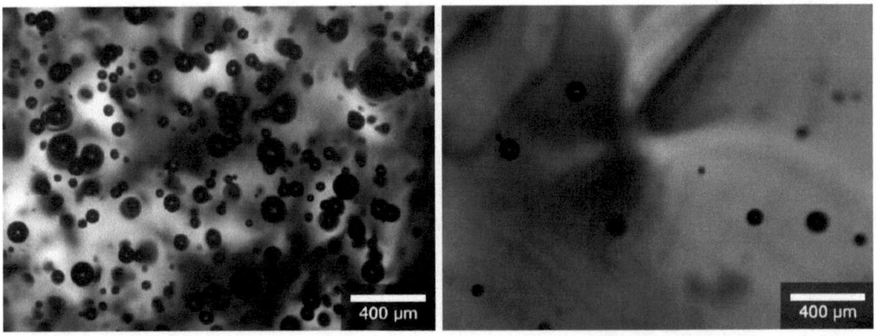


**Fig. 2.15** Comparison of melting results of raw quartz, fraction 0.1–0.3 mm (*left*) and quartz after thermal treatment (*right*)

investment costs and quality requirements. Engineering services need to cover all the technology and economic details such as estimation of investment costs, details of main plant and equipment, calculation of mass balances, energy requirements, and specific production costs before the final investment decision can be made. A precise and sustainable definition of the product split, in terms of relevant chemistry and appropriate physical characteristics, and end-user acceptance are two prerequisites for the success of any minerals-based project. End-user approval of processed samples at potential qualities is the ultimate, and necessary, risk management process.

## References

Andres U, Jirestig J, Timoshkin I (1999) Liberation of minerals by high voltage electrical pulses. Powder Technol 104:37–49

Blankenburg H-J, Götze J, Schulz H (1994) Quarzrohstoffe. Deutscher Verlag für Grundstoffindustrie, Leipzig, p 296

Dal Martello E, Bernardis S, Larsen RB, Tranell G, Di Sabatino M, Arnberg L (2011a) Electrical fragmentation as a novel refining route for hydrothermal quartz for SoG-Si production. Miner Eng. doi: 10.1016/j.powtec.2012.02.055

Dal Martello E, Tranell G, Gaal S, Raaness OS, Tang SK, Arnberg L (2011b) Study of pellets and lumps as raw materials in silicon production from quartz and silicon carbide. Metall Mater Trans B, p 12

Donovan JJ, Lowers HA, Rusk BG (2011) Improved electron probe microanalysis of trace elements in quartz. Am Mineral 96:274–282

Flem B, Larsen RB, Grimstvedt A, Mansfeld J (2002) In situ analysis of trace elements in quartz by using laser ablation inductively coupled plasma mass spectrometry. Chem Geol 182:237–247

Geerligs LJ, Wyers GP, Jensen R, Raaness O, Waernes AN, Santen S, Reinink A, Wiersma B (2002) Solar grade silicon by a direct route based on carbothermic reduction of silica: requirements and production technology. Energy Research Centre of the Netherlands (ECN) publication ECN-RX–02-042

Götze J (2009) Chemistry, textures and physical properties of quartz—geological interpretation and technical application. Mineral Mag 73(4):645–671
Götze J, Plötze M (1997) Investigation of trace element distribution in detrital quartz by electron paramagnetic resonance. Eur J Mineral 9:529–537
Götze J, Plötze M, Graupner T, Hallbauer DK, Bray CJ (2004) Trace element incorporation into quartz: a combined study by ICP-MS, electron spin resonance, cathodoluminescence, capillary ion analysis, and gas chromatography. Geochimica et Cosmochimica Acta 68(18):3741–3759
Haus R (2005) High demands on high purity. Ind Miner 10:62–67
Haus R (2010) High-purity quartz resources. In: PHOTON′s 8th solar silicon conference, Stuttgart
Heaney PJ, Prewitt CT, Gibbs GV (eds) (1994) Silica—physical behavior, geochemistry and materials applications. Rev Mineral (29), Mineralogical Society of America
Hyrsl J, Niedermayr G (2003) Magic world: inclusions in quartz. Bode Verlag, Haltern, p 240
Moore P (2005) High purity quartz. Ind Miner 8:54–57
Nuttall RHD, Weil JA (1981) The magnetic properties of the oxygen-hole aluminium centers in crystalline $SiO_2$. I. $[AlO_4]^0$. Canadian J Phys 59:1696–1708
Okada M, Rinneberg H, Weil JA (1971) EPR of Ti3+ centers in alpha quartz. Chem Phys Lett 11:275–276
Roedder E (1984) Fluid inclusions. Reviews in Mineralogy, vol 12. Mineralogical Society of America, Washington, p 644
Rykardt R (1995) Quarz-monographie, 2nd edn. Ott Verlag, Thun
Tlili A, Smith DC, Beny J-M, Boyer H (1989) A Raman microprobe study of natural micas. Mineral Mag 53:165–179
Van den Kerkhof AM, Hein UF (2001) Fluid inclusion petrography. Lithos 55:27–47

# Chapter 3
# Quality Requirements of Quartz Sand in the Building Industry

**Hartmut B. Walther**

**Abstract** Quartz sand is a valuable raw material for the building and construction industry. Individual branches demand very different sand qualities. The requirements of the following products are discussed: autoclaved aerated concrete, calcium silicate units, cement, clay bricks and roof tiles, concrete, mortar and render. Not only very pure sand qualities with >98 wt% $SiO_2$ are required, some applications need only 95, 88 or 70 wt% $SiO_2$. The requirements for the individual uses are, however, strict. New market developments and changes in legislature force the construction industry to improve its products. These developments cause modifications in the production process as well as new requirements to the raw materials used. Consequently, traditional branches may have to change their raw material deposits. There is a strong need for data concerning the chemical and granulometrical composition of near surface sand deposits.

## 3.1 Introduction

### *3.1.1 The Idea Behind*

The aim of this chapter is to increase the understanding of little researched but important commercial raw materials that suffer from some kind of ignorance. Although sand deposits form the basis of many important industrial activities, it is very difficult to learn something about their existence and it is almost impossible to find mineralogical or chemical data. If data are available they mainly cover special

H. B. Walther (✉)
Xella Technologie- und Forschungsgesellschaft mbH, Kloster Lehnin, Germany
e-mail: Hartmut.Walther@xella.com

J. Götze and R. Möckel (eds.), *Quartz: Deposits, Mineralogy and Analytics*, Springer Geology, DOI: 10.1007/978-3-642-22161-3_3,

**Table 3.1** Industrial uses of sand and gravel (Anonymous 2011)

| | | |
|---|---|---|
| Civil engineering 95% | Structural engineering 65% | Construction industry (not quantified) |
| | | Autoclaved aerated concrete |
| | | Calcium silicate units |
| | | Cement |
| | | Clay bricks/roof tiles |
| | | Concrete products |
| | | Precast concrete blocks |
| | | Prefabricated concrete units |
| | | Ready mixed concrete |
| | | Mortar & render products |
| | Underground use 35% | |
| Special uses 5% | | Ceramic industry |
| | | Chemical industry |
| | | Filter sands |
| | | Foundry sands |
| | | Glass industry |
| | | Silicon and silicon alloys |

sand of high purity—but 95% of sand and gravel are used in civil engineering and only 5% for special uses (Table 3.1).

New market requirements and changes in legislature force the construction industry to improve its products—building materials for the safe use in sustainable buildings. This development causes changes in the production process as well as new requirements to the raw materials used. Unfortunately, most geologists and even raw material specialists have little concern about the needs of the construction industry and its requirements for raw materials. This chapter intends to reduce this gap of knowledge. It will inform about the needs of building material producers and will encourage people to have a closer look to standard materials. Quartz sand according to the definition of the German law contains minimum 80 wt% quartz (Anonymus 1985). The needs from the industry differ from that value (Blankenburg et al. 1994; Lorenz and Gwosdz 1999). There is a strong need for data about sand and gravel everywhere in the world.

### *3.1.2 General Development*

A building constructed in the year 2010 is quite different from another one built in the 1960s. New requirements for instance on thermal insulation, earth quake resistance or wind suction, demand new construction types and advanced building materials. Modern wall forming materials are often much lighter than those used 20 years ago and show either a lower raw density such as autoclaved aerated concrete (AAC) or display a honeycomb structure such as clay bricks. Leading brick products display both: a honeycomb structure and cell walls with a considerably high porosity.

On the other hand, the demand for high sound insulation requires massive walls of high density produced from calcium silicate blocks. Wall blocks of high density are needed on streets with heavy traffic or between semi-detached houses to ensure privacy.

In addition, it is the aim of product research to allow a short construction time of the buildings and to avoid possible sources of mistakes by craftsmen. As a consequence, individual elements increased in size. Traditional bricks of "normal format" NF have a size of 7.1 × 11.5 × 24 cm but modern blocks have sizes like 25 × 30 × 36.5 cm or 50 × 100 × 36.5 cm. Even storey-high panels of different materials are available today.

The whole production process was optimized and developed to produce construction elements of sizes as described above and with reduced or increased raw densities. This process resulted in new requirements to the raw materials including quartz sand.

## 3.2 Special Developments of Individual Construction Products and Increasing Requirements for Siliceous Raw Materials

### *3.2.1 Autoclaved Aerated Concrete Industry*

AAC is produced in large moulds of several cubic metres in volume. Sand based AAC is produced from slurry of milled sand, water, cement and lime containing a very small amount of dispersed aluminium metal. The aluminium reacts with the hydrated lime and releases small hydrogen bubbles. After a first stiffening of this mass the blocks are cut with metal wires. Finally, the material is cured in autoclaves for several hours at a pressure of about 12 bar and a temperature of 180°C. During autoclaving, quartz dissolves and reacts together with the calcium component released from burnt lime and cement into tobermorite. The primary production of several cubic metre sized blocks with subsequent cutting to sizes of wall forming elements provides the basis for the production of large AAC blocks since inception. The industrial production in Sweden started in the 1920s, in Germany in the 1950s.

The main development in this time was an increase of the rising height, old blocks had a height of about 25 cm, nowadays 62.5 cm or even 80 cm (for wall panels) can be achieved. Yet, in the 1970s some factories could produce 62.5 cm blocks only by a trick. 35 cm AAC were cast and stiffened and then a second 35 cm layer was cast and stiffened directly on top of the first layer. Cutting and curing were performed with these combined blocks. Technical improvements and high quality raw materials today allow continuous production of the full height.

The developments of the last decades mainly focused on the reduction of raw density, increase of compressive strength, reduction of shrinkage due to loss of

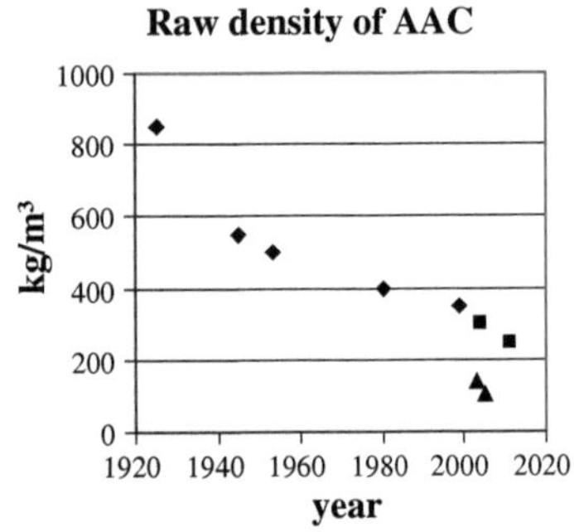


**Fig. 3.1** Raw densities of AAC as produced—existing products: *black diamonds*; new products of lower raw densities: *squares*; mineral insulation board: *triangles*

moisture and improvements in the phase composition. Figure 3.1 illustrates the development of the raw density. Beside an ongoing reduction of the raw density of traditional products new products of lower raw densities were developed. Lightweight mineral insulation board of a raw density of about 100 kg/m$^3$ is state of the art. Its brand name is YTONG-Multipor and it is used for insulation purposes.

The compressive strength of a material with a given recipe/composition is of course directly related to the density of this material. Thus, the compressive strength of YTONG-Multipor is lower than the compressive strength of normal wall materials. However, an identical compressive strength of 2 N/mm$^2$ can be achieved today with a raw density of 300 kg/m$^3$ which is about half the value of that from the 1950s.

This development caused a remarkable increase in the required quality of the raw sand (Table 3.2). Products of high raw density and low strength can be produced with sand of lower quality. However, a high content of alkaline elements can lead to efflorescing salt crystals forming a white covering of the wall.

### *3.2.2 Calcium Silicate Unit Industry*

Calcium silicate units were traditionally called sand lime bricks. They are mainly produced in Germany, Poland, the Netherlands and Belgium. Extended areas of these countries are covered with glacial quartz sand deposits. Obviously, these sediments provided the basis for the development of this building material.

Calcium silicate units were produced from a mixture of sand and burnt lime, mechanically pressed into the final block format and cured in autoclaves like AAC but at up to 16 bar pressure and 200 °C, where tobermorite is formed from quartz and burnt lime. The production process is consequently similar to that of AAC but the sand is used in its original form and the grains support each other in the structure. The grain size of the sand is very important and the producers optimize their products mainly by influencing the grain size distribution (Table 3.3). Furthermore, the grain shape is very important. Sharp edged grains interlock in the structure and give an optimal support to each other. The smoother the surface of a

**Table 3.2** Requirements of siliceous raw materials for the use in modern AAC

| Property | Requirement | Comments |
|---|---|---|
| Chemical composition | | |
| $SiO_2$ | >80 wt% | High end products >98 wt% |
| $K_2O + Na_2O$ | <1 wt% | |
| Content of humic acids | <0.5 wt% | Light yellow at NaOH-test |
| Maximum grain size | No performance required | Normally <4 mm for well grinding |
| Fraction <63 μm | <4 wt% | Higher contents possible if quartz |
| Critical components | Montmorillonite | Binds water and requires binder |
| | Coal and wood | Cause decrease in compressive strength |

**Table 3.3** Grain size distribution of sand for the use in calcium silicate blocks according to Quincke and Eden (2002)

Unfavourable grain size distributions

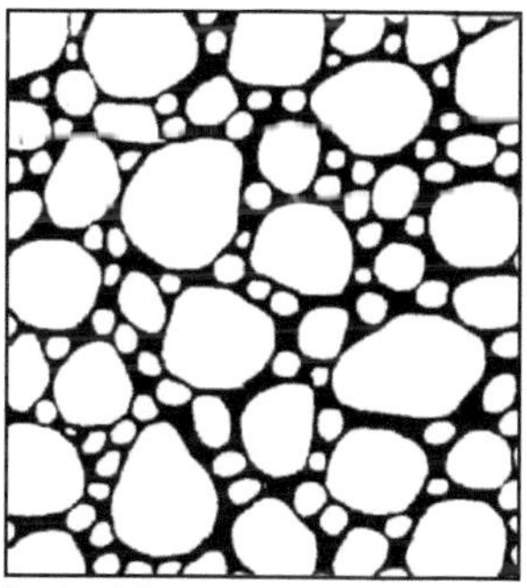

Lack of fine fraction

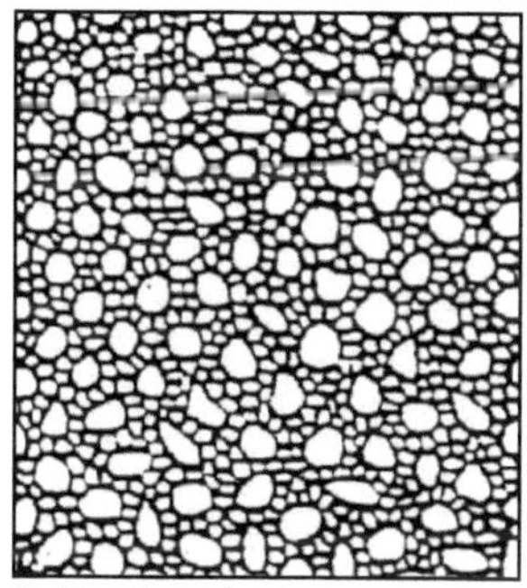

Uniformly grained

Favourable grain size distributions

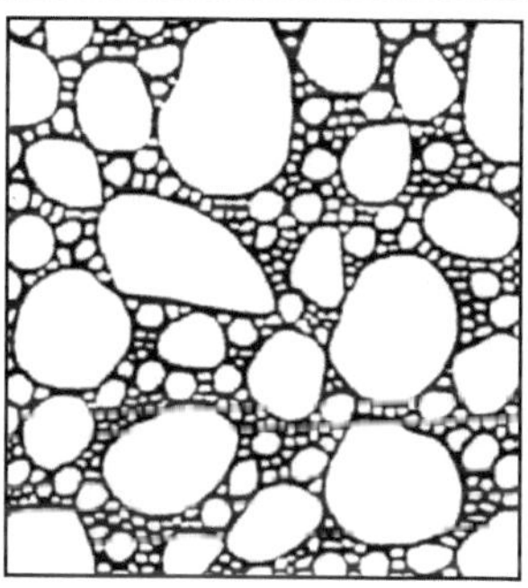

Consistent grain size distribution

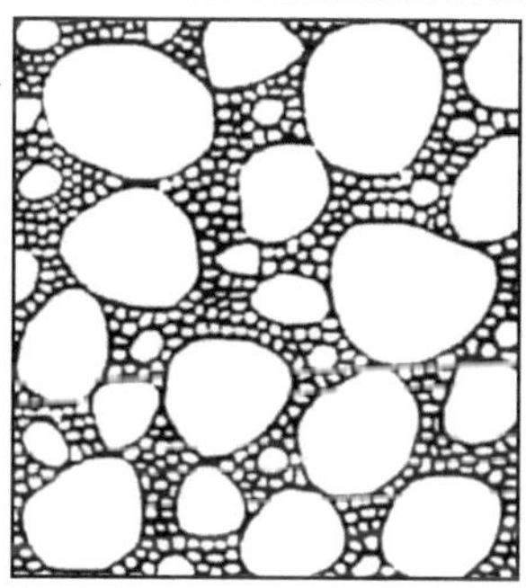

Gap grading

grain is the smaller is its geometrical resistance against pressure (Table 3.4). Optimal raw sand allows a reduction of the lime without any influence on the mechanical strength. Single grained raw sands restrict the possible variety of products and require higher amounts of binder.

**Table 3.4** Grain shape groups and its importance for calcium silicate units according to Gundlach (1973)

| Grain shapes | Group / suitability | Description of the outline |
|---|---|---|
|  | sharp edged<br>very well suitable | sharp edges, sharp profiled big and small recesses, protruding peaks |
|  | edged<br>well suitable | start of round edges, big recesses still visible, small recesses rather weak and less frequent |
|  | moderately rounded<br>suitable | rounded edges, big recesses rather weak, little small recesses, no protruding peaks |
| 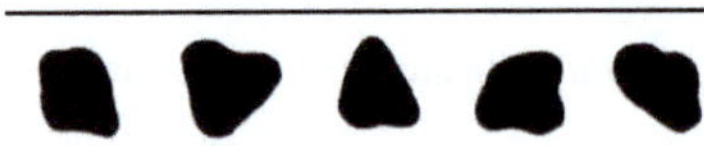 | rounded<br>sufficient | primary edges rounded, big recesses only allusively, no small recesses |
|  | well rounded<br>restricted use | original edges and big recesses not visible any more, clearly convexly shaped, scarcely flat areas |

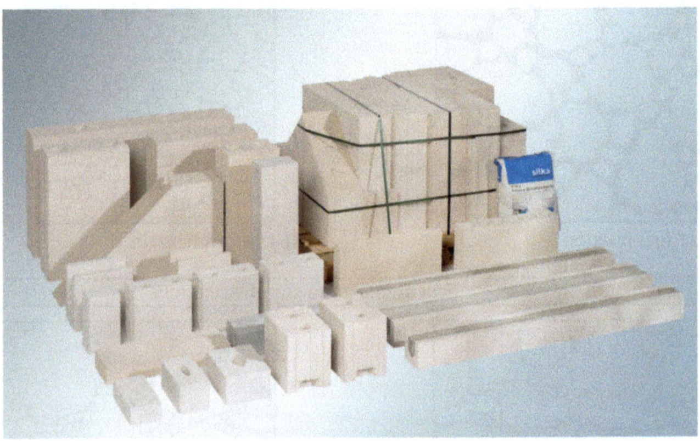

**Fig. 3.2** Production program of a calcium silicate unit producer in 2010, 50 years ago only block sizes of the lower corner were produced

At the beginning, more than 100 years ago, calcium silicate units were produced in the traditional brick formats like 240 × 115 × 71 mm ("Normalformat", NF, in Germany) or 250 × 125 × 65 mm ("Reichsformat", RF, in Germany) or 240 × 115 × 52 mm ("Dünnformat", DF, in Germany). Later the height of the blocks increased and in the 1950s a large portion of the production was delivered in 2DF i.e. 240 × 115 × 113 mm. Later improvements in the moulding press technology allowed the production of larger and even higher blocks of 6 DF or 12 DF. A new technological era started in 1973 (Kendel 1973). Calcium silicate units reached a height of 50 cm (Fig. 3.2) and at this point the compressibility of the raw materials and the internal structure of the compressed and hardened product obtained a new importance. Single grained raw sands could not be used any longer

Table 3.5 Requirements of siliceous raw materials for the use in calcium silicate units

| Property | Requirement | Comment |
|---|---|---|
| Chemical composition | | |
| $SiO_2$ | >50 wt% | Lower contents possible according to Gundlach (1973) |
| $K_2O + Na_2O$ | <1 wt% | |
| Content of humic acids | <0.8 wt% | Yellow at NaOH-test |
| Maximum grain size | About 20 mm | |
| Fraction <125 μm | >5 wt% | For example: for strength class 20 N/mm$^2$ |
| Critical components | Mica | Cause decrease in compressive strength |
| | Coal and wood | Cause decrease in compressive strength |
| | Pyrite/ marcasite | Can cause rust spots on surface |
| | Clay minerals | If covering the quartz grain surface they may influence the phase transformation in autoclave process |

for these large products. Their technological disadvantages could not be compensated even by using larger amounts of lime.

A second development influenced the requirements of the mineralogical composition of the raw sand. Increasing legal demands concerning sound insulation require walls of high weight and consequently higher raw density.

Increasing block heights require raw material with sufficient fine fraction which will be well compacted (Table 3.5). The fine fraction has not necessarily to consist of quartz only, but may contain also clay minerals. For example: In one case a minimum content of about 5 wt% of quartz <125 μm is needed to produce 50 cm high elements of the strength class 20 N/mm$^2$. The sand for the use in the production is sometimes mixed from 3 up to 5 different qualities of sand. In collaboration with the University of Kassel the German Research Center for the Calcium Silicate Industry developed a computer program to optimize the grain size distribution from up to 5 individual sand qualities (Eden et al. 2008).

High sound insulation of a wall requires a high density of the building material. This can be achieved by adding coarse particles into the production mixture, particles up to 20 mm are possible. To increase the density, rocks of higher density like basalt can be added. High end products contain heavy mineral fractions (Bertran 1995, Table 3.8 "heavy aggregates") and the remaining quartz grains assure the stability of the building material. Its structure complies with that of a gap grading (Table 3.3).

### 3.2.3 Cement Industry

The use of quartz sand in the production of cement is not in the focus of people who are not directly linked to the cement industry. It is generally known that cement is produced from marl or lime and clay, but the natural given purity and

**Table 3.6** Requirements of siliceous raw materials for the use in the cement industry

| Property | Requirement | Comment |
|---|---|---|
| Chemical composition | | |
| $SiO_2$ | >95 wt% | The higher the better |
| $K_2O + Na_2O$ | <1 wt% | |
| Maximum grain size | No performance required | Normally <4 mm for well grinding |
| Critical components | Alkaline elements | |

stability in composition of these materials is inadequate to ensure a constantly high level of cement quality. It is common knowledge that cement factories use highly advanced laboratories to ensure a high standard of their products. These labs are often fully automated, thus sampling, preparation and measurement are carried out automatically, sometimes using robots. Several analyses have to be carried out every hour, 24 h a day and 7 days a week.

These analytical efforts only make sense, if correction methods exist, which influence the clinker composition. This can be achieved by changing the mixing proportions of the standard raw materials. However, this may not be enough and correction materials have to be used to compensate inadequacies of raw materials. Main correction materials are e.g. quartz sand (silica), magnetite (iron) and "oxiton" (aluminium), the latter being a by-product of aluminium recycling.

The grain size of the quartz sand is of minor significance as the raw materials are ground before being fed into the rotary kiln. The sand is used as a chemical agent and is assimilated completely into the cement clinker. The sand should contain only very small amounts of feldspar, because of the sensitivity of cement to alkaline elements.

Quartz sand used in cement production should not contain critical components for the cement composition such as alkaline elements (Table 3.6). Moreover, its $SiO_2$ content should be as high as possible because it has only the purpose to increase this content in the clinker.

### *3.2.4 Clay Brick Industry/Roof Tile Industry*

Clay bricks are one of the oldest artificial building materials and have been produced for thousands of years. The old Romans already produced durable bricks from clay as it is still visible at many places all over Europe. At that time wooden moulds were filled with clay by hand. Today, the blocks are produced by extrusion moulding with subsequent wire cutting, roof tiles are pressed using turntable presses. Afterwards, the production of bricks and roof tiles is nearly identical and includes two important steps: drying and burning. The mass has to be ductile during shape forming and contains a considerable amount of water at this step, too much for the burning process. Thus, intermediate drying is required. This drying step was formerly carried out for several weeks under normal ambient conditions and large

**Table 3.7** Requirements of siliceous raw materials for the use in the clay brick/roof tile industry

| Property | Requirement | Comment |
|---|---|---|
| Chemical composition | | |
| $SiO_2$ | >85 wt% | |
| $Al_2O_3$ | No performance required | |
| $K_2O + Na_2O$ | No performance required | |
| Content of humic acids | No performance required | |
| Maximum grain size | <4 mm | Grains crushed in rolling mills <0.6 mm |
| Fraction <63 μm | No performance required | |
| Critical components | Coarse carbonate grains | Grains crushed in rolling mills <0.6 mm, destructed in brick kiln, destroying body |

drying sheds were typical for old brick factories. The drying is accompanied by shrinkage of the clay material. During burning the final material properties are generated and a second volume reduction occurs. Modern factories include a drying plant for well defined drying with minimal stress to the structure of the material.

The stringent requirements for building materials introduced in recent decades also affected the traditional brick industry. The height of the bricks increased remarkably. The Romans used formats only a few centimetres high, subsequently the height was increased to 5.2 cm and today, most blocks are about 24 cm high. New thermal requirements demanded a reduction of raw density of the bricks. This goal is mainly reached by a honeycomb or cellular structure of the bricks. In some cases the cell walls are also porous.

The final dimensions of the block/tile have to be reached after two shrinkage steps of several millimetres each. Consequently, the producers take special care to limit shrinkage. One approach is to reduce the moisture of the ductile material, a second one is to reduce the content of clay minerals in the initial raw material by adding fine grained quartz sand. However, this approach introduces a new source of defects—at burning temperature quartz occurs in its hexagonal high temperature modification and its transformation into the trigonal low temperature modification during cooling results in a volume decrease of about 0.8% at 573 °C. This volume change causes cracks in the material and is one of the most important scrap sources, especially in roof tile production. Consequently, a small grain size is needed to reduce quartz stress in the ceramic material. Because of mixing of sand and clay before processing the ceramic raw mass, the grain size of quartz is of minor interest. The material is passing two or three rolling mills and the final one has an opening of 0.6–5 mm and all larger grains are crushed.

The quartz grains in the ceramic mass for the production of clay bricks and roof tiles should be very small because of the volume change from the high temperature modification to the low temperature modification. Coarse grains are broken in rolling mills during processing of the raw materials. Coarse carbonate grains must not be contained within the sand because of their thermal decomposition during the burning process (Table 3.7).

**Table 3.8** Natural and artificial aggregates for use in concrete

| Type of aggregate | Limit of grain density (kg/dm$^3$) | Natural aggregates | Artificial aggregates |
|---|---|---|---|
| Heavy aggregates | $\geq$3.0 | Barite | Steel sand |
| | | Magnetite | Heavy metal slag |
| | | Ilmenite | Ferrophosphorous |
| | | Hematite | |
| Normal aggregates | $\geq$2.0 … <3.0 | Sand | Blast furnace slag |
| | | Gravel | Corundum |
| | | Crushed stone | Crushed concrete |
| Light weight Aggregates | 0.4 … 2.0 | Pumice | Expanded shale |
| | | Perlite | Expanded clay |
| | | | Expanded perlite |
| | | | Crushed bricks |

### *3.2.5 Concrete Industry*

The concrete industry covers a variety of individual branches like pre-cast concrete blocks, prefabricated concrete units and ready mixed concrete. Like other building materials concrete went through tremendous developments within the last decades. The general principles still remained the same and the well known Fuller curves (Fuller and Thomson 1907) are still valid. However, large efforts have been made to improve the homogeneity of the material, its compressive and tensile strength, thermal properties, density, water demand and shrinkage.

The density and also the thermal properties are mainly influenced by the use of different aggregates as shown in Table 3.8.

Special concrete for the absorption of radioactivity or X-ray has been created using large amounts of barite, thermal insulation blocks containing pumice, expanded clay or expanded perlite. This, however, influences the requirements to quartz sand and gravel only concerning grain size distribution. To obtain an optimal Fuller curve is more or less trivial.

Other developments are rather sophisticated. To avoid even the smallest cracks in the final concrete block, the temperature within the body has to be controlled and maximum values up to 60°C have to be kept even within the centre of large elements. This can be achieved only by homogeneous raw material properties, an efficient temperature management and intelligent mixing programs (Weber and Riechers 2003).

Concrete can be improved against dissolution. In this case, a very dense structure has to be achieved and, instead of water, chemicals enhance the flowability of the ready mixed concrete; the slurry flows but is almost dry. This requires special actions at the building site, such as a silo for predried quartz sand, storage of all aggregates in closed boxes or under a shelter and some technical measures as well. Concrete like this is called "Ultra High Performance Concrete"—UHPC (Weber and Riechers 2003).

UHPC can reach compressive strength values of more than 120 N/mm$^2$, mainly by optimized aggregate mixtures, low water content, chemicals and a silica suspension. The latter is the joker of the concrete specialists: dispersed silica tremendously increases the strength of concrete. Using these ingredients and natural aggregates with a high content of quartz grains and pebbles, a concrete can be produced having a compressive strength corresponding to quartzitic sandstone or dense limestone. Such building projects can be a challenge in terms of homogeneity of material and continuity of operation. A construction site of a cooling tower in Niederaußem/Germany required 17,650 m$^3$ UHPC permanently delivered in a constant quality of aggregates and mixture during several weeks without any interruption (Weber and Riechers 2003).

UHPC is a material providing a new level of stability and strength compared to regular concrete. This offers the possibility to produce products using UHPC that were formerly produced from metal. An example is given by Niemann et al. (2008) in the German patent application DE 10 2007 016 719 A1 concerning calendar rolls in the production of paper and cardboard. Even compressive strength values between 150 and 200 N/mm$^2$ can be reached by using Reactive Powder Concrete (RPC).

The requirements for the use of sand and gravel in concrete are mandatory regulated by the European standard DIN EN 12620. This standard as well as the German standard DIN 1045-2 must be considered in Germany together with some other regulations. Amongst them, the "Alkali-Richtlinie" is of outstanding importance. Many damages on German highways (Autobahnen) were caused by the long term reaction of alkaline elements released from the cement together with reactive silica from sand and gravel (ASR—alkali-silica reaction). It is the aim of this regulation to prevent further damages based on this reaction.

These standards and regulations restrict the use of sand and gravel in concrete. They are generally available and can not be reduced down to a few paragraphs and a single table. Therefore, this chapter will not describe the quality requirements of aggregates for the use in concrete.

### *3.2.6 Mortar and Render Industry*

Mortar is normally used to connect the blocks within masonry. Traditionally, horizontal and vertical joints are solidified with mortar of about 12 mm thickness. Modern masonry quite often displays mortar only in the horizontal joints and, beyond that the thermal conductivity of the mortar is reduced by lowering its raw density. This is achieved by the reduction of the quartz content in the mortar. The heat flow across the mortar joint can also be reduced by decreasing its thickness. Modern production procedures allow block tolerances below 1 mm and the mortar layer can be reduced to 1–2 mm. Coarse sands used in former times are not suitable for such thin bed mortars. Sand composed of well rounded quartz grains instead of broken fragments assures the smoothness of the fresh mortar and is a prerequisite for the mortar sliding technique.

Render is also prepared using quartz sand. However, in this case well rounded grains are now a disadvantage. Splintery broken grains display a much better support of the fresh mortar structure, thus the use of natural quartz sand is limited. Nevertheless, quartz sand and broken limestone are equal in this application. Coarse quartz grains cause structural effects in the render and special products contain grains up to 6 mm size.

Floor pavements on the basis of cement also belong to the mortar group.

The requirements of this industry in context of grain size distribution are manifold but for the single product quite detailed. The relevant European standard is DIN EN 13139. The preparation of mortar and render on site by the craftsmen mixing sand and cement by volume in a mortar mixer is decreasing year by year. Plants producing premixed dry mortar and render have been established also in Eastern Europe.

Sand fractions used in render and mortar are characterized by narrow grain size distributions like 0.3–0.5 mm, especially in thin bed mortar. Mortar layers on exposed masonry have a normal thickness of 12 mm and require grains of a size up to 4 mm. Sand fractions used for render are rather fine grained, render with structural effects contains several grains of 2–3 mm diameter, sometimes even grains of 6 mm diameter may be used. Floor screeds require sand with a main fraction of 0–8 mm.

However, all of these materials must be dried and should have moisture contents clearly below 1 wt%. The material should be free of clay minerals because they will cause high rates of shrinkage resulting in cracks.

A special requirement of this industry is the increasing demand on very light coloured white sands. These very pure sands are not only used for render, but are requested increasingly for white mortar. Market leader of white masonry blocks demand white mortar for combined supply with the blocks.

No tabulated values were given, because of the wide variety of requirements and the existence of a standard for the raw materials.

## 3.3 Recent Developments in Mining and Processing

Two opposing processes are currently noticeable in the economic system. There is the increasing demand of well processed raw materials with improved and very stable production characteristics at one side. This leads to highly advanced processing plants and technological progress. On the other side, the heavy burden of cost imposed on industry results in the use of simply processed raw materials. Furthermore, due to the economic crisis in the construction industry, the demand of raw materials has decreased in some regions. As costs for truck transport have decreased because of free logistic capacities, high quality raw materials can be transported over large distances. Thus, medium quality sand producers are faced with growing difficulties and geological parameters of the deposit gain importance. At present times, it is uneconomic to process low quality source materials to produce mediocre products.

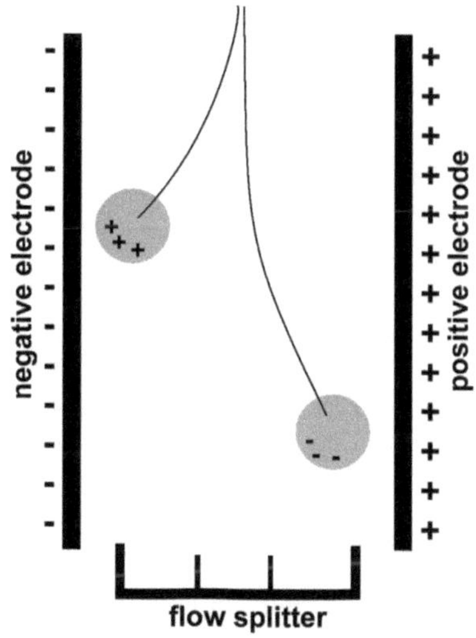


Fig. 3.3 Principle of sorting in a static electrical field

### 3.3.1 Processing of Quartz Sand

The "traditional" treatment processes are generally known. They cover washing to remove the clay fraction, dry or wet sieving and mineralogical separation by magnetic sorting or flotation. These methods are well described in several textbooks, e.g. Schubert (1986, 1989, Ney (1986)).

However, a well known technique seems to be coming up again: sorting in a static electrical field (Fig. 3.3). Traditionally, this method was mainly used in separation of heavy minerals, in the processing of iron ores and for the separation of quartz and feldspar (Schubert 1986). In former times this separation method required the quartz grains to be activated using hydrofluoric acid (HF). The favoured technique now is the use of the pyroelectric properties of the quartz grains. Heating of the sand shows a similar effect as activation by HF. This method is relatively simple and can be used for processing fine sand. Ideally, the sand is washed and has a grain size below 0.6 mm (30 mesh). The sorting equipment should be positioned directly after the sand dryer because the activation of the grains requires temperatures of 100–150°C.

The Outotec company in the USA produces this kind of sorting equipment. Available devices have a capacity of 6–8 t/h using 1.5 m working width. However, no regular use of the equipment could be named by the general agent of the company in Germany.

A test of this method was successfully performed in the pilot plant station of the producer. A low quality quartz sand was treated containing 64 wt% quartz, 10 wt% feldspar, 21 wt% calcite + dolomite and 4 wt% of mica + clay. The result of one test series is given in Fig. 3.4. A commercial device does not use 12 slots but only one splitter dividing output from losses. However, the quartz content

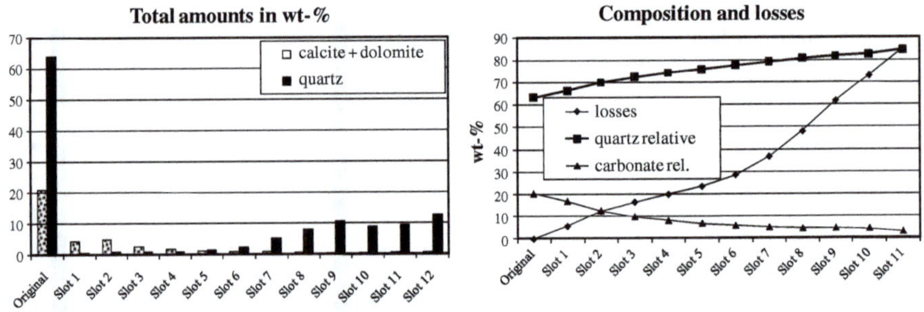


**Fig. 3.4** *Left*: total amounts of carbonates and quartz in the single fractions. *Right*: slot 1 until the named slot were discarded, contents of the processed material and theoretical losses

**Fig. 3.5** Feeding material for autogenous milling; coarse fraction *left* and fine fraction (wet) *right*

in the dry processed material was increased from 64 to 76%, accepting losses of ¼. This was achieved without the use of any chemicals.

### *3.3.2 Tendencies in Grinding of Quartz*

Ball mills, like in cement industry, are used for the grinding of quartz sand. Steel balls destroy the grains but introduce iron as impurity into the product. In cases where quartzite was used as source material, so called autogenous milling was performed for decades. Autogenous milling implies that quartzite boulders act like crushing balls and destroy the small pebbles. This technique has two main advantages: First, there are no costs for grinding balls and second, no impurities from the balls become admixed.

Autogenous milling requires close sampling of the product and intermittent feeding of the mill by finely crushed quartzite and coarse boulders respectively (Fig. 3.5). As soon as the final product starts to become comparatively fine, new grinding boulders have to be added. This results in an increase of the coarse fraction within the fine material. The main disadvantage of this method is that it is not possible to achieve a constantly fine grained product, which is the aim of milling. Step by step the milling systems are now changed into ball mills.

Some applications, for example in the chemical industry, require low iron contents in the milled sand. Consequently, steel balls must not be used in grinding. Alternatively, balls from corundum or even $ZrO_2$ are used. Very high requirements concerning the chemical purity of quartz sand require grinding with silica balls, normally flint balls. These "grinding balls" are commonly no spheres but rather ellipsoids. This use of flint in the mill is also called autogenous milling, because of the chemical and mineralogical identity of quartz sand and flint—however, "true autogenous milling" requires the same quartzite for grinding-boulders and fine material as well.

Two main principles are applied in grinding: dry and wet milling. For dry milling, the material has to be dried before the grinding process. After grinding, the fine material can be stored in silos and can easily be shipped. Wet milling saves the drying energy but instead of powder a slurry is produced which has to be stirred continuously. Wet mills are normally plated with rubber and are not as noisy as steel plated dry mills. Furthermore, there is no quartz dust problem. Main disadvantage is the necessity of stirring the slurry and the impossibility of shipping. The ground slurry has to be used on site.

The finest material can be achieved by wet milling. But use of economically prized steel balls for fine grinding introduces a new problem. If ground to long, the metallic iron abrasion oxidizes and takes the oxygen out of the water releasing hydrogen. Consequently, only balls from corundum or $ZrO_2$ can be used.

## 3.4 Quartz Sand Deposits for the Use in Construction Industry

Northern regions of Germany and Poland as well as Belgium and the Netherlands are covered with glacio-fluvial sand deposits of extraordinary sizes. The construction industry there is well positioned to acquire sufficient amounts of raw material. Simple processing by washing and sieving provides the sand quality needed. Sometimes, only dry sieving is sufficient. However, there is a huge demand for gravel, mainly of the concrete industry. Unfortunately, these sediments contain only small amounts of gravel frequently containing flint. This restricts their use as concrete aggregate, because of the ASR, causing long term destruction of the product.

The German glass sand deposits of Tertiary age are purified in processes connected with the formation of lignite. Most minerals are dissolved by humic acids and only the stable quartz remained, as frequently described (Wopfner 1983; Götze and Blankenburg 1992).

**Fig. 3.6** Quartzite quarry from SIBELCO near Robilante/Italy

**Fig. 3.7** Underground mine for sand near Pula/Croatia in 2000

Southern Europe was affected by the Alpine orogenesis causing the formation of numerous mountains and massive plains of debris. Neither valleys in mountain regions do exhibit quartz sand deposits nor river plains in their surroundings. If sand deposits exist, their phase composition is controlled by the weak source rocks in the mountains and by weathering products in the river plains. Only deposits of Tertiary and older age contain suitable sand. The efforts to produce high silica products are immense; for instance the company Sibelco mines massive quartzite in the northwest of Italy, close to the French border (Fig. 3.6). The material has to be crushed, ground and purified by grain size fractionation and magnetic sorting before selling. Even glass sand quality can be produced in that way.

Istria is dominated by massive lime sediments but containing one sand layer of up to 5 m thickness, which is sometimes solidified by calcareous cement. This sand material was mined, processed by flotation and exported from Yugoslavia to Italy. The raw material was mined in the last underground mine of Croatia until 2000 (Fig. 3.7). After the political change, the material was not flotated any longer but only broken and solely used for the production of AAC in the YTONG factory of Pula.

**Table 3.9** Use characteristics in different branches

| Physical use | Physical and chemical use | Chemical use |
|---|---|---|
| Clay bricks | Calcium silicate units | Autoclaved aerated concrete |
| Concrete products | Ceramic products (pottery) | Cement |
| Foundry sand | | Chemicals |
| Grit | | Glass |
| Mortar and render | | |

## 3.5 Conclusions

The different scenarios of usage within individual branches call for different raw material characteristics. As long as the grain is used in its original state, its size and shape is a characteristic feature. The grain size is of minor importance, if the sand is ground and used chemically. An overview about the characteristic uses is given in the following Table 3.9.

Quartz sands are valuable raw materials for the building and construction industry. The individual branches demand very different sand qualities and the requirements grow. Consequently, there is a strong need for information concerning potential raw materials. The acquisition of this data exceeds the capability of individual producers. Any data about mineralogical and chemical composition as well as granulometric data about sand deposits and occurrences are worthy of publication.

**Acknowledgments** I would like to thank numerous colleagues for their help, information and stimulating discussions. Special support was given by W. Eden, U. Jakobs, M. Kanig, W. Krcmar, H.-J. Riechers, T. Schoch, C. Wertel and, last but not least, H. Wopfner. Comments by T. Götte and an anonymous reviewer led to a considerable improvement of the original manuscript.

## References

Anonymous (1985) Gemeinsamem Runderlass vom 23.09.1985 des Umwelt- und des Wirtschaftsministeriums NRW

Anonymous (2011) Verwendung von Kies und Sand im Überblick. http://www.bks-info.de/images/ks-baust/verwendung_ueberblick.gif. Accessed 30 May 2011

Bertran A (1995) DE 43 39 916 A1, published 24.05.1995, Zusammensetzung zur Herstellung von Baustoffen mit schalldämmender Wirkung

Blankenburg H-J, Götze J, Schulz H (1994) Quarzrohstoffe. Leipzig and Stuttgart

DAfStb Alkali-Richtlinie 2007-02 Vorbeugende Maßnahmen gegen schädigende Alkalireaktion im Beton (Alkali-Richtlinie)—Teil 1: Allgemeines—Teil 2: Gesteinskörnungen mit Opalsandstein und Flint—Teil 3: Andere alkaliempfindliche Gesteinskörnungen

DIN 1045-2 2008-08 Tragwerke aus Beton, Stahlbeton und Spannbeton- Teil 2: Beton—Festlegung, Eigenschaften, Herstellung und Konformität—Anwendungsregeln zu DIN EN 206-1

DIN EN 12620 2008-07 Gesteinskörnungen für Beton, Deutsche Fassung EN 12620:2002 + A1:2008

DIN EN 13139:2002-08 Gesteinskörnungen für Mörtel, Deutsche Fassung EN 13139:2002

DIN EN 13139:2011-03 Gesteinskörnungen für Mörtel, Deutsche Fassung prEN 13139:2011

Eden W, Geisenhanslüke C, Glaubitt A, Middendorf B, Schmidt M (2008) Verdichtung von Kalk-Sand-Rohmassen, Forschungsvereinigung Kalk-Sand eV. Forschungsbericht Nr. 107, Hannover

Fuller WB, Thomson SE (1907) The laws of proportioning concrete. Trans Am Soc Civ Eng 59:67–143

Götze, J. and Blankenburg, H-J (1992) The genesis of the Hohenbocka quartz sand (Eastern Germany)—new results of mineralogical and geochemical investigations. Neues Jahrbuch für Geologie und Paläontologie Teil I, H. 11/12:pp 1217–1231

Gundlach H (1973) Dampfgehärtete Baustoffe. Bauverlag, Wiesbaden and Berlin

Kendel F (1973) Die Eigenschaften großformatiger Kalksandblöcke, Forschungsvereinigung Kalk-Sand eV. Forschungsbericht Nr. 35, Hannover

Lorenz W, Gwosdz W (1999) Bewertungskriterien für Industrieminerale, Steine und Erden Teil 3: Quarzrohstoffe. Geologisches Jahrbuch Reihe H. Heft 6, Hannover

Ney P (1986) Gesteinsaufbereitung im Labor. Stuttgart

Niemann J, Krüger L, Conrad H-R, Wiemer P, and Schnyder E (2008) DE 10 2007 016 719 A1, published 09.10.2008, Walze

Quincke JE, Eden W (2002) Handbuch zur Herstellung von Kalksandsteinen, Hannover

Schubert H (1986) Aufbereitung fester mineralischer Rohstoffe Bd. 2. Leipzig

Schubert H (1989) Aufbereitung fester mineralischer Rohstoffe Bd. 1. Leipzig

Weber R, Riechers H-J (2003) Kies und Sand für Beton. Düsseldorf

Wopfner H (1983) Environment of silcrete formation: a comparison of examples from Australia and the Cologne Embayment, West Germany. In: Wilson RCL (ed) Residual deposits. Special Publication No. 11, Geological Society of London, London, pp 151–158

# Chapter 4
# Petrological and Chemical Characterisation of High-Purity Quartz Deposits with Examples from Norway

**Axel Müller, Jan Egil Wanvik and Peter M. Ihlen**

**Abstract** Demand for high-purity quartz (HPQ) is strongly increasing worldwide owing to growing consumption and an increasing range of high-technology applications. This study includes: (1) a refined definition of HPQ (2) a discussion of the impurities controlling the chemical quality of HPQ products and (3) descriptions of selected HPQ deposits in Norway, both economic and potentially economic examples. The suggested definition of HPQ proposes concentration limits for the most important detrimental elements. The maximum content of each element should be: Al <30 μg $g^{-1}$, Ti <10 μg $g^{-1}$, Na <8 μg $g^{-1}$, K <8 μg $g^{-1}$, Li <5 μg $g^{-1}$, Ca <5 μg $g^{-1}$, Fe <3 μg $g^{-1}$, P <2 μg $g^{-1}$ and B <1 μg $g^{-1}$ whereby the sum of all elements should not exceed 50 μg $g^{-1}$.Impurities within quartz crystals (intracrystalline impurities) control the quality of HPQ products because they cannot be removed by conventional processing. These impurities include (i) lattice-bound trace elements, (ii) submicron inclusions <1 μm, and (iii) mineral and fluid micro inclusions (>1 μm). Present knowledge about intracrystalline impurities in natural quartz is described. The methods used here for identification and analysis of impurities are backscattered electron (BSE) and cathodolumine-sence (SEM-CL) imaging and laser ablation inductively coupled plasma mass spectrometry (LA-ICP-MS). The HPQ deposits discussed include the Melkfjell quartzite, several kyanite quartzites, the Nedre Øyvollen pegmatite and the Kvalvik, Nesodden and Svanvik hydrothermal quartz veins. The discussion focuses on the content of lattice-bound trace elements and the micro-inclusion inventory because these are the major parameters which determine the quality of HPQ products. Finally, processes leading to HPQ formation are discussed.

---

A. Müller (✉) · J. E. Wanvik · P. M. Ihlen
Geological Survey of Norway, Sluppen 6315,
7491 Trondheim, Norway
e-mail: axel.muller@ngu.no

J. Götze and R. Möckel (eds.), *Quartz: Deposits, Mineralogy and Analytics*,
Springer Geology, DOI: 10.1007/978-3-642-22161-3_4,

## 4.1 Introduction

High-purity quartz (HPQ), which is generally defined as quartz containing less than 50 μg $g^{-1}$ of contaminating elements (Harben 2002; Fig. 4.1), is a valuable commodity used in a wide range of high-technology products. Demand for HPQ is increasing strongly due to the rapid development and expansion of the HPQ-consuming industry. Security of supply necessitates the identification and characterization of new HPQ deposits in more countries, particularly in Europe, possibly including deposits of a different kind compared to those currently in production. Prerequisites for developing exploration tools for such deposits are the application of state-of-the-art microanalytical methods for appropriate petrological and chemical characterisation of potential deposits in order to achieve a better understanding of the environment and conditions of HPQ formation.

Industrial HPQ products are commonly delivered as processed fine-grained sand. Therefore, all types of medium- to coarse-grained quartz-rich (>20 wt.%) rocks possess, theoretically, the potential for being HPQ deposits, given that the deposit is large enough (>80,000 t; Fig. 4.1) and the other minerals of the rock can be separated by conventional processing. Impurities within quartz crystals (intracrystalline impurities) are the principal control on the quality of quartz because they can be removed by processing to some extent only. The impurities which are discussed in the first part of this study include: (i) lattice-bound trace elements, (ii) submicron inclusions <1 μm, and (iii) mineral and fluid micro-inclusions (>1 μm). In addition, intercrystalline impurities may occur along grain boundaries in the form of mineral coatings or micro-crystals. This type of impurities are not discussed further because they can be eliminated if the raw material is crushed down to its average crystal size, allowing the removal of these impurities by processing.

High-purity quartz (HPQ) is often thought of as a relatively easily available commodity. In fact, HPQ is very rare in economic and near-economic quantities (>80,000 t), but at microscopic scale (<1 mm) it is rather common and occurs as recrystallised and neocrystallised micro-domains (secondary quartz) within quartz crystals (e.g. Müller et al. 2000; Van den Kerkhof and Hein 2001; Van den Kerkhof et al. 2004; Müller et al. 2008a). Another reason for this misconception is uncertainty about the definition of HPQ. A revised definition of HPQ based on previous classifications by Harben (2002) and Müller et al. (2007) is therefore suggested in the second part of this study.

In the third part of the paper examples of potential economic and economic Norwegian HPQ deposits are introduced. Some of the deposits are in operation, others are under investigation. The deposits described represent a wide range of genetic environments, including hydrothermal, igneous and metamorphic settings. Attention is paid to the micro-inclusion inventories of the quartz and how they can be described, since micro-inclusions are the major contaminants of HPQ products. Finally, aspects of HPQ formation in the Norwegian deposits are discussed.

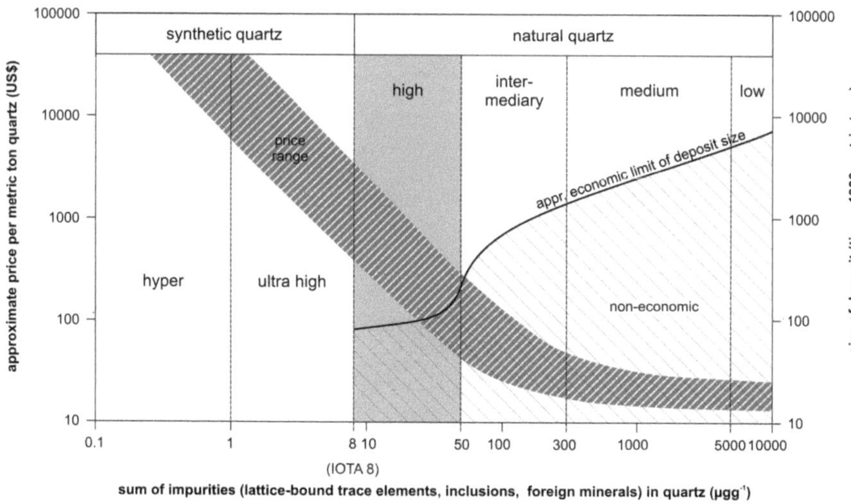


**Fig. 4.1** Classification of chemical quartz qualities (hyper to low) and their approximate price range according to Harben (2002). High-purity quartz is defined containing 8–50 µg $g^{-1}$ impuritites (*grey shaded field*). The lower limit of 8 µg $g^{-1}$ is represented by the quartz product IOTA 8 which is the puriest quartz product on the market produced from natural quartz (IOTA® 2011). The approximate economic limit of the deposit size bases on data of active quartz mines

## 4.2 Impurities in Quartz

### *4.2.1 Lattice-Bound Trace Elements*

Lattice-bound trace elements in quartz either substitute for $Si^{4+}$ ($Al^{3+}$, $Fe^{3+}$, $B^{3+}$, $Ti^{4+}$, $Ge^{4+}$, $P^{5+}$) or occur at interstitial channel positions ($Li^{+}$, $K^{+}$, $Na^{+}$, $H^{+}$, $Fe^{2+}$) where the ions act predominantly as charge compensators for the trivalent and pentavalent substitutional ions $Al^{3+}$, $Fe^{3+}$ and $P^{5+}$ (e.g. Weil 1984, 1993). Coupled substitution of $Al^{3+} + P^{5+} = Si^{4+} + Si^{4+}$, known as the berlinite substitution from feldspars (e.g., Simpson 1977), has been suggested to be the lattice configuration of P in natural quartz (e.g., Maschmeyer and Lehmann 1983). Recent work suggests that $B^{3+}$ may also substitute for $Si^{4+}$ (Thomas 2008; Müller and Koch-Müller 2009). Figure 4.2 illustrates the theoretical configuration of these trace elements in the quartz lattice. Based on the lattice configurations of these elements Dennen (1966) suggested that the atomic ratio of ($Al^{3+} + Fe^{3+}$) to ($H^{+} + Li^{+} + Na^{+} + K^{+}$) should correspond to 1 for some natural quartz crystals, thus achieving an approximately charge-neutral condition. Muller and Koch-Müller (2009) suggested a refinement of the hypothetical charge neutrality equation, where the atomic ratio of ($Al^{3+} + Fe^{3+} + B^{3+}$) to ($P^{5+} + H^{+} + Li^{+} + Na^{+} + K^{+}$) should correspond to 1 for natural quartz crystals. In theory, it is possible that electron defects, e.g. caused by vacancies, may also contribute to the charge balance. However, determinations of the ($Al^{3+} + Fe^{3+} + B^{3+}$)/($P^{5+} + H^{+} + Li^{+} + Na^{+} + K^{+}$) ratio in natural quartz

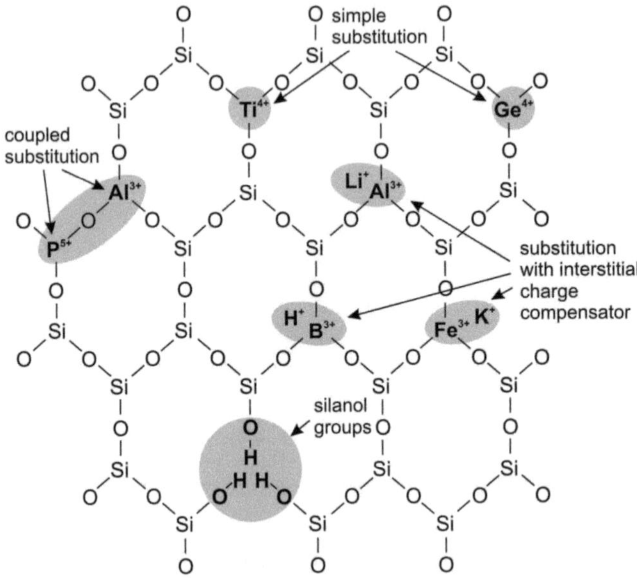


**Fig. 4.2** Schematic quartz structure showing the configuration of trace elements in the quartz lattice (modified from Götze 2009). McLaren et al. (1983) proposed the substitution of $Si^{4+}$ by four $H^+$ is possible (silanol groups). Because of the two-dimensional illustration the fourth $H^+$ is not shown on the figure

have shown that electron defects have only a minor or no contribution to the charge balance (Müller and Koch-Müller 2009). The equation is important for the determination of the chemical quality of quartz. Aluminium is the most common trace element in natural quartz, in concentrations of up to several thousand μg $g^{-1}$ and can thus be more easily determined by analytical methods than the other elements. If high Al concentrations are detected then concentrations of Li, K, Na and H will be high as well and possibly also the concentrations of B and P. The Al concentration in quartz is thus an important quality indicator.

Aluminium, Fe, and alkali metals tend to form minute atomic clusters which are incorporated along specific growth axes: this is especially true for low-temperature quartz (Pfenninger 1961; Flicstein and Schieber 1974; Siebers 1986; Ramseyer and Mullis 1990; Brouard et al. 1995). Investigations of hydrogen isotopes by Simon (2001) documented two H-reservoirs in quartz: (i) fluid inclusions and (ii) structurally-bound water in small, homogeneously distributed micro-clusters and bubbles. According to Simon's (2001) model, such micro-clusters should be accompanied by high $Al^{3+}$ and $K^+$, $Na^+$, $Li^+$ abundances. The possible configuration of such a cluster is illustrated in Fig. 4.3. The relatively high abundance of H-compensated Al defects and structurally bound water molecules observed by the Fourier transform infrared spectroscopy (FTIR) may indicate the existence of such micro-clusters (e.g. Müller et al. 2003a; Müller and Koch-Müller 2009).

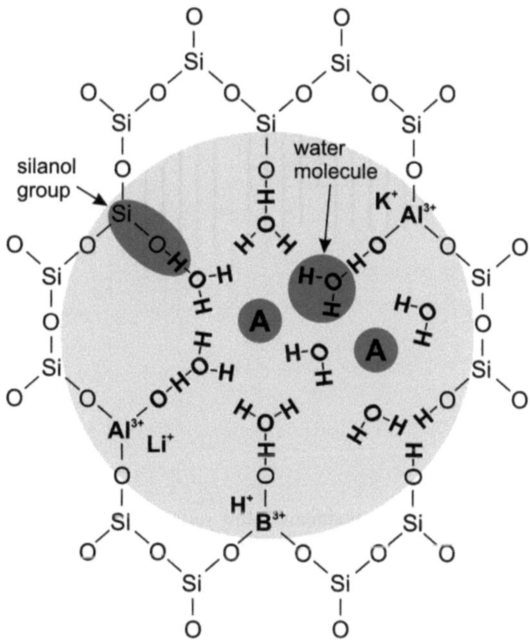


**Fig. 4.3** Theoretical configuration of atomic micro-cluster comprising predominantly Al, alkali ions and H in the form of $OH^-$ and molecular water. The A position can be occupied by hydroxyl-attracting ions such as $P^{5+}$, $B^{3+}$ and $Fe^{3+}$. Such a cluster would represent a transition between lattice defects and nanometer-sized inclusions (modified from Müller et al. 2003a)

A number of additional elements with concentrations >0.01 μg $g^{-1}$ have been detected in natural quartz (e.g. Gerler 1990; Blankenburg et al. 1994; Monecke et al. 2002). Figure 4.4 shows the average abundance of these elements. Hydrogen is the only common trace element which is not listed in Fig. 4.4 because limited data exist. Existing sources document H concentrations, bound as $OH^-$, in the range of 0.5–1,500 μg $g^{-1}$ (Bambauer et al. 1962; Luckscheiter and Morteani 1981; Hertweck et al. 2003; Miyoshi et al. 2005; Jourdan et al. 2009; Müller and Koch-Müller 2009). Hydrogen is not considered to be an important contaminant. However, the presence of structural water lowers the melting temperature of quartz which might be important in processing the quartz. The chart shown in Fig. 4.4 was introduced by Gerler (1990) and afterwards refined by Götze (2009) using data mainly from Blankenburg et al. (1994). In this study a further refinement is suggested, using 2,117 laser ablation inductively coupled plasma mass spectrometry (LA-ICP-MS) analyses from the data base of the Geological Survey of Norway (NGU). The analyses comprise quartz samples from all over the world and from all common geological environments, including sedimentary, diagenetic, hydrothermal, igneous and metamorphic settings. The LA-ICP-MS data reveal a lower average abundance of all determined elements (shown in *black*) compared to the original data provided by Gerler (1990) and Götze (2009) (shown in *grey*). State-of-the-art laser ablation equipment allows high-resolution optical control of the analysed quartz domain to avoid analysis of contaminating micro fluid- and mineral inclusions down to 0.5 μm. Conventional bulk quartz analyses applied by Gerler (1990) and Blankenburg et al. (1994 and data therein) such as atomic

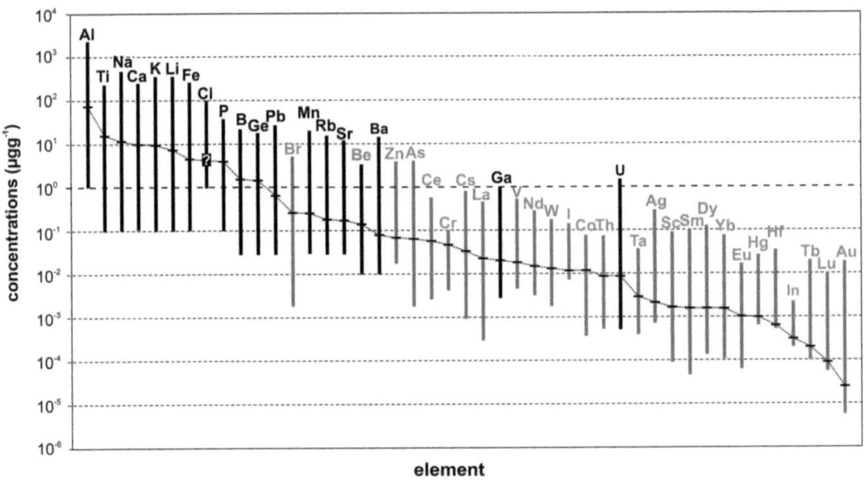


**Fig. 4.4** Average abundance and variations of trace elements in natural quartz. Data shown in *grey* are from Gerler (1990), Blankenburg et al. (1994) and Götze (2009). Data shown in *black* represent 2117 LA-ICP-MS analyses of quartz carried out at NGU over the last 6 years. Concentrations of Cl are uncertain due to the very high detection limit of about 100 μg $g^{-1}$. Cl is presented because Gerler (1990), Blankenburg et al. (1994), and Götze (2009) suggested that it occurs in considerable amounts in quartz

emission spectrometry (AES), atomic absorption spectrometry (AES), total reflection X-ray fluorescence (XRF), instrumental neutron activation analysis (INAA) and solution ICP-MS bear a higher risk of analyzing inclusions, thus superimposing additional concentrations to those of lattice-bound trace elements. However, it is assumed that concentrations of Na, K and Ca are still overestimated because these are the most common trace elements in fluid inclusions. Concentrations of Cl in quartz remain uncertain due to the high detection limit of about 100 μg $g^{-1}$ for the LA-ICP-MS. The abundance of Cl shown in Fig. 4.4 is based on data by Gerler (1990), Blankenburg et al. (1994), and Götze (2009), who suggested that Cl occurs in considerable amounts in quartz. Elements, for which a structural incorporation in the quartz lattice have not been proven, occur presumably as inclusions (Götze et al. 2004 and reference therein).

With modern in situ micro beam techniques such as electron probe micro analysis (EPMA), secondary ion mass spectrometry (SIMS), LA-ICP-MS and particle-induced X-ray emission (PIXE) (e.g., Bruhn et al. 1996; Flem et al. 2002; Müller et al. 2003a) the analysis of micro-inclusions (1,000–1 μm) can be avoided by optical control of the analysed quartz domain prior to, and during analysis. However, sub-micron (1–0.1 μm) and nano-inclusions (<0.1 μm) cannot be identified by these methods and, thus, the superimposition of concentrations of lattice-bound trace elements by concentrations of elements bound in inclusions cannot be completely excluded.

Concentrations of trace elements are mainly controlled by the crystallization temperature (e.g., Wark and Watson 2006), the extent of melt fractionation (for igneous and pegmatitic quartz; e.g., Müller et al. 2002a; Larsen et al. 2004; Breiter and Müller 2009; Jacamon and Larsen 2009; Müller et al. 2010a; Beurlen et al. 2011; Fig. 4.5) and the chemistry and acidity of the fluid from which the quartz crystallized (for diagenetic and hydrothermal quartz; e.g., Rusk et al. 2008; Jourdan et al. 2009; Müller et al. 2010b). Metamorphism, metasomatism and fluid-driven alteration, causing the recrystallisation and neocrystallisation of secondary quartz by replacing primary quartz, commonly cause the redistribution of incorporated trace elements (e.g., Van den Kerkhof et al. 2004; Müller et al. 2008a) and the formation of submicron inclusions of, e.g., mica-like composition (Müller et al. 2002b). The origin of submicron-scale rutile needles is considered to be the result of Ti exsolution from Ti-rich quartz (e.g., Frondel 1962, Meinhold 2010 and references therein). Post-crystallisation diffusion of ions may also occur. Significant diffusion has been observed for Fe (e.g., Penniston-Dorland 2001; Müller et al. 2005c; Fig. 4.6). Penniston-Dorland (2001) pointed out that the diffusion coefficient of Fe in quartz is probably much higher than $10^{12}$ $cm^2$ $s^{-1}$, and the migrating ion is presumably $Fe^{3+}$. However, Götze and Plötze (1997) showed that because of the relatively large ion radius Fe seems to be incorporated in the marginal parts of quartz crystals. Titanium, for example, has a much lower diffusion coefficient, and diffusion occurring in a million years would be of the order of 500 μm at 800°C and ~15 μm at 600°C (Cherniak et al. 2007).

### 4.2.2 Submicron and Nano-scale Inclusions

Quartz crystals may contain submicron-sized solid inclusions. The term "submicron" is used here for dimensions smaller than one micrometer, while the term "nano" refers to dimensions <100 nm. According to Seifert et al. (2011) the nanometre size is usually applicable only to one dimension, because the particles are generally needle-shaped or occur as thin strips. The lengths of nano-crystallites often exceed 100 nm, whereas diameters and thicknesses of needles and strips, respectively, correspond to dimensions <100 nm. Inclusions much larger than one micrometer, usually of zircon, apatite, monazite, feldspar, mica and other minerals, are considered as micro-inclusions or simply inclusions.

Only a few studies deal with submicron inclusions in natural quartz: these studies are predominantly focused on coloured quartz, in particular, blue igneous quartz (Parker 1962; Frazier and Gobel 1982; Zolensky et al. 1988; Bartovic and Beane 2007; Seifert et al. 2011). Therefore, it remains uncertain as to how common submicron inclusions in quartz in general are. Submicron inclusions observed so far include rutile, ilmenite, mica, tourmaline and Al–Si phases which could represent either $Al_2SiO_5$ polymorphs, AlOOH phases, or corundum (Zolensky et al. 1988; Seifert et al. 2011; Fig. 4.7). The study of blue quartz by Seifert et al. (2011) demonstrated that mica inclusions represent the largest group, but not the

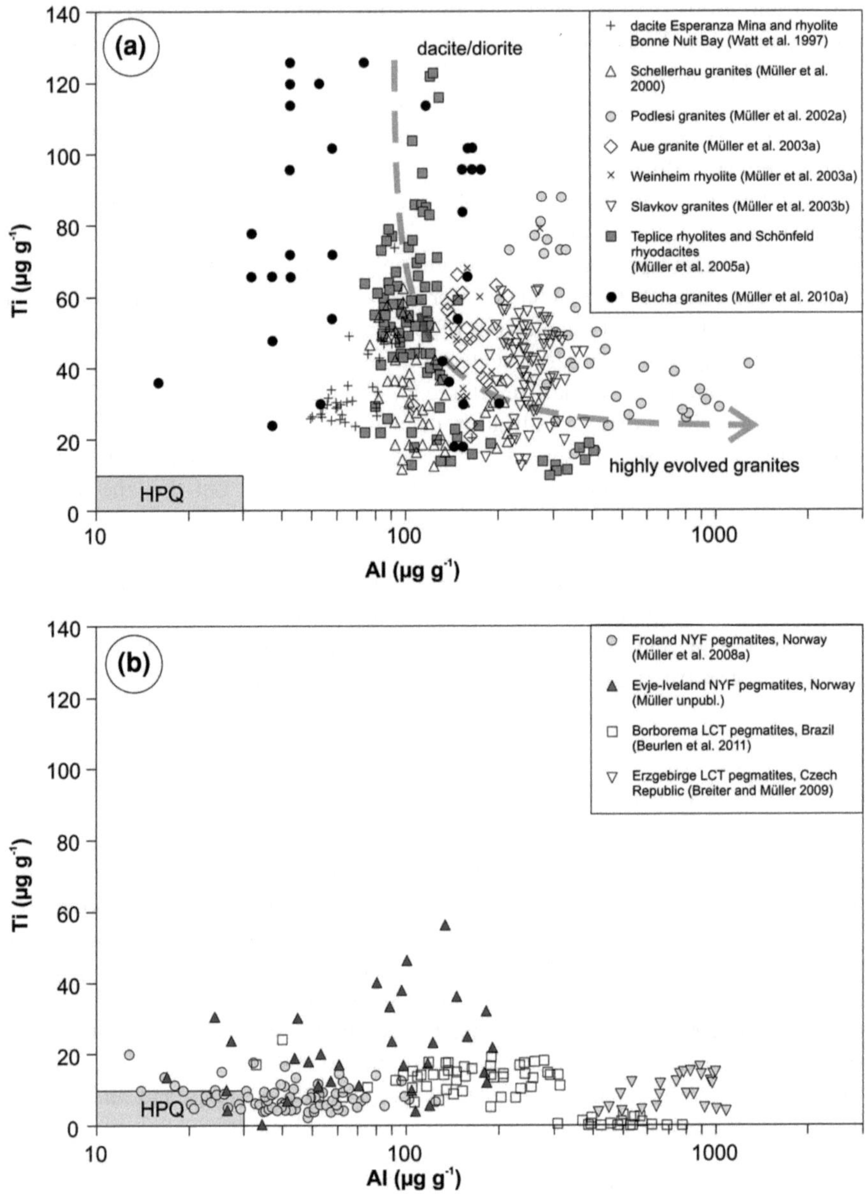


**Fig. 4.5** **a** Trace element evolution of quartz in granitic rocks. Quartz found in dacitic/dioritic rocks is characterised by high Ti and moderate Al. During further magmatic evolution Ti in quartz decreases and Al increases. Aluminium and Ti contents of igneous quartz plot far from the HPQ field (modified from Müller et al. 2010a). Data were acquired using EPMA and SIMS. **b** Compared to igneous quartz shown in **a**, pegmatite quartz has low Ti due to the generally low crystallization temperatures. Some of the Niobium-Yttrium-Fluorine (NYF)-type pegmatites plot in the HPQ field. Lithium-Cesium-Tantalum(LCT)-type pegmatites are characterized by high Al. Data were acquired using LA-ICP-MS

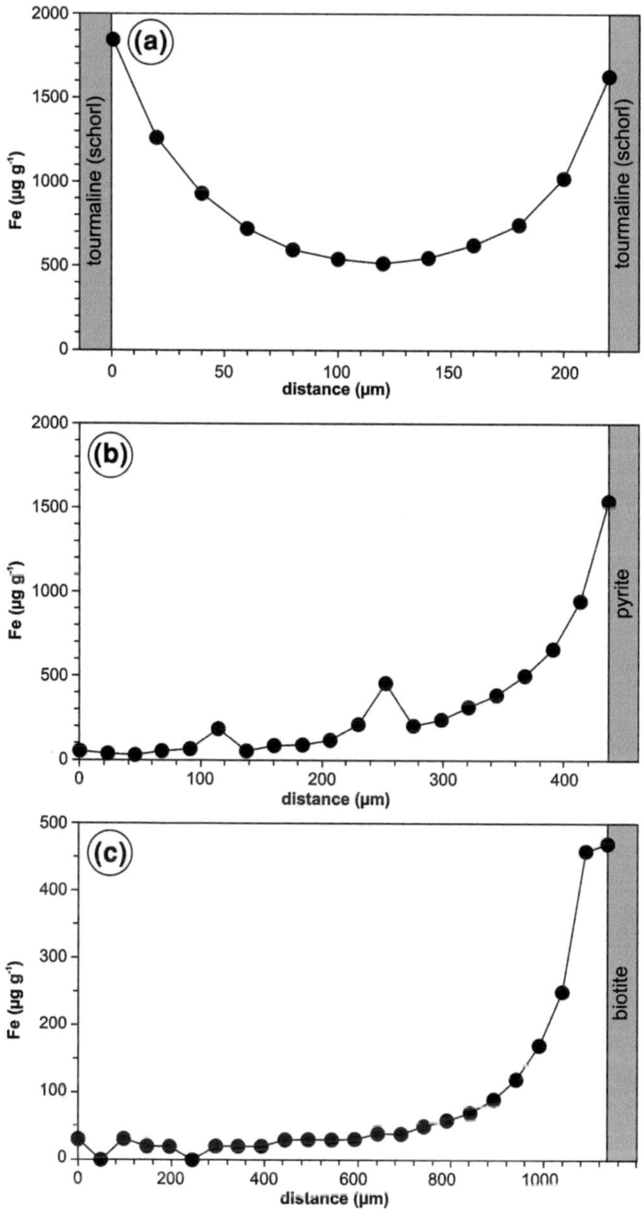


**Fig. 4.6** Iron concentration profiles of quartz crystals at the contact of Fe-rich minerals. The profiles illustrate that high Fe concentrations in quartz are predominantly diffusion-controlled. **a** Quartz crystal enclosed in tourmaline (schorl) (modified from Müller et al. 2005c). **b** Hydrothermal quartz in the contact with pyrite (modified from Müller et al. 2010b). **c** Igneous quartz in contact with biotite (modified from Müller et al. 2002b)

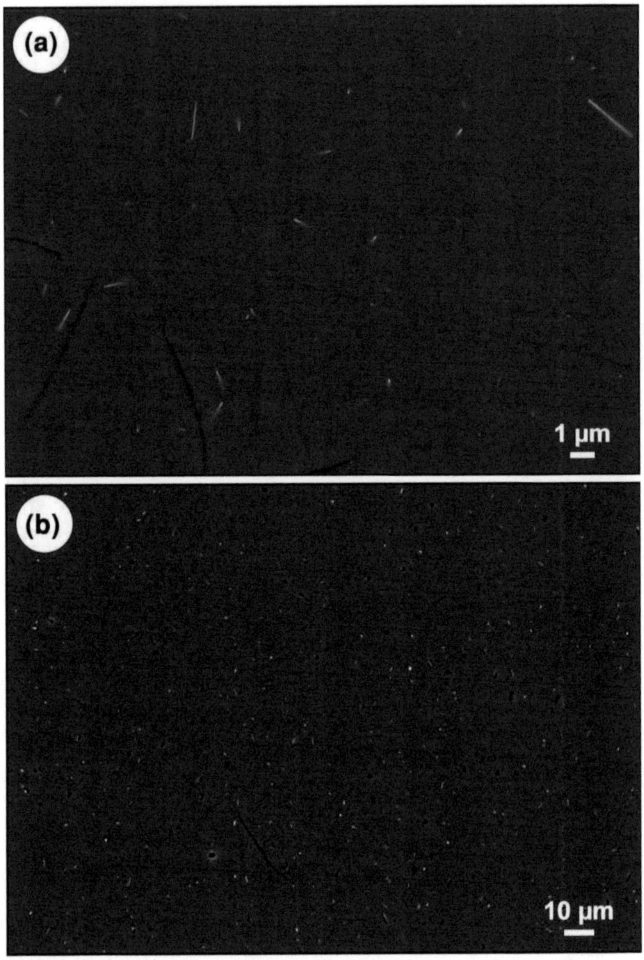


**Fig. 4.7** BSE images of submicron inclusions in quartz (with permission from Seifert et al. 2011). **a** Submicron rutile needles (whiskers) in blue quartz from Broken Hill, Australia. **b** Distribution of submicron inclusions mainly comprising mica and minor rutile in blue quartz from the Faja de Eruptiva Oriental rapakivi granite, NW Argentina. The density of submicrometre inclusions corresponds to 8,910 particles per $mm^2$

majority of submicron particles. The length of mica "needles" often exceeded 1 μm, while the width is normally <100 nm. The bulk of inclusions consist of needle- to ribbon-shaped ilmenite, on average ~500 nm in length and 50–100 nm across. The smallest identifiable inclusions are euhedral rutile crystallites of 25 × 50 nm in lowest dimension (50 × 150 nm on average). BSE-based counting of submicron inclusions (regardless of type) in blue quartz yielded 5,000–41,500 particles per square millimetre. TEM-based counting of nano-inclusions in several zoned blue-quartz grains resulted in 1.1–1.7 particles per $\mu m^3$ in the cores and 0.4–0.6 in the rims.

Müller et al. (2002b) interpreted Al–K spikes observed in EPMA profiles across quartz grains in deformed granites from the Lachlan Fold Belt, Australia, as submicron inclusions of mica-like composition. They concluded that multiple deformation of quartz caused the redistribution of Al and K in the quartz lattice, which results in the accumulation of these elements in submicron inclusions (<0.5 μm) of muscovite-like composition.

In summary, submicron inclusions may occur in high densities in quartz crystallised in specific environments: where these inclusions are present, they are the major source of contamination. Further systematic studies of submicron and nano-inclusions in quartz are needed.

### 4.2.3 *Mineral, Melt and Fluid Inclusions*

Natural quartz commonly contains micro-inclusions (>1 μm) of minerals, silicate melt and fluids. The kind and abundance of the inclusions depends on the crystallization environment and post-crystallisation alteration and deformation. If inclusions are present in large quantities they have a strong influence on the chemical quality of quartz raw materials. Because they are enclosed in quartz crystals they can be hardly removed by beneficiation processes.

Fluid inclusions which are by far the most common inclusions in quartz are formed during crystal growth (primary fluid inclusions) and also during the re-sealing of cracks in a pre-existing crystal (secondary fluid inclusions) when mineralizing fluids penetrate the quartz (e.g., Roedder 1984; Leeder et al. 1987; Van den Kerkhof and Hein 2001). Water is the most common fluid present, but carbon dioxide, methane, heavier hydrocarbons, and nitrogen can also occur. If the fluid carried high amount of dissolved material it may precipitate as daughter crystals during cooling to surface conditions. Halite is the most common daughter mineral, but many other salts and some silicate minerals also occur in the same way (e.g., Van den Kerkhof and Hein 2001). If bulk quartz samples are analysed by conventional methods Na, K, Cl and (Ca) from fluid inclusions are commonly the major contamination source. However, fluid inclusions and their dissolved components can be removed during the beneficiation process by thermal treatment and calcination (e.g., Haus 2005).

Silicate melt inclusions are small blebs (~1–300 μm) of silicate melt that are trapped within igneous and pegmatitic quartz. They are glassy or crystalline and relatively rare compared to fluid inclusions (e.g., Sheperd et al. 1985; Leeder et al. 1987; Frezzotti 2001; Webster 2006). They are hard to identify in intrusive rocks (granites and pegmatites) because they are crystallised and often overprinted or hidden by fluid inclusions. The composition of melt inclusions corresponds to the silicate melt composition at the time of entrapment, comprising the major elements Si, Al, Fe, Ca, Na and K. In pegmatitic quartz, melt inclusions can host significant concentrations of F, Cl, B, P, Li, Cs, and Rb, up to several weight percent (e.g., Thomas et al. 2006) and, thus can be the major source of contamination, e.g. of B and P in quartz raw materials produced from pegmatitic quartz.

In theory, all mineral phases which occur in the host rock may also occur as micro-inclusions in quartz. Mineral species included in igneous quartz are mainly feldspar, mica, rutile, zircon, apatite, Fe oxides, etc. (Roedder 1984; Leeder et al. 1987). The spectrum of mineral inclusions in metamorphic quartz depends on the conditions of metamorphism. Whereas mineral inclusions of chlorite, muscovite or amphibole are more characteristic of low-grade metamorphic rocks, kyanite, staurolite or garnet occur especially in high-grade metamorphic rocks. Inclusions of anhydrite, gypsum, polyhalite, calcite, several salt minerals, organic matter, etc. have been described in sedimentary authigenic quartz (e.g. Richter 1971; Fruth and Blankenburg 1992; Hyrsl and Niedermayr 2003; Götze 2012). The formation of mineral inclusions in quartz is manifold. They were enclosed during growth from melt and fluids or by solid-state grain boundary migration during metamorphism and subsequent crystal lattice recovery. The latter process seems to be very common in metamorphic rocks since mineral inclusions are most common in quartz of quartzites. Exsolution of rutile needles as a result of cooling or decompression of Ti-rich quartz is the third process leading to the formation of mineral inclusions (e.g., Adachi et al. 2010).

## 4.3 Quality Definition of High-Purity Quartz

The commercial definition of quartz qualities by Harben (2002), including high-, ultra-high- and hyper-qualities, is presumably the most widely used commercial definition (Fig. 4.1). His classification is based on qualities of processed quartz products available on the market. For the classification of high- and ultra-high-qualities he utilized the trace-element concentrations of the IOTA quartz products from Unimin (Jung 1992; IOTA 2011). According to Harben (2002) HPQ concentrates are designated to contain less than 50 μg $g^{-1}$ of contaminating trace elements.

As Fig. 4.4 illustrates, concentrations of different elements vary over several orders of magnitude. The commercial concentration requirements for critical elements which are detrimental to the HPQ products vary considerably as well. Thus, the definition of HPQ from a commercial point of view should be based on the concentration limits of elements which interfere with the quality of melted HPQ products, which are predominantly silica and quartz glass products. Most of the HPQ products are used in applications where the temperature reaches up to 1,400°C. Other applications utilise the excellent optical properties of silica glass. At high temperature, elements such as Na, K and Li diffuse easy through the glass and Al changes the viscosity of the glass. At high temperature Ca will cause unwanted crystal nucleation in the quartz glass over time. Ti reduces the UV-light transmission in silica glass. Oxides of heavy elements such as Fe, Mn, Cr, Co, Cu

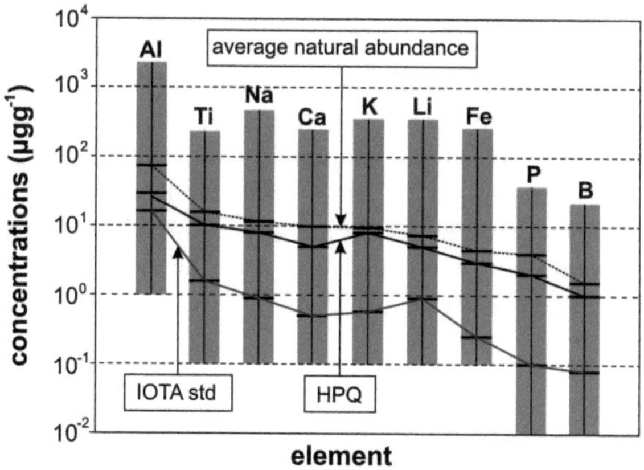


**Fig. 4.8** Average abundance and variations (*bars*) of nine detrimental trace elements in natural quartz compared with the suggested upper concentration limits of HPQ and the IOTA standard quartz (IOTA 2011)

and Ni give a colouration of the silica glass that reduces its transmission properties. Both P and B are unwanted in photovoltaic or semiconductor products manufactured from quartz. For these reasons upper concentration limits for the most critical trace elements should be included in a revised HPQ definition. The classification should be applicable not only to processed quartz raw material but also to unprocessed natural quartz in order to allow useful classification of potential quartz deposits. For the exploration of HPQ deposits it is important to characterise unprocessed quartz by in situ analysis of quartz crystals.

Müller et al. (2007) made a first attempt at HPQ classification based on the concentration limits of certain elements. They suggested upper concentration limits of 25 $\mu g\ g^{-1}$ for Al and 10 $\mu g\ g^{-1}$ for Ti. With the background knowledge mentioned above, together with trace element concentrations of HPQ products available on the market, a more comprehensive definition of HPQ is suggested. The sum of the nine elements Na, K, Li, Al, Ca, Fe, Ti, B and P analysed on quartz crystals (single grain analysis) or processed quartz sand (bulk product analyses) should be <50 $\mu g\ g^{-1}$. The maximum content of each element is suggested as: Al <30 $\mu g\ g^{-1}$, Ti <10 $\mu g\ g^{-1}$, Na <8 $\mu g\ g^{-1}$, K <8 $\mu g\ g^{-1}$, Li <5 $\mu g\ g^{-1}$, Ca <5 $\mu g\ g^{-1}$, Fe <3 $\mu g\ g^{-1}$, P <2 $\mu g\ g^{-1}$ and B <1 $\mu g\ g^{-1}$ whereby the sum of all elements should not exceed 50 $\mu g\ g^{-1}$. Figure 4.8 illustrates the suggested upper concentration limits of HPQ compared with average abundances in natural quartz and the concentrations of the IOTA standard of these nine elements. The proposed upper concentrations limits for Na, K and Ca are relatively high compared to their average abundance in natural quartz, because in processed quartz the concentrations are superimposed by "contributions" from fluid inclusions (containing NaCl and KCl), and mica and feldspar micro-inclusions (containing K, Na and Al) which are the most common intracrystalline impurities. Other elements that also occur in the HPQ products in quantities of up to a few $\mu g\ g^{-1}$ (e.g. Mg, Zr and Ge) are rarely given attention, because they have only minor effects on silica glass properties. The content of lattice-bond H might be several tens of $\mu g\ g^{-1}$ but is not included in the definition because there are no published requirements and because of the analytical challenges in quantifying the H concentration.

It should be mentioned that trace element concentrations in HPQ products do not necessarily represent hundred percent lattice-bound elements if processed HPQ sand is analysed, for example by solution ICP-MS. Portions of the element concentrations might originate from inclusions or other minerals which were not completely removed during processing. The analysis of inclusions or foreign minerals can be avoided by the application of in situ micro-beam techniques on single quartz crystals such as EPMA, LA-ICP-MS or SIMS.

## 4.4 Methods

### *4.4.1 Laser Ablation Inductively Coupled Plasma Mass Spectrometry (LA-ICP-MS)*

Concentrations of Li, Be, B, Al, P, K, Ti, Mn, Fe, and Ge were analysed in situ by laser-ablation inductively-coupled plasma mass spectrometry (LA-ICP-MS). The analyses were performed on the double-focusing sector field mass spectrometer model ELEMENT 1 from Finnigan MAT which is combined with the NewWave 193-nm laser probe. The laser had a repetition rate of 15 Hz, a speed of 15 $\mu m\ s^{-1}$, a spot size of 75 μm, and energy fluence of about 14 mJ $cm^{-2}$ on the sample surface. Raster ablation was applied in the centre of quartz crystals on an area of approximately 225 × 300 μm. The approximate depth of ablation was between 40 and 100 μm depending on the crystallographic orientation and absorption behaviour of the individual quartz crystals. The carrier gas for transport of the ablated material to the ICP-MS was He mixed with Ar. External calibration was performed using four silicate glass reference materials (NIST SRM 610, 612, 614, 616), the NIST SRM 1,830 soda-lime float glass, the BAM No.1 amorphous $SiO_2$ glass and the synthetic pure quartz monocrystal Qz-Tu. Certified, recommended and proposed values for these reference materials were taken from the certificates of analysis where available, or otherwise from the web site Geological and Environmental Reference Materials (GeoReM 2011). The isotope $^{29}Si$ was used as the internal standard. A linear regression model, including several measurements of the different reference materials, was used to define the calibration curve for each element. For the calculation of P concentrations, the procedure of Müller et al. (2008b) was applied. Ten sequential measurements on the "$SiO_2$ blank" crystal were used to estimate the limits of detection (LOD) which were based on 3 × standard deviation (3$\sigma$) of the ten measurements. LODs vary for each analysis sequence (measurement day). Examples of LODs are given in Table 4.2. The analytical error ranges within 10% of the absolute concentration of the element. More details of the measurement procedure are provided by Flem et al. (2002).

Petrographic examination of the samples using optical microscopy prior to analysis and optical control during laser ablation commonly allowed the selection of ablation areas in quartz crystal centres free of mineral and fluid micro-inclusions (>0.1 μm). However, quartz grains may be very rich in micro-inclusions or may have small grain size (<500 μm). In these cases micro-inclusions and impurities at grain boundaries may superimpose concentrations of lattice-bound trace elements.

### *4.4.2 Backscattered Electron and Cathodoluminescence Imaging*

Backscattered electron (BSE) and cathodoluminescence (SEM-CL) images of quartz and micro-inclusions in quartz were obtained from polished thin sections coated with carbon using the LEO 1450VP analytical SEM with an attached Centaurus BS Bialkali CL detector. The applied acceleration voltage and current at the sample surface were 20 kV and ~2 nA, respectively. The Bialkali tube has a CL response range from 300 (violet) to 650 nm (red). The detector sensitivity peaks in the violet spectrum range around 400 nm. The BSE and SEM-CL images were collected from one scan of 43 s photo speed and a processing resolution of 1,024 × 768 pixels and 256 grey levels. The brightness and contrast of the collected CL images were improved with PhotoShop software.

SEM-CL imaging reveals micro-scale (<1 mm) growth zoning, alteration structures and different quartz generations which are not visible with other methods. Grey-scale contrasts visualised by SEM-CL are caused by the heterogeneous distribution of lattice defects (e.g., oxygen and silicon vacancies, broken bonds) and lattice-bound trace elements (e.g., Götze et al. 2001). Although the physical background of the quartz CL is not fully understood, the structures revealed by CL give information about crystallisation, deformation and fluid-driven overprint.

## 4.5 Norwegian HPQ Deposits

### *4.5.1 The Melkfjell Quartzite*

*Location*. The Melkfjell quartzite is situated about 30 km SE of Mo i Rana in Nordland county, approximately 3 km south of Kaldvatn lake (Figs. 4.9, 4.10a). The quartzite is exposed at an altitude of 700–850 m.

*Geology*. The quartzite is part of the Kjerringfjell Group forming the base of the Caledonian Rødingsfjell Nappe Complex where layers of quartzite are present together with mica gneisses and mica schists. The Melkfjell quartzite extends about 7 km in a WSW-ENE direction and is generally 200 m wide. At its ENE end

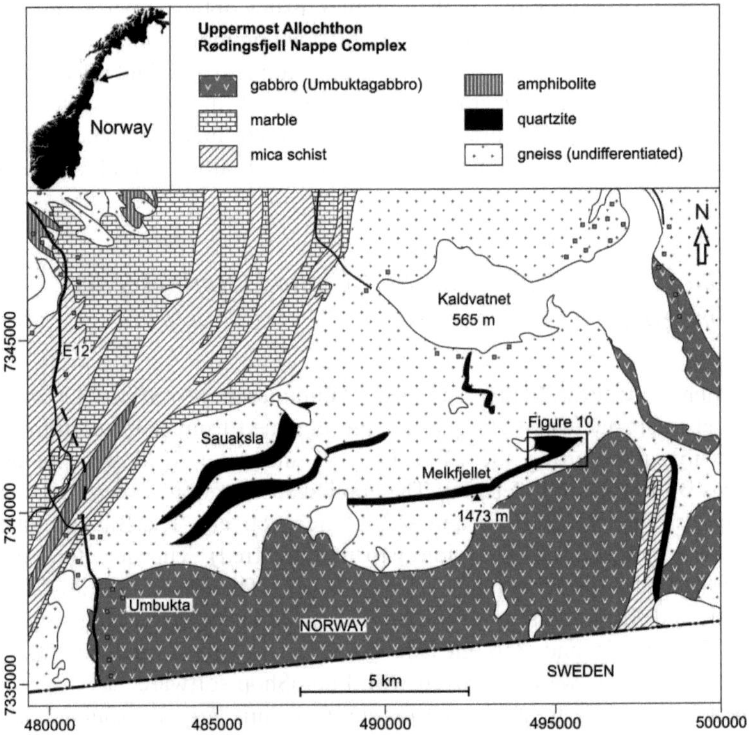


**Fig. 4.9** Simplified geological map of the Melkfjell area (modified from Gustavson and Gjelle 1991). The outline of the Melkfjell quartzite is according to Wanvik (2009)

the quartzite belt turns eastwards and widens to 600 m, forming the most interesting section from an economic point of view (Fig. 4.11). Several quartzite bodies of similar lithology and tectonostratigraphic position occur in the area (Wanvik 2001). Wanvik (2009) distinguished white-weathering and brownish-weathering quartzites. The latter is characterized by a higher Fe content. The average bulk composition of the Melkfjell quartzite is given in Table 4.1. The quartzite contains minor muscovite and accessory biotite, feldspar and graphite. The quartzite is commonly cross-cut by pegmatite and amphibolite veins. The amphibolite veins are between 0.1 and 10 m wide end extend up to lengths of 100 m. Irregular lenses and veins of pegmatite are 0.1–1 m in size and have quartz-dioritic to trondhjemitic compositions. The crystal sizes of quartz and feldspar are 1–2 cm. The pegmatites become dominant in the quartzite at the eastern end of the quartzite body, east of the Melkfjelltjønna (Fig. 4.11). The high pegmatite proportion makes this part of the Melkfjell quartzite uneconomic. However, several million tons of quartzite are present in the more pure and homogeneous parts of the deposit (Wanvik 2009).

**Fig. 4.10** **a** Outcrops of the Melkfjell quartzite in the foreground with Melkfjell to the left. View towards W. **b** Outcrop of kyanite quartzite at Juovvačorrú in Skjomen. The *bluish shade* of the rock is caused by fine-grained kyanite. **c** Contact of the massive quartz core and the feldspar-dominated wall zone of the Nedre Øyvollen pegmatite exposed in the underground mine. The short edge of the photograph corresponds to 2 m. **d** Exposure of the Nesodden quartz vein with the Hardangerfjord in the background. The *yellow dashed* line marks the outline of the Nesodden vein. View towards NE. **e** Outcrop of the Kvalvik quartzite with irregular, cross-cutting quartz veins of high-purity quality. **f** Temporary outcrop of the Svanvik quartz vein. The quartz block in the foreground is about 1 m high

*Quartz petrography.* The quartz crystals range in size from 10 to 2,000 μm with an average of about 500–800 μm. The quartz is strongly recrystallised and large grains show undulatory extinction. Inclusions in quartz are very common.

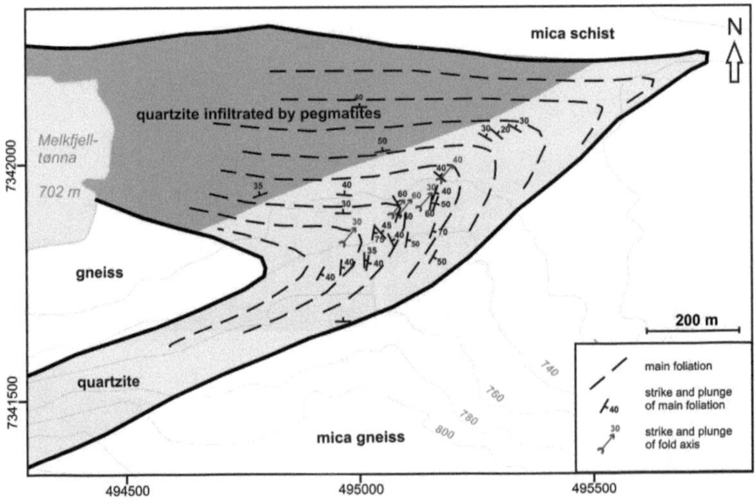


**Fig. 4.11** Detailed geological map of the Melkfjell quartzite according to Wanvik (2009)

**Table 4.1** Average whole rock composition of the Melkfjell quartzite according to Wanvik (2009)

| $SiO_2$ | $Al_2O_3$ | $Fe_2O_3$ | CaO | MgO | $Na_2O$ | $K_2O$ | $TiO_2$ | $P_2O_5$ |
|---|---|---|---|---|---|---|---|---|
| 98.05 | 0.67 | <0.41 | 0.02 | 0.11 | 0.09 | 0.11 | 0.02 | 0.02 |

The concentrations represent the average of 24 XRF analyses

The inclusions observed are (in order of frequency) biotite and muscovite (20–2,000 μm), albite (100–2,000 μm), plagioclase (20–2,000 μm), apatite (10–50 μm), rutile (<1–10 μm), monazite (10–100 μm), zircon (10–40 μm), pyrite (10–50 μm), Fe-oxides (5–20 μm) and arsenopyrite (10 μm; Figs. 4.12a, 4.13a). CL imaging revealed that grain-boundary migration is a widespread phenomenon, whereby grains with low CL intensity replacing large grains with higher CL intensity (Fig. 4.14).

*Quartz crystal chemistry.* The quartz of the Melkfjell quartzite is characterized by low concentrations of Li (mean 1.1 μg $g^{-1}$), B (mean 0.8 μg $g^{-1}$) and Al (mean 7.7 μg $g^{-1}$; Table 4.2; Fig. 4.15). The Ti concentration is moderately high due to high-temperature Caledonian metamorphism. Application of the Ti-in-quartz geothermometer by Wark and Watson (2006) reveals the peak metamorphic temperature of about 520 ± 8°C (Ti = 8.76 μg $g^{-1}$) which is somewhat higher than the maximum temperature of 506°C determined by Müller and Koch-Müller (2009). Ti saturation is assumed due to the common exsolution of rutile needles (Fig. 4.12). The analysis with the low Ti content of 2.09 μg $g^{-1}$ was from neo-crystallized quartz with low CL intensity, reflecting retrograde crystallization conditions. This low-Ti quartz shows also the lowest Al and Li contents.

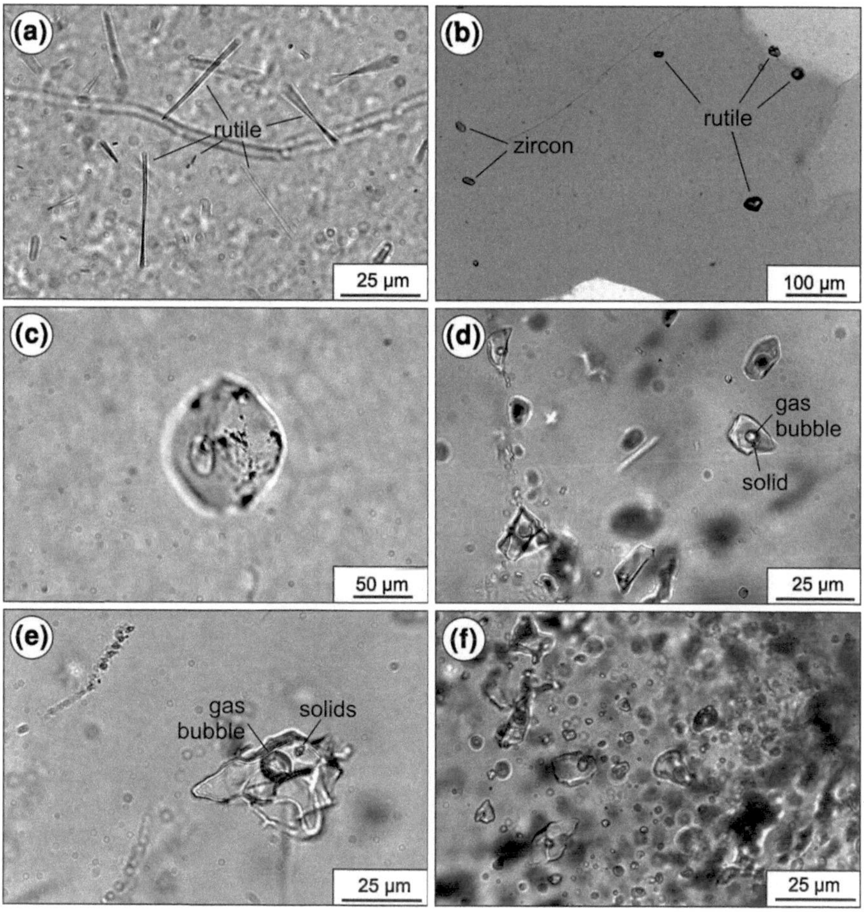


**Fig. 4.12** Optical microscope images of micro inclusions in quartz. **a** Rutile needles in quartz from Melkfjell. **b** Rutile and zircon inclusions in quartz from the Gullsteinberg kyanite quartzite. **c** Silicate melt inclusion in pegmatite quartz from Nedre Øyvollen. **d** Fluid inclusions with gas bubble and solids (presumably halite) in quartz from the Nesodden vein. **e** Fluid inclusion with gas bubble and solids (presumably halite and sylvite) in quartz from the Kvalvik deposit. **f** Liquid-rich fluid inclusions in quartz from the Svanvik vein

*Economic assessment and remarks on HPQ formation.* The common occurrence of rutile needles in the Melkfjell quartz will make it very hard to achieve a low Ti content in the final, processed product and, thus, only applications with low-quality requirements for Ti content are feasible. The common occurrence of apatite and monazite as interstitial accessories and inclusions in quartz will cause elevated P concentrations in the product, limiting possible photovoltaic applications. However, the quality of the quartzite is adequate for ferrosilicon applications if contaminating rocks of amphibolite and the pegmatite dykes are properly handled (Wanvik 2009). Apart from its Ti and P contents the quartz might be of HPQ

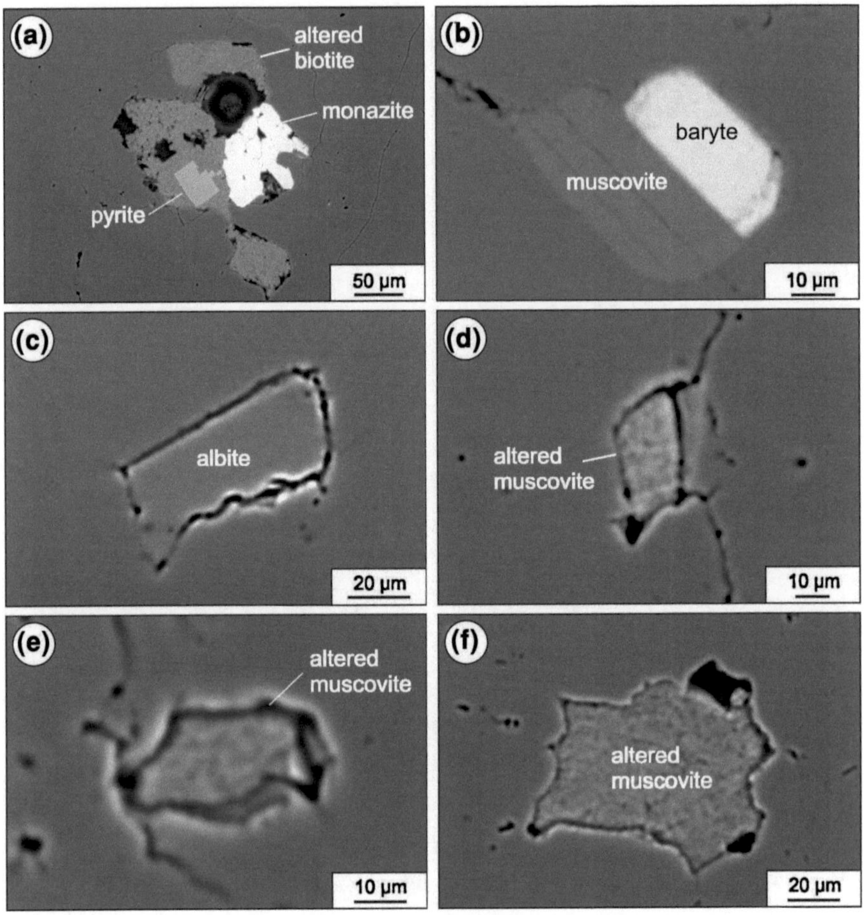


**Fig. 4.13** Backscattered electron images of micro inclusions in quartz. **a** Polyphase inclusion in quartz of the Melkfjell quartzite comprising weathered biotite, monazite and pyrite. **b** Muscovite and baryte inclusion in quartz of the Tverrådal kyanite quartzite. **c** Albite inclusion in the Nedre Øyvollen pegmatite quartz. **d** Inclusion of altered muscovite in the Nesodden quartz. **e** Inclusion of altered muscovite in the Kvalvik quartz. **f** Inclusion of altered muscovite in the Svanvik quartz

quality in a processed condition, because the content of lattice-bound Al and Li is exceptionally low (Fig. 4.15).

The metamorphic conditions of the Caledonian orogenesis produced quartz with very low Al and Li and moderate Ti. The protolith was relatively pure sandstone as indicated by the low content of minor and accessory minerals. However, some of the accessory minerals, such as monazite and arsenopyrite, are rather uncommon for quartzite. Müller et al. (2002b) showed that moderate pressure (~3–4 kbars) and temperatures >350°C cause the expulsion of Al and alkalis from the quartz lattice during metamorphic overprinting, resulting in low

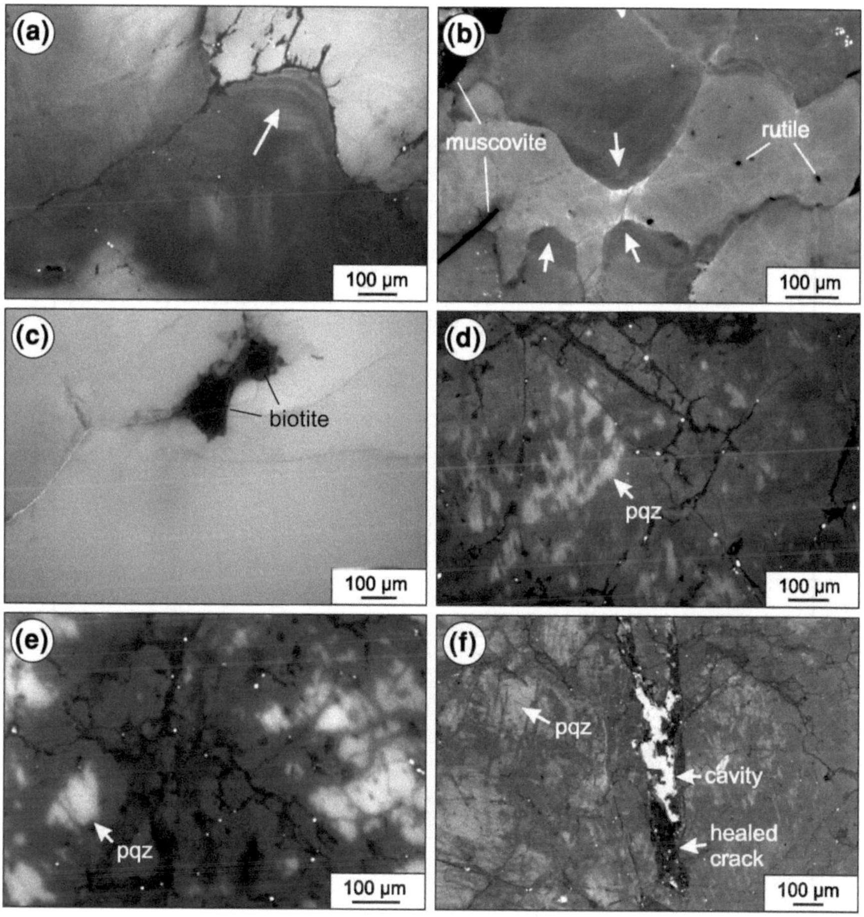


**Fig. 4.14** Cathodoluminescence images of quartz. **a** Dull luminescent, weakly zoned quartz grain bulged into and replaced bright luminescent quartz by grain boundary migration. The *arrow* indicates the growth direction. Melkfjell quartzite. **b** Dull luminescent, weakly zoned quartz grains replaced bright luminescent quartz by grain boundary migration. Tverrådal kyanite quartzite. **c** Biotite inclusions in bright and homogeneous luminescent quartz in the Nedre Øyvollen pegmatite. **d** Strongly overprinted quartz from the Nesodden vein. The bright luminescent areas are remains of the primary, unaltered hydrothermal quartz (pqz). The tiny, bright dots are inclusions of mica-like composition. **e** Strongly overprinted quartz from the Kvalvik deposit. The bright luminescent areas are remains of the primary, unaltered hydrothermal quartz (pqz). The white dots are tiny (<1 µm) mica inclusions. The CL structures are similar to those in the Nesodden quartz. **f** Strongly altered quartz from the Svanvik vein with relic domains of primary hydrothermal quartz (pqz). The altered quartz is cross-cut by a younger partially healed quartz vein

Al and Li contents in the newly formed quartz. The Melkfjell quartzite was exposed to pressures between 3 and 5 kbars (Stephens et al. 1985) and temperatures of 466–528°C (Müller and Koch-Müller 2009 and this study) and, thus, the

Table 4.2 Trace element concentrations (μg $g^{-1}$) of quartz crystals determined by LA-ICP-MS

| | Li | Be | B | Al | P | K | Ti | Mn | Fe | Ge |
|---|---|---|---|---|---|---|---|---|---|---|
| LOD (example) | 0.80 | 0.10 | 1.3 | 6.0 | 3.1 | 0.5 | 0.9 | 0.2 | 0.40 | 0.10 |
| *Melkfjell quartzite* | | | | | | | | | | |
| | <0.16 | 0.25 | 1.95 | 6.9 | 4.5 | <2.3 | 7.87 | 0.12 | <1.20 | 0.54 |
| | 1.18 | <0.13 | 0.72 | 11.2 | <1.8 | <2.3 | 8.76 | 0.32 | <1.20 | 0.81 |
| | 0.89 | <0.13 | 0.50 | <6.6 | 6.1 | <2.3 | 2.09 | 0.32 | <1.20 | 0.82 |
| | 2.19 | <0.13 | 0.47 | 7.1 | 2.5 | <2.3 | 5.59 | 0.34 | <1.20 | 0.99 |
| | 1.25 | <0.13 | 0.35 | <6.6 | 3.4 | <2.3 | 7.75 | 0.19 | <1.20 | 1.00 |
| | (0.51) | (0.21) | (0.43) | (25.3) | (3.3) | (<2.3) | (28.86) | (0.11) | (<1.20) | (0.70) |
| Average | 1.13 ± 0.73 | <0.15 | 0.80 ± 0.66 | <7.7 | <3.7 | <2.3 | 6.41 ± 2.68 | 0.26 ± 0.1 | <1.20 | 0.83 ± 0.19 |
| *Kyanite quartzites* | | | | | | | | | | |
| Gullsteinberg | 0.81 | <0.10 | <1.3 | 25.2 | <3.1 | 1.0 | 10.49 | <0.20 | <0.40 | 0.29 |
| | <0.8 | <0.10 | <1.3 | 18.6 | <3.1 | <0.5 | 10.22 | 0.24 | <0.40 | 0.24 |
| | <0.8 | <0.10 | <1.3 | 27.7 | <3.1 | 0.7 | 9.64 | <0.20 | <0.40 | 0.27 |
| Average | <0.80 | <0.10 | <1.3 | 23.8 ± 4.7 | <3.1 | <0.7 | 10.12 ± 0.44 | <0.21 | <0.40 | 0.27 ± 0.03 |
| Skjomen | 1.84 | <0.10 | <1.3 | 13.8 | 3.8 | <0.5 | 3.55 | <0.20 | <0.40 | 0.40 |
| | 1.80 | <0.10 | <1.3 | 29.2 | 5.3 | 17.4 | 4.72 | <0.20 | 2.07 | 0.52 |
| | 1.70 | <0.10 | <1.3 | 13.1 | 3.9 | <0.5 | 9.30 | <0.20 | <0.40 | 0.53 |
| Average | 1.78 ± 0.07 | <0.10 | <1.3 | 18.7 ± 9.1 | 4.3 ± 0.9 | <6.1 | 5.86 ± 3.04 | <0.20 | <0.96 | 0.48 ± 0.07 |
| Tverrådal | <0.80 | <0.10 | <1.3 | 15.0 | <3.1 | <0.5 | 13.61 | <0.20 | <0.40 | 0.44 |
| | 0.86 | <0.10 | <1.3 | 17.5 | <3.1 | <0.5 | 10.28 | <0.20 | <0.40 | 0.48 |
| | <0.80 | <0.10 | <1.3 | 10.0 | <3.1 | <0.5 | 8.02 | <0.20 | <0.40 | 0.38 |
| Average | <0.82 | <0.10 | <1.3 | 14.2 ± 3.5 | <3.1 | <0.5 | 10.64 ± 2.81 | <0.20 | <0.40 | 0.44 ± 0.05 |
| Nasafjell | <0.80 | <0.10 | <1.3 | 10.2 | <3.1 | <0.5 | 2.70 | <0.20 | <0.40 | 0.36 |
| | <0.80 | <0.10 | <1.3 | 9.9 | <3.1 | 0.8 | 2.07 | <0.20 | <0.40 | 0.48 |
| | <0.80 | <0.10 | <1.3 | 15.1 | <3.1 | <0.5 | 2.60 | 0.24 | 2.51 | 0.57 |

(continued)

**Table 4.2** (continued)

| | Li | Be | B | Al | P | K | Ti | Mn | Fe | Ge |
|---|---|---|---|---|---|---|---|---|---|---|
| Average | <0.80 | <0.10 | <1.3 | 11.7 ± 2.9 | <3.1 | <0.6 | 2.46 ± 0.34 | <0.21 | <1.10 | 0.47 ± 0.11 |
| *Nedre Øyvollen pegmatite quartz* | | | | | | | | | | |
| | 4.89 | 0.36 | 1.82 | 14.4 | <2.4 | <3.2 | 5.46 | <0.50 | <1.00 | 1.25 |
| | 4.08 | 0.24 | 1.50 | 30.4 | <2.4 | <3.2 | 1.04 | <0.50 | <1.00 | 0.47 |
| | 5.87 | 0.33 | 1.44 | 21.4 | <2.4 | <3.2 | 4.47 | <0.50 | <1.00 | 0.70 |
| | 4.83 | 0.23 | 1.52 | 16.4 | <2.4 | <3.2 | <1.00 | <0.50 | <1.00 | 0.36 |
| | 4.89 | 0.36 | 1.82 | 14.4 | <2.4 | <3.2 | 5.46 | <0.50 | <1.00 | 1.25 |
| | 4.08 | 0.24 | 1.50 | 30.4 | <2.4 | <3.2 | 1.04 | <0.50 | <1.00 | 0.47 |
| Average | 4.92 ± 0.74 | 0.29 ± 0.06 | 1.57 ± 0.17 | 20.6 ± 7.1 | <2.4 | <3.2 | <2.99 | <0.50 | <1.00 | 0.70 ± 0.40 |
| *Nesodden hydrothermal quartz vein* | | | | | | | | | | |
| | 4.26 | <0.10 | 1.42 | 16.7 | <5.1 | <2.8 | 1.59 | <0.40 | <1.20 | 1.02 |
| | 3.52 | <0.10 | 1.25 | 14.6 | <5.1 | <2.8 | 2.18 | <0.40 | <1.20 | 0.99 |
| | 6.91 | <0.10 | 1.64 | 29.8 | <5.1 | <2.8 | 2.02 | <0.40 | <1.20 | 0.86 |
| | 7.44 | <0.10 | 1.25 | 31.4 | <5.1 | <2.8 | 2.03 | <0.40 | <1.20 | 1.04 |
| | 3.25 | <0.10 | 0.81 | 18.4 | <5.1 | <2.8 | 1.74 | <0.40 | <1.20 | 1.08 |
| | 5.58 | <0.10 | 1.47 | 13.7 | <5.1 | <2.8 | 2.42 | 0.44 | <1.20 | 0.97 |
| | 7.29 | <0.10 | 1.40 | 23.4 | <5.1 | <2.8 | 3.62 | 0.47 | <1.20 | 1.14 |
| | 5.32 | <0.10 | 1.46 | 11.0 | 5.4 | <2.8 | <1.60 | 0.70 | <1.20 | 0.89 |
| | 7.34 | 0.14 | 1.28 | 12.1 | <5.1 | <2.8 | 1.66 | <0.40 | <1.20 | 0.91 |
| | 6.34 | <0.10 | 1.55 | 15.0 | <5.1 | <2.8 | 1.76 | <0.40 | <1.20 | 0.92 |
| | 7.83 | <0.10 | 1.39 | 18.4 | <5.1 | <2.8 | 1.94 | <0.40 | <1.20 | 1.00 |
| | 4.66 | <0.10 | 1.40 | 28.9 | <5.1 | 12.9 | 3.45 | 0.43 | <1.20 | 0.94 |
| | 5.31 | <0.10 | 2.23 | 6.9 | <5.1 | <2.8 | 1.96 | 0.47 | <1.20 | 0.92 |
| | 5.22 | <0.10 | 1.51 | 12.0 | <5.1 | <2.8 | 2.02 | <0.40 | <1.20 | 1.16 |
| | (3.53) | (< 0.10) | (1.80) | (61.3) | (< 5.1) | (23.8) | (2.32) | (< 0.40) | (1.26) | (1.13) |
| | (4.16) | (< 0.10) | (1.53) | (149.2) | (< 5.1) | (91.5) | (4.45) | (< 0.40) | (11.41) | (1.05) |

(continued)

**Table 4.2** (continued)

| | Li | Be | B | Al | P | K | Ti | Mn | Fe | Ge |
|---|---|---|---|---|---|---|---|---|---|---|
| | (4.55) | (< 0.10) | (1.99) | (101.2) | (< 5.1) | (13.1) | (3.24) | (0.59) | (1.20) | (1.01) |
| | (3.56) | (< 0.10) | (1.94) | (134.8) | (< 5.1) | (35.5) | (3.34) | (0.50) | (1.20) | (1.13) |
| Average | 5.74 ± 1.48 | <0.10 | 1.43 ± 0.3 | 18.02 ± 7.6 | <5.1 | <3.5 | 2.14 ± 0.64 | <0.44 | <1.20 | 0.99 ± 0.09 |
| *Kvalvik hydrothermal quartz* | | | | | | | | | | |
| | 3.32 | <0.10 | 1.71 | 16.5 | <5.1 | <2.7 | 2.69 | <0.40 | <1.23 | 0.79 |
| | 3.33 | <0.10 | 1.59 | 12.8 | <5.1 | <2.7 | 3.09 | <0.40 | <1.23 | 0.91 |
| | 3.22 | <0.10 | 1.47 | 15.1 | <5.1 | <2.7 | <1.60 | <0.40 | <1.23 | 0.77 |
| | 3.63 | <0.10 | 1.20 | 14.3 | <5.1 | <2.7 | <1.6 0 | <0.40 | <1.23 | 0.84 |
| | 4.12 | <0.10 | 1.14 | 16.7 | <5.1 | <2.7 | 1.77 | <0.40 | <1.23 | 0.91 |
| | 7.06 | <0.10 | 1.70 | 20.8 | <5.1 | <2.7 | 1.92 | 0.53 | <1.23 | 0.83 |
| | 5.47 | <0.10 | 2.06 | 13.6 | <5.1 | <2.7 | 2.63 | <0.40 | <1.23 | 0.64 |
| | (3.69) | (< 0.10) | (1.73) | (59.7) | (< 5.1) | (21.0) | (3.30) | (< 0.40) | (< 1.23) | (1.00) |
| Average | 4.23 ± 1.45 | <0.10 | 1.55 ± 0.32 | 15.7 ± 2.7 | <5.1 | <2.7 | <2.19 | <0.40 | <1.23 | 0.81 ± 0.09 |
| *Svanvik hydrothermal quartz vein* | | | | | | | | | | |
| | 2.12 | <0.10 | <1.3 | 1.9 | <3.1 | <0.5 | 1.10 | <0.20 | <0.40 | 0.18 |
| | 2.11 | <0.10 | <1.3 | 8.3 | <3.1 | <0.5 | <0.90 | <0.20 | <0.40 | 0.15 |
| | 2.17 | <0.10 | <1.3 | 5.7 | 4.7 | <0.5 | 1.36 | <0.20 | <0.40 | 0.23 |
| Average | 2.14 ± 0.03 | <0.10 | <1.3 | 5.3 ± 3.2 | <3.6 | <0.5 | <1.12 | <0.20 | <0.40 | 0.19 ± 0.04 |

Data on kyanite quartzites and Svanvik are from Müller and Koch-Müller (2009) and from Melkfjell from Wanvik (2009). Note that the limits of detection (LOD) are different for different analytical sequences. LOD's given are examples of the kyanite quartzite measurement sequence. Analyses in parenthesis are not considered in the calculation of the average value because these analyses are superimposed by mineral inclusions, such as rutile (Ti) and mica (Al, K)

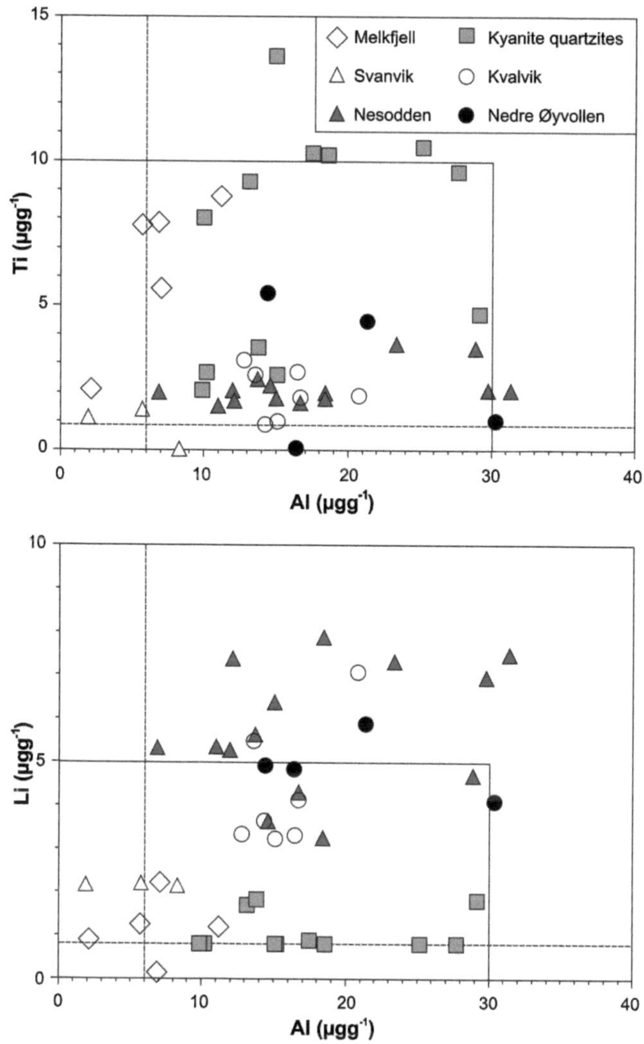


**Fig. 4.15** Diagrams of Al versus Li in quartz. Concentrations were determined by LA-ICP-MS

metamorphic conditions coincide with the pressure–temperature window in which quartz purification occurs (Müller et al. 2002b). The moderate Ti content reflects the relatively high crystallization temperature during metamorphism.

CL imaging showed that grain-boundary migration due to grain-stress relaxation is a widespread phenomenon in the quartzite, whereby dull luminescent grains replaced bright luminescent grains. Dull luminescent quartz is generally considered to be quartz with a low content of lattice defects and trace elements (e.g., Götze et al. 2001). Therefore, the grain-boundary migration caused a further purification of the quartz grains. In particular, the Ti content was lowered

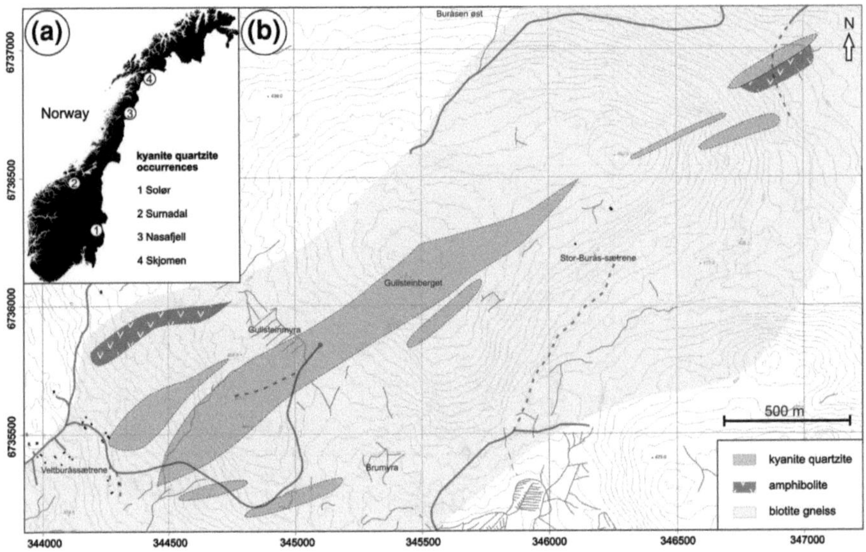


**Fig. 4.16** **a** Locations of kyanite quartzite occurrences in Norway. **b** Geological map of the kyanite quartzite occurrence at Gullsteinberg in Solør according to Jakobsen and Nielsen (1977)

considerably by this process, from about 7–2 μg $g^{-1}$. The expelled Ti may have crystallised as sub-micron rutile needles which are common in the Melkfjell quartz (Fig. 4.12a). From an economic point of view the disadvantage of the widespread grain-boundary migration is the enclosure of minor and accessory minerals by quartz, resulting in a high density of mineral inclusions in the quartz.

## 4.5.2 Kyanite Quartzites

*Location.* Kyanite quartzites occur in four areas in Norway: Solør, Surnadalen, Nasafjell and Skjomen (Ihlen 2000; Müller et al. 2007; Fig. 4.16). Four large kyanite quartzite bodies are located in Solør, 15–35 km SW of Elverum in SE Norway, at Gullsteinberg, Knøsberg, Kjeksberg and Sormbru. The kyanite quartzite at Surnadal is exposed in the Tverrådal valley about 8 km SE of the municipal centre Skei in Møre-og-Romsdal county in central Norway. In Skjomen the kyanite quartzite bodies are situated 20 km SSE of the village, Elvegård, S of Narvik in Nordland county in northern Norway. The rocks are exposed on a mountain plateau called Juovvačorrú at an altitude of about 1,000 m (Fig. 4.10b). The kyanite quartzites at Nasafjell form three WNW-ESE elongated lenses at the SW slope of the Nasafjell (1,210 m a.s.l.) in the Saltfjellet region, close to the Arctic Circle in Nordland county.

*Geology.* Kyanite quartzites form strata-bound lens-shaped bodies, which may extend for several kilometres as trains of lenses. They are commonly associated with metamorphosed acid and intermediate volcanic and sub-volcanic rocks of island-arc setting (e.g., Ihlen 2000; Larsson 2001). The Al-rich rocks were affected by upper greenschist to amphibolite facies metamorphism (>420°C and >2.8 kbar). Kyanite quartzites have been mined for kyanite in USA and Sweden. The rocks are fine-grained, laminated rocks which contain on average 70–85 vol.% quartz and >15 vol.% kyanite. Their composition changes gradually into kyanite-poor or kyanite-free pyrophyllite/muscovite-bearing quartzites between the layers and towards the contact with the host rocks. The muscovite content in the rock commonly increases from the central part of the kyanite quartzite bodies towards the margin, from zero to up to 30 vol.%. The kyanite quartzites do not contain feldspar or biotite, which are common minor minerals in quartzites of sedimentary origin. Common accessories are rutile, muscovite, pyrophyllite, zircon, apatite, pyrite, and barytes. Sporadic accessories are topaz, fluorite, lazulite, wavellite, Al–Sr phosphates and Cr mica (Müller et al. 2005b). Representative whole-rock analyses of kyanite quartzites in Norway are given in Table 4.3. Metamorphosed advanced argillic alteration zones associated with felsic and intermediate volcanites and sub-volcanic intrusions of island-arc settings are suggested to be the pre-metamorphic protoliths of the kyanite quartzites (Larsson 2001; Bibikova et al. 2001; Müller et al. 2005b, 2007).

The Solør kyanite quartzites are hosted in Middle to Upper Proterozoic (1.6–1.8 Ga; Nordgulen 1999), fine- to medium-grained granitic gneisses of the Solør Complex. All four deposits are in close relationship with metagabbros and fine-grained amphibolites (hyperites) which are common in the granitic gneisses of the Solør Complex. The largest kyanite quartzite occurrence is the Gullsteinberg body, 2 km long and up to 200 m thick (Fig. 4.16), which makes it the largest deposit known in Norway.

A small kyanite quartzite body (5 × 10 m) occurs at Tverrådal valley in the Surnadal area (Wanvik 1998). The quartzite is situated in the border zone between basal, autochthonous Proterozoic gneiss units and allochthonous Caledonian nappes that extend from the Trondheim region along Surnadal valley and further westwards to the Molde area. The kyanite quartzite lens is hosted in autochthonous basement granitic gneiss of Mesoproterozoic age (1.6–1.0 Ga; Tveten et al. 1998). Due to the small size the occurrence is non-economically nevertheless it is mentioned here for completeness.

A nearly 70 km-long folded zone of muscovite-rich rocks with lenses of kyanite quartzites is known from Nasafjell (Dahl 1980). The folded zone forms a part of the Caledonian Gargatis nappe, which consists of Mesoproterozoic (1.6–1.9 Ga) granitic gneisses, metarhyolites, metagabbros, muscovite and biotite schists (Gjelle 1988). The four known kyanite quartzite deposits are 150, 400, 500 and 600 m long and up to 30 m wide.

About 16 small bodies (20–140 m in length) are exposed in Skjomen, belonging to the Sørdalen Supracrustal Belt, part of the Palaeoproterozoic Rombak basement window (Korneliussen and Sawyer 1989; Fig 4.10b). They are embedded in muscovite schist. The sequence of the Sørdalen Supracrustal Belt is mainly composed of porphyritic, mafic, intermediate and felsic volcanites. Lenses of kyanite

**Table 4.3** Represenstative whole rock analyses (XRF) of kyanite quartzites in Norway

| | Gullstein-berg | Knøsberg | Kjeksberg | Sormbru | Tverrådal | Skjomen | Nasafjell |
|---|---|---|---|---|---|---|---|
| *Major elements (wt.%)* | | | | | | | |
| $SiO_2$ | 76.09 | 75.89 | 82.54 | 77.37 | 82.67 | 77.68 | 76.27 |
| $Al_2O_3$ | 17.7 | 19.39 | 12.76 | 13.78 | 14.28 | 20.8 | 20.84 |
| $Fe_2O_3$ | 1.42 | 0.71 | 1.67 | 1.41 | 0.13 | 0.09 | 0.54 |
| $TiO_2$ | 0.32 | 0.23 | 0.48 | 0.32 | 0.44 | 0.18 | 0.18 |
| MgO | 0.07 | 0.05 | <0.01 | 0.07 | 0.06 | 0.03 | 0.05 |
| CaO | 0.01 | <0.01 | <0.01 | 0.02 | 0.01 | <0.01 | 0.01 |
| $Na_2O$ | <0.1 | <0.1 | <0.1 | 0.2 | <0.1 | <0.1 | <0.1 |
| $K_2O$ | 0.63 | 0.05 | <0.01 | 3.64 | 0.15 | 0.05 | 0.32 |
| MnO | <0.01 | <0.01 | <0.01 | <0.01 | <0.01 | <0.01 | <0.01 |
| $P_2O_5$ | 0.02 | 0.04 | 0.05 | 0.04 | 0.03 | 0.02 | 0.07 |
| LOI | 2.32 | 1.74 | 1.17 | 2.38 | 0.62 | 0.21 | 0.56 |
| total | 98.58 | 98.11 | 98.63 | 99.24 | 98.33 | 99.05 | 98.78 |
| *Trace elements ($\mu g\ g^{-1}$)* | | | | | | | |
| Ba | 422 | 74 | 1722 | 646 | 399 | <10 | 285 |
| Ga | 11 | <10 | <10 | <10 | <10 | 31 | 11 |
| Zn | 17 | 8 | 10 | 13 | 7 | 15 | 7 |
| Cu | <10 | <10 | 10 | <10 | <10 | <10 | <10 |
| Ni | <5 | <5 | <5 | 5 | <5 | <5 | <5 |
| Co | 5 | 8 | <5 | <5 | 8 | <5 | 6 |
| Ce | <10 | <10 | <10 | 104 | 10 | <10 | 84 |
| La | <10 | <10 | 11 | 63 | 13 | <10 | 65 |
| Nd | <10 | <10 | <10 | 50 | <10 | <10 | 30 |
| W | 47 | 24 | 57 | 35 | 31 | 48 | 27 |
| Pr | <10 | <10 | <10 | 14 | <10 | <10 | 11 |
| Mo | <5 | <5 | <5 | 7 | <5 | <5 | 5 |
| Nb | 21 | 15 | 16 | 13 | 16 | 33 | 22 |
| Zr | 213 | 119 | 251 | 227 | 310 | 73 | 221 |
| Y | <5 | <5 | <5 | 7 | <5 | <5 | 8 |
| Sr | 26 | 47 | 92 | 73 | 26 | 5 | 69 |
| Rb | 22 | <5 | <5 | 78 | 5 | <5 | 10 |
| Th | <5 | 7 | <5 | 12 | 8 | <5 | 24 |
| Pb | <10 | <10 | <10 | 21 | <10 | <10 | 10 |
| Cr | 68 | 17 | 27 | 65 | <10 | 42 | 19 |
| V | <10 | 19 | 29 | 23 | 25 | 21 | 27 |
| As | 8 | <5 | 5 | <5 | <5 | 5 | 17 |
| Hf | <10 | <10 | <10 | <10 | 14 | <10 | <10 |
| S | 0.32 | 0.32 | 0.36 | 0.32 | <0.1 | <0.1 | <0.1 |

Data from Müller et al. (2005b)

quartzite and metasediments are tectonically intercalated within the volcanic rocks. The rocks are assumed to have ages between 1.91 and 1.88 Ga (Korneliussen and Sawyer 1989) and were metamorphosed at amphibolite facies conditions, at 6 kbar and ca. 575°C (Sawyer 1986). The age of the prograde metamorphism is unknown.

Locally, the rocks were affected by greenschist-facies metamorphism related to Palaeozoic Caledonian deformation.

*Quartz petrography.* The kyanite quartzites are commonly fine-grained. The quartz grains in kyanite-rich layers are smaller (100–300 μm) than in kyanite-poor layers (300–1,000 μm). Average grain sizes vary between 100 and 600 μm for the different deposits. The quartz textures consist either of elongated, sutured and recrystallised grains or a granoblastic–polygonal texture depending on the local deformation history. Some of the larger grains (>500 μm) have undulatory extinction with sub-grains forming a stripe-like pattern. Textures are common, which are related to grain-boundary area reduction due to lattice recovery and strain-induced grain boundary migration formed during retrograde deformation.

Quartz forms a close intergrowth assemblage with kyanite and muscovite. Inclusions of kyanite (5–50 μm), muscovite (10–200 μm), pyrophyllite (5–100 μm), rutile (5–50 μm), zircon (10–200 μm), apatite (10–400 μm), pyrite (2–50 μm) and baryte (2–50 μm) are common in quartz (Figs. 4.12b, 4.13b). Rutile inclusions occur as tiny needles (exsolution product) and as prismatic crystals (2–50 μm). Tiny accessory minerals, in particular sericite, are preferentially enriched along grain boundaries and trans-granular micro cracks. Fluid inclusions are extremely rare.

Quartz generally has a weak luminescence. Some quartz grains show slight variations in CL intensity. In the samples from Tverrådal and Nasafjell, tongues of dull luminescent quartz replaced quartz with brighter CL due to grain boundary migration (Fig. 4.14b). The tongues show weak contrasted growth banding.

*Quartz crystal chemistry.* The quartz in the kyanite quartzites is characterised by very low Li, K, Fe and Mn, moderately high Ti, and variable Al concentrations. Concentrations of Li are mostly below the limit of detection (<0.8 μg $g^{-1}$) with the exception of the Skjomen quartz which contains 1.8 μg $g^{-1}$ Li on average. Al content ranges from 9.9 to 29.2 μg $g^{-1}$. The average Ti content differs for the different deposits and varies from 2.5 μg $g^{-1}$ (Nasafjell) to 10.9 μg $g^{-1}$ (Tverrådal) reflecting different deformation temperatures. Average Ti concentrations of quartz from Tverrådal and Gullsteinberg are slightly above the upper concentration limit of HPQ and, thus, this quartz cannot be considered as HPQ in a strict sense.

*Economic assessment and remarks on HPQ formation.* The quartz found in the kyanite quartzites generally has very low Li and variable Al and Ti concentrations. The quartz from the Gullsteinberg and Tverrådal deposits plots at the boundary between the high and medium quality fields due to the relatively high Ti content of about 10 μg $g^{-1}$. Kyanite can presumably be easily separated after milling by density separation due to the high density contrast between quartz (2.62 $gcm^{-1}$) and kyanite (3.61 $gcm^{-1}$). Kyanite concentrates can be used for refractory products. The major challenges from an economic point of view are the small average grain size of the quartz and the high content of mineral inclusions, namely kyanite, muscovite, pyrophyllite, rutile, zircon and apatite. These inclusions will cause elevated concentrations of Al, K, Ti, Zr and P in the product. On the other hand, fluid inclusions are almost absent. Taking all the advantages and disadvantages into account, the kyanite quartzites may be considered as potential resources of HPQ. However, their small grain size, the intimate intergrowth of HPQ, kyanite and other mineral constituents and the high content of mineral inclusions in quartz will be a challenge for mineral processing engineers.

The metamorphic history of the kyanite quartzite is responsible for the specific trace element signature of the quartz. As discussed for the Melkfjell quartzite, moderate pressure ($\sim$3–4 kbars) and temperatures >350°C cause the expulsion of Al and alkalis from the protolithic quartz during metamorphism, resulting in low Al and Li contents in the newly formed quartz. Peak metamorphic conditions were between 3–6 kbar and 420–530°C (Müller and Koch-Müller 2009 and references therein).

Widespread grain-boundary migration related to retrograde lattice recovery may be caused by the further local lowering of certain trace element concentrations, particularly of Ti and Al. This process explains the variation of Al and Ti within one sample (Table 4.2). Grain-boundary migration resulted in the healing of crystal defects, expelling trace elements to the grain boundaries and/or concentrating them in micro-inclusions (e.g., Passchier and Trouw 2006). It is generally accepted that incipient recrystallisation leads to strain reduction by slight boundary movements into neighbouring grains that are more strained. Tongues of newly formed, low-stress quartz domains grow into neighbouring grains of higher strain and defect concentration. The process is frozen and visualised by CL in Fig. 4.14b, in which dull luminescent (defect-poor) quartz is seen to have grown at the cost of the brighter luminescent (relative defect-rich) crystal. Thus, the trace element signature of the quartz has been slightly modified by retrograde recrystallisation after peak metamorphism.

### 4.5.3 The Nedre Øyvollen Pegmatite

*Location.* The Nedre Øyvollen pegmatite is located ca. 1.5 km WNW of the village Drag, on the west side of Tysfjord in Nordland county, northern Norway (Fig. 4.17).

*Geology.* The pegmatite is hosted by the Tysfjord granitic gneiss which is considered to have an emplacement age of about 1.74 Ga (Andresen and Tull 1986). The deformational fabric of the granitic gneiss is considered to be of Caledonian age although we do not exclude the possibility that some of the foliations seen in the Tysfjord area can be of pre-Caledonian age (Andresen and Tull 1986). The Caledonian amphibolite-facies peak metamorphism took place around 432 Ma ago (Northrup 1997). The deformation temperature was presumably in the range of 420–450°C and the pressure between 2 and 3 kbar (Björklund 1989).

The pegmatite is part of the Drag pegmatite cluster which comprises ca. Thirty major pegmatite bodies in the vicinity of Drag which are genetically related to the Tysfjord granite emplacement. The Nedre Øyvollen pegmatite has a vertical, cigar-shaped quartz core enclosed by a wall zone consisting of plagioclase,

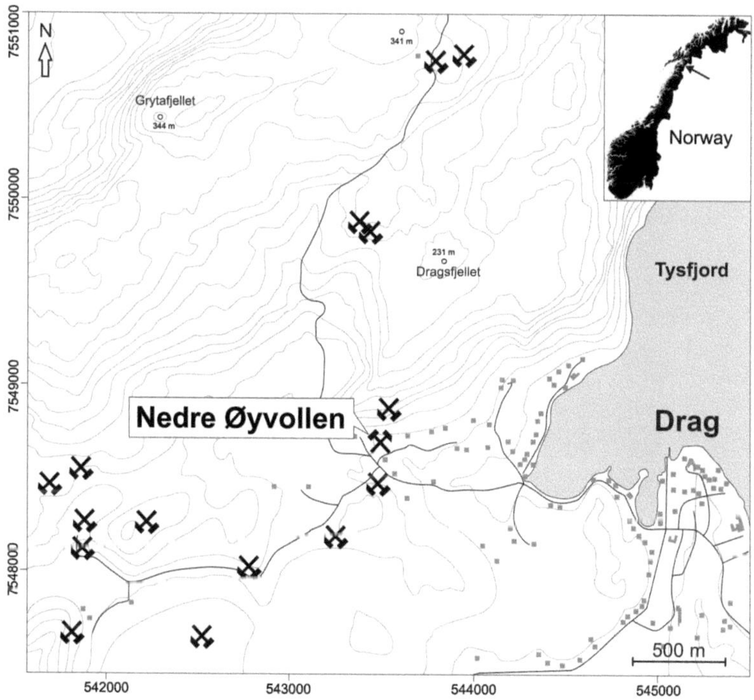


**Fig. 4.17** Location of the Nedre Øyvollen pegmatite mine, 1.5 km WNW of Drag. The *hammer and pick symbols* indicate locations of former feldspar mines in the Drag pegmatite cluster. The area shown in the map is within the Proterozoic Tysfjord granite

K-feldspar and mica (Figs. 4.18, 4.10c). The top of the pegmatite consisted of a plagioclase-rich cupola which has been mined (Neumann 1952). The quartz core has a marginal zone of smoky quartz. The body is about 30 × 60 m in horizontal section and approximately 80 m deep. The occurrence of yttrofluorite masses (up to 2 m in size) within the marginal part of the quartz core is a feature of the pegmatite. Yttrofluorite is a greenish white fluorite with variable contents of REE. Other accessories are native bismuth, chalcopyrite, galenite, cassiterite, euxenite-(Y) and calcite (Husdal 2008). According to the pegmatite classification by Černý and Ercit (2005) the pegmatite is a Niobium-Yttrium-Fluorine (NYF)-type pegmatite of the rare element REE class. Despite the amphibolite-facies overprint, the pegmatite body shows no obvious signs of deformation at the macro scale.

*Mining history*. Mining of feldspar in the Drag area started in 1907. The Nedre Øyvollen pegmatite was discovered during construction of a mine railway in 1909 and in the same year the mining of the pegmatite started in an open pit. Feldspar mining continued, with interruptions, until the 1930s. The total feldspar production was about 12,000 tons, resulting in a 30-m deep open pit. The deposit was drilled in 1975 and 1979 by NGU and about 175,000 tons of massive quartz was proven

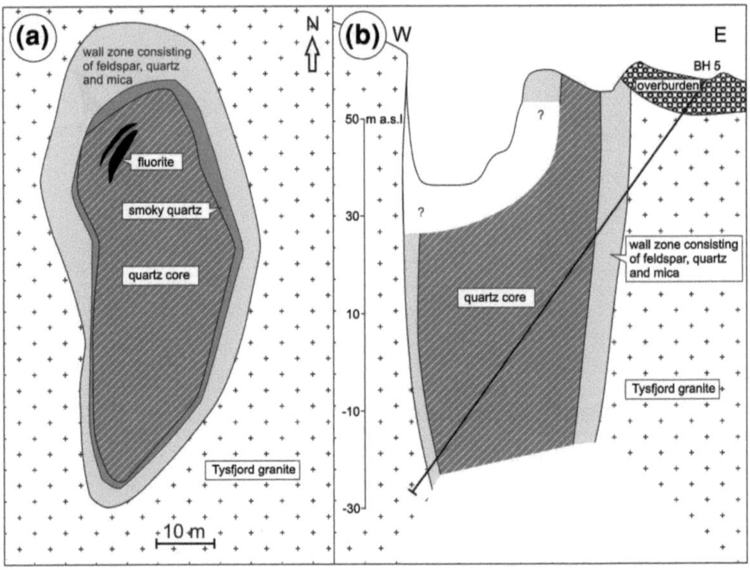


**Fig. 4.18** The Nedre Øyvollen pegmatite. **a** Horizontal section through the central part of the Nedre Øyvollen pegmatite (modified from Norwegian Crystallites 2011). **b** Vertical E-W section of the pegmatite according to Åmli and Lund (1979)

below the open pit (Åmli and Lund 1979). Mining of quartz, underground, started in 1987 and continued, with interruptions, until Norwegian Crystallites AS overtook the mine in 1996. Since then the pegmatitic quartz has been the major source of the HPQ produced by Norwegian Crystallites AS.

*Quartz petrography*. The average quartz crystal size in the pegmatite core is about 6 mm. The crystals are clear, and fluid and mineral inclusions are very rare. The few inclusions identified comprise (in order of frequency): biotite (10–200 μm), albite (20–50 μm), K-feldspar (10–50 μm), silicate melt inclusions (5–50 μm) and calcite (10–50 μm; Figs. 4.12c, 4.13c, 4.14c). The quartz grains commonly show no indications of metamorphic overprint such as undulatory extinction or sub-grain formation (recrystallisation). The shear stress which affected the pegmatite core was focused in thin shear zones resulting in cleavage-like shear planes cross-cutting the quartz body. The CL of quartz is very homogeneous and very intense at initial electron exposure (Fig. 4.14c). The CL intensity decreases rapidly during the first seconds of electron beam exposure. Micro cracks healed by secondary dull-luminescent quartz are very rare.

*Quartz crystal chemistry*. The quartz is characterised by low Ti (mean 3.0 $\mu g\ g^{-1}$) and moderately low Al (mean 20.6 $\mu g\ g^{-1}$). The average Li content is 4.9 $\mu g\ g^{-1}$ and B is 1.6 $\mu g\ g^{-1}$. It is the only investigated deposit which contains quartz with Be concentrations above the limit of detection (mean 0.29 $\mu g\ g^{-1}$).

The average quartz composition suggests high-purity quality despite the fact that individual analyses plot close to the concentration limits for HPQ.

*Economic assessment and remarks on HPQ formation*. The Nedre Øyvollen pegmatite is one of the few deposits in the world from which HPQ has been produced for more than a decade. The large, massive quartz body, the large crystal size and the homogeneous chemistry with very low trace element concentrations have made it a highly economic, world-class HPQ deposit.

The pegmatitic quartz core of the Nedre Øyvollen pegmatite consists of primary magmatic HPQ which was generally not or only weakly affected by secondary recrystallisation due to shearing related to the Caledonian metamorphism. However, it cannot be ruled out that the metamorphic temperature, which was presumably in the range of 420–450°C and the pressure of 2–3 kbar (Björklund 1989), may have had an effect on the trace element distribution in the quartz.

Figure 4.5b shows that NYF-type pegmatite melts, such as the Nedre Øyvollen pegmatite, have the potential to crystallise HPQ directly from silicate melts. Thus, the process leading to HPQ crystallisation in the Nedre Øyvollen quartz is very different from HPQ genesis in the quartzites discussed above.

### 4.5.4 The Nesodden Quartz Vein

*Locality*. The Nesodden hydrothermal quartz vein is exposed on the steep, forested mountainside SE of Hardangerfjord at 210–320 m a.s.1. It is close to the settlement of Lyngstranda in Kvinnherad municipality, Hordaland county in W Norway (Figs. 4.19, 4.10d). The quartz vein is cut by the narrow, steep Klubbdal valley which causes an altitude difference of about 150 m along the strike of the vein.

*Geology*. The vein is situated in Proterozoic basement rocks, south of the Hardanger Fault Zone (HFZ). The HFZ is a 600 km long, SW-NE striking late Caledonian ductile shear zone (Fossen and Hurich 2005). Fluids mobilised during the extensional shearing crystallised as hydrothermal quartz veins, such as the Nesodden and Kvalvik veins, in the vicinity of the fault. The Nesodden quartz vein is 580 m long, averages 12 m wide and plunges 80–85° to the NW (Fig. 4.20). The vein is situated in a 5 km long, 20–50 m wide fault zone which strikes parallel to the Nordlifjell mountain ridge in a NE-SW direction, parallel to Hardangerfjord and the HFZ. Several small quartz lenses up to 10 m long and 5 m wide occur in parallel at the NW side of the major vein (Fig. 4.20 ). The massive quartz contains enclaves and inclusions of granitic gneiss particularly close to the contacts. The inclusion size ranges from several meters to micro-meter scale.

The deposit was investigated by Christiana Spigerverk, who indicated that it contained about 0.5 million tons of quartz of Si- and SiC-quality (low quality according to Harben 2002) to a depth of 100 m (Geis 1964a, b, 1965a, b). New mapping carried out by NGU in 2007 revealed that the inferred resources of the deposit are at least 1 million m$^3$ (580 × 12 × 150 m), which corresponds to 2.7 million tons of quartz (Ihlen and Müller 2011).

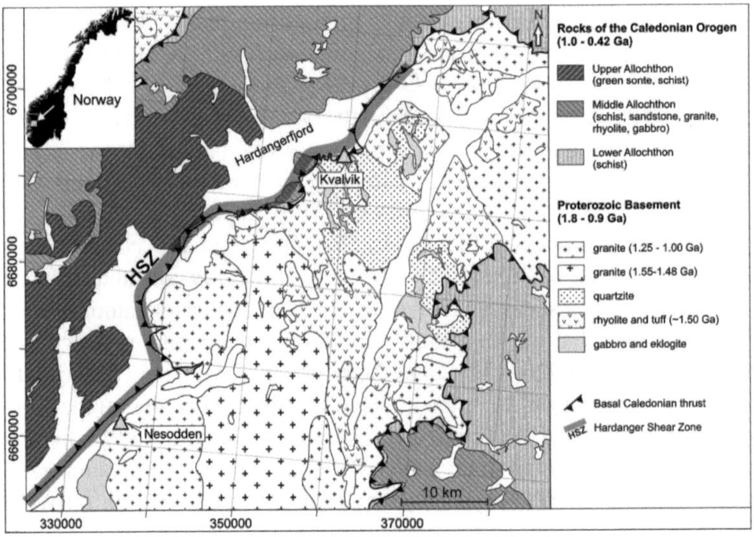


**Fig. 4.19** Simplified geological map according to Solli and Nordgulen (2006) with locations of the Nesodden and Kvalvik quartz occurrences in the Hardangerfjord area. The extension of the Hardanger Shear Zone (HSZ) is according to Fossen and Hurich (2005)

*Quartz petrography*. The crystal sizes of the Nesodden quartz are highly variable, ranging from <1 μm to about 3 cm with an average of about 3 mm. The quartz is locally strongly recrystallised due to post-crystallisation deformation related to late Caledonian tectonics. The quartz shows a high porosity at microscale partially due to the high content of fluid inclusions. The fluid inclusions are liquid-rich and occasionally contain precipitated solids, presumably crystals of halite and sylvite (Fig. 4.12d). Inclusions of muscovite (1–200 μm) are very common (Fig. 4.13d): occasionally, inclusions of calcite are present (1–20 μm).

CL imaging shows that the Nesodden quartz is strongly altered due to recrystallisation and fluid-driven overprint. Domains of primary, unaltered quartz occur as bright luminescent relicts and comprise less than 25 vol.% of the investigated quartz samples (Fig. 4.14d). CL reveals tiny inclusions of calcite and feldspar-like composition along healed micro cracks (Fig. 4.14d).

*Quartz crystal chemistry*. The quartz from Nesodden has low Ti (mean 2.1 μg $g^{-1}$), moderately low Al (mean 18.0 μg $g^{-1}$) and Ge (mean 1.0 μg $g^{-1}$) and relatively high Li (mean 5.7 μg $g^{-1}$) contents. The average B content is about 1.4 μg $g^{-1}$. Four of the quartz analyses listed in Table 4.2 show high Al (>60 μg $g^{-1}$) together with high K (>13 μg $g^{-1}$) indicating that mica-like micro- inclusions contaminated the analyses. These analyses are not included in the calculation of the average values, because they do not reflect the content of lattice-bound trace elements. Because of the high average Li content, the Nesodden quartz is not a HPQ in the strict sense. However, the experience of the authors in processing of

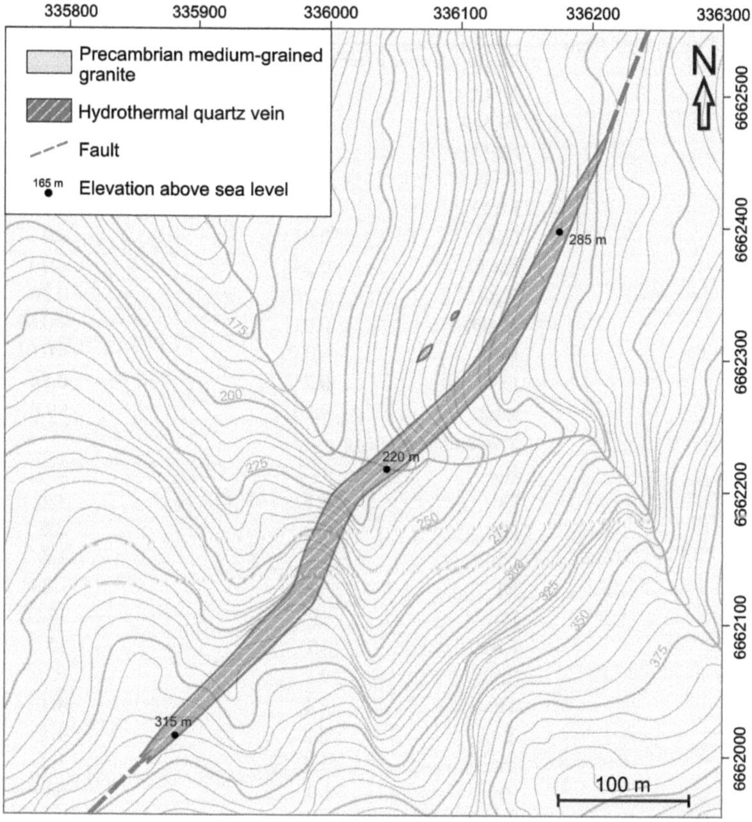


**Fig. 4.20** Detailed geological map of the Nesodden quartz vein (modified from Ihlen and Müller 2011)

other quartz deposits, suggests that parts of the interstitially bound $Li^+$ could possibly be removed during chemical quartz treatment.

*Economic assessment and remarks on HPQ formation.* The Nesodden quartz vein is a moderate-sized, massive quartz deposit. Relatively high Li and B concentrations, a high content of high-salinity fluid inclusions, common enclaves of granitic gneiss and micro-inclusions of muscovite, feldspar and calcite are the challenges in processing HPQ from the deposit.

Fluids responsible for the quartz vein crystallisation were presumably mobilised during Caledonian extensional shearing related to the motion of the nearby HFZ. The LA-ICP-MS analyses suggest that the primary crystallised hydrothermal quartz contained low concentrations of lattice-bound trace elements. Subsequent alteration and recrystallisation associated with shearing and fluid flux may have contributed to further purification of the quartz as indicated by the low CL intensity of the secondary quartz.

**Table 4.4** XRF bulk rock analyses of quartzites from Kvalvik

| | Sample 58906 | Sample 58908 |
|---|---|---|
| *Major elements (wt.%)* | | |
| $SiO_2$ | 93.9 | 96.9 |
| $Al_2O_3$ | 2.44 | 0.98 |
| $Fe_2O_3$ | 0.36 | 0.29 |
| $TiO_2$ | 0.06 | 0.03 |
| MgO | 0.06 | 0.17 |
| CaO | 0.01 | <0.01 |
| $Na_2O$ | 0.24 | <0.1 |
| $K_2O$ | 1.16 | 0.36 |
| MnO | 0.01 | <0.01 |
| $P_2O_5$ | <0.01 | <0.01 |
| LOI | 0.21 | 0.24 |
| total | 98.5 | 99.0 |
| *Trace elements* ($\mu g\ g^{-1}$) | | |
| Ba | 112 | 28 |
| Cr | 8.5 | <4 |
| Ga | 3.7 | 2.6 |
| Nb | 1.7 | 1.3 |
| Ni | <2 | <2 |
| Pb | 19 | 17 |
| Rb | 36 | 15 |
| Sr | 9 | <1 |
| Th | <4 | 4.1 |
| Y | 8.2 | 4.1 |
| Zn | 1.4 | 1.3 |
| Zr | 60 | 68 |

KvalvikData from Ihlen and Müller (2011)

### *4.5.5 The Kvalvik Deposit*

*Location.* The Kvalvik quartz deposit is exposed on the hillside SE of Hardangerfjord at an altitude between 160 and 340 m a.s.l., S of the settlement Alsåker in Ullensvang municipality, Hordaland county, W Norway (Fig. 4.19).

*Geology.* Like the Nesodden vein, the Kvalvik quartz deposit is situated in Proterozoic basement rocks south of the HFZ (Fig. 4.19). The deposit is characterised by a dense network of snow-white quartz veins up to several deca-metres long and 5 m wide (Fig. 4.10e). The veins are hosted by fine-grained, grey feldspar-bearing quartzite of the Aga formation (Sigmond 1998, Ingdal et al. 2001) and occur immediately below the Caledonian nappe units exposed several hundred metres NW of the deposit. The formation of the veins is presumably related to fluid mobilisation during late Caledonian extension when the HFZ developed (Ihlen and Müller 2011). The composition of the quartzitic host rock is given in Table 4.4. The quartz veins tend to be sub-parallel to the NW–SE striking foliation of the

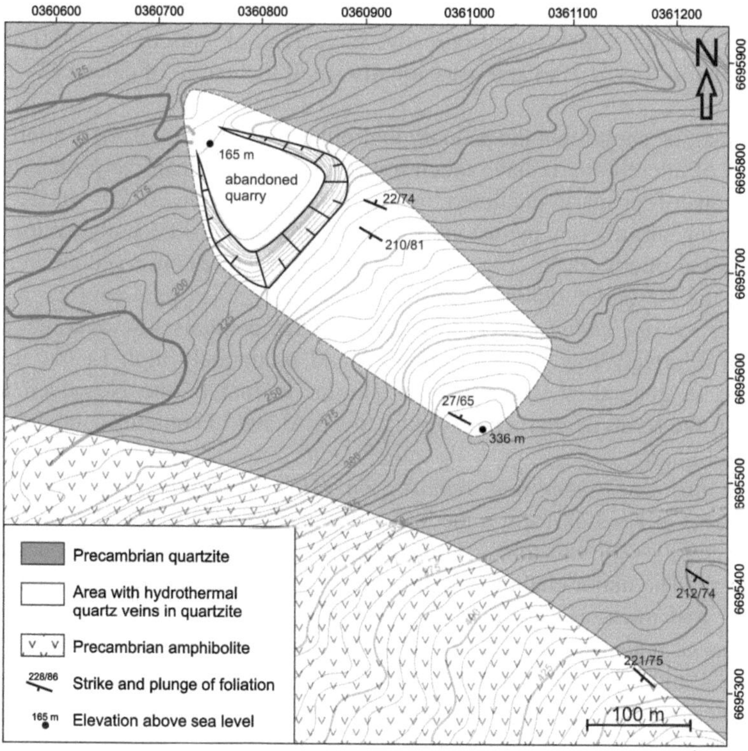


**Fig. 4.21** Detailed geological map of the Kvalvik quartz deposit (modified from Ihlen and Müller 2011)

quartzite, but cross-cutting, irregular vein structures are common as well. The zone of vein-infiltrated quartzite extends about 350 m in NW–SE direction and up to 120 NE-SW (Fig. 4.21). It is approximately 40 m thick and plunges 20–25° N to the NE. The volume of the zone corresponds to 1.4 mill $m^3$ (350 × 120 × 40 m). Ca. 0.3 million $m^3$ (120 × 120 × 20 m) have been mined. The average percentage of hydrothermal veins in this volume is about 25 vol.% resulting in inferred remaining resources of about 0.7 million tons snow-white quartz. The percentage increases from ca. 10–15 vol.% in the SW to 60–80 vol.% in the NE. The NW part of the deposit was mined for quartz to supply the ferrosilicon smelter at Bjølvefossen. The open pit mine is about 120 × 120 m in size and the SE wall is about 40 m high.

*Quartz petrography*. The crystal size of the Kvalvik quartz ranges from <1 μm to about 2 cm with an average of about 2 mm. The quartz is cut by micro-scale shear zones consisting of micro-crystalline, recrystallised quartz. Locally the quartz is completely recrystallised due to late Caledonian deformation. The content of liquid-rich fluid inclusions (2–50 μm) is very high. The fluid inclusions often contain a gas bubble and occasionally precipitated salt crystals, presumably halite (Fig. 4.12e). The high fluid inclusion content and common micro-scale cavities result in a high micro-porosity. Inclusions of muscovite (1–50 μm) are very common (Fig. 4.13e) and occasionally small inclusions of calcite (2–20 μm) occur.

CL imaging shows that the primary hydrothermal quartz is strongly altered. The domains of altered quartz have much lower CL intensity than the primary quartz. The relicts of primary unaltered quartz comprise about 20–30 vol.%. Bright luminescent sub-micron inclusions (<1 μm) of feldspar-like composition occur along healed micro cracks (Fig. 4.14e). The CL structures observed are similar to those in the Nesodden quartz.

*Quartz crystal chemistry.* The Kvalvik quartz contains low average Ti (2.2 $\mu g\ g^{-1}$), relatively low Al (15.7 $\mu g\ g^{-1}$) and moderately high Li (4.2 $\mu g\ g^{-1}$) contents. In general, the chemistry is similar to that of the Nesodden quartz. However, the Kvalvik quartz has slightly lower Al and Li contents compared to the Nesodden quartz.

*Economic assessment and remarks on HPQ formation.* The hydrothermal quartz veins at Kvalvik form a medium-size quartz deposit. The chemistry of the quartz crystals suggests that a HPQ product could possible be produced from the deposit. The major challenge will be the separation of the vein quartz from the feldspar- and mica-bearing quartzitic host. The mica inclusions and sub-micron inclusions of feldspar will presumably add to the concentrations of lattice-bound Al, K, Na, and Ca of the quartz.

Quartz chemistry, mineral and fluid inclusion inventory and CL intensity and structures are similar to those of the Nesodden quartz, suggesting a similar genesis and source for the quartz–forming fluids. The processes of HPQ formation are analogous to those of the Nesodden deposit as well.

### 4.5.6 The Svanvik Quartz Vein

*Location.* The Svanvik quartz vein is situated ca. 35 km south of Kirkenes in Finnmatk county in northern Norway, close to border with Russia. The new quarry is situated ca. 2 km NW of the village Svanvik and 150 m E of the state road 885 between Bjørnevatn and Svanvik (Fig. 4.10f).

*Geology.* The quartz vein at Svanvik comprises an up to 20 m wide and 500 m long sub-vertical and E-W striking hydrothermal vein. The eastern part of the vein is covered by Pleistocene and Holocene sediments which are up to 20 m thick. The vein is hosted by grey granitic to granodioritic gneiss at the SE edge of the late Archean Svanvik Complex (Siedlecka and Nordgulen 1996; Fig. 4.22). The gneiss has an age of 2825 ± 34 Ma (Levchenkov et al. 1995). The Svanvik Complex is framed in the north by the late Archean Brannfjell Complex, in the west by the Neiden pluton (2483 ± 28 Ma; Levchenkov et al. 1995) and in the south and east by the Early Proterozoic Pechenga–Varzuga Greenstone Belt (Melezhik and Sturt 1994). The

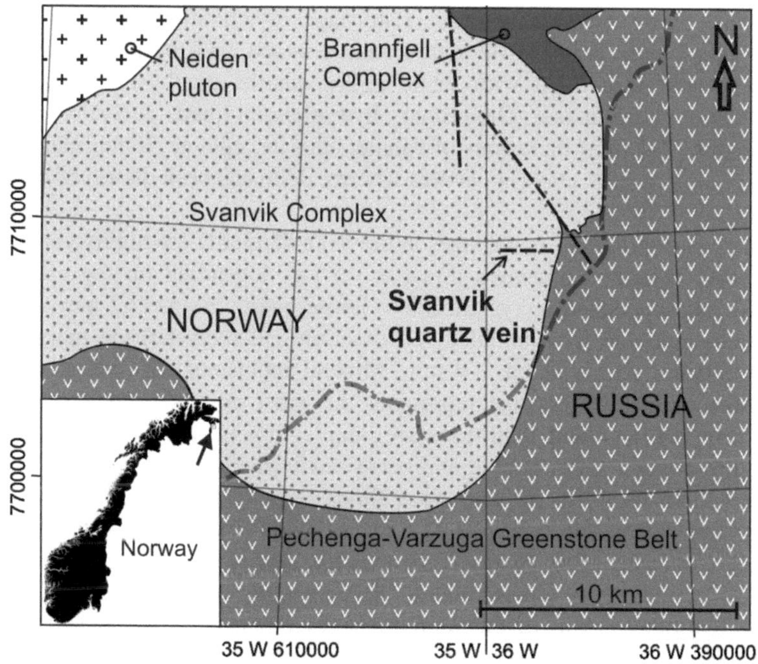


**Fig. 4.22** Simplified geological map of the Pasvik area with the location of the Svanvik quartz vein (modified from Siedlecka and Nordgulen 1996)

contact between the Svanvik Complex and the Pechenga–Varzuga Greenstone Belt is just 2.2 km E of the new quartz mine at Svanvik.

The hydrothermal quartz vein at Svanvik was discovered in 1984 by the geologist Mogens Marker (University of Copenhagen) during regional mapping of the Pasvik area. Between 1985 and 1987 the deposit was sampled, mapped and drilled by NGU (Wanvik 1988, 1989a, 1989b). Strongly deformed chlorite schist, with a thickness of up to 10 m, occurs on both sides of the quartz vein (Fig. 4.23). The vein and the chlorite schist are part of an E-W striking shear zone within the granitic Svanvik gneiss. The vein was emplaced at an initial stage of shear zone formation, presumably during the Early Proterozoic (~1.75 Ga), when the Pechenga–Varzuga Greenstone Belt was sandwiched between two colliding continents (Melezhik and Sturt 1994). Wanvik (1989a) suggested that the Svanvik quartz vein is up to 20 m wide and at least 500 m long and 50 m deep, which corresponds to about 1 million tons of quartz.

Formation of the vein is associated with strong chloritisation of the host rocks. During vertical movements in the shear zone, calcic (carbon-rich) fluids entered extensional fractures crossing the vein and calcite crystallised together with minor chlorite and epidote. The veins are up to 0.5 m wide. In addition there are small K-feldspar-quartz veins, up to 2 cm in width. Contact-parallel sheets and layers (shear planes) of chlorite schist, 0.5 mm to 10 cm in thickness, occur in a 0.5–2 m wide zone along the vein margins. Late-stage re-activation of the shear zone caused minor brecciation at the quartz vein contacts and the formation of open cracks, which were filled with white milky quartz, amethyst, and clear rock crystals up to 4 cm in size. The primary hydrothermal quartz at Svanvik contains macroscopic (>1 mm) inclusions of calcite, chlorite, epidote, K-feldspar, chlorite schist, granitic gneiss, clay (in cavities where calcite has been dissolved), and iron oxides.

*Quartz petrography*. Crystal sizes in the Svanvik quartz are highly variable and range from several centimetres down to <1 μm depending on the intensity of shearing and associated recrystallisation. Large quartz grains (>2 mm) are commonly elongate and have undulatory extinction and recrystallised margins. The average crystal size is about 2–3 mm. Mineral inclusions >5 μm are relatively rare. Muscovite (10–100 μm) represents the most common mineral inclusion phase (Fig. 4.13f). Fluid inclusions are very common. The aqueous inclusions (2–60 μm) contain a gas bubble and occasionally halite crystals, indicating high salinities (>15 wt.%) in the entrapped fluids (Fig. 4.12f). In addition, traces of sylvite were detected in decrepitated fluid inclusions by EDX.

The CL intensity of the Svanvik quartz is low (Fig. 4.14f). Structures visualised by CL indicated a very intense alteration, recrystallisation and brecciation of the primary hydrothermal quartz due to multiply reactivated shearing. In addition, tiny (<1 μm) inclusions of calcitic and feldspar-like composition are revealed by CL in healed micro cracks and recrystallised domains.

*Quartz crystal chemistry*. Quartz from Svanvik has very low Li (mean 2.1 $\mu g\ g^{-1}$), Al (mean 5.3 $\mu g\ g^{-1}$), Ti (mean 1.1 $\mu g\ g^{-1}$) and Fe (mean 0.2 $\mu g\ g^{-1}$) concentrations. In general, the Svanvik quartz has the lowest total content of lattice-bound trace elements ($\sim$15 $\mu g\ g^{-1}$) compared to the other deposits discussed in this study.

*Economic assessment and remarks on HPQ formation*. The Svanvik deposit is a medium-sized, massive quartz deposit with outstanding pure quartz chemistry, with a total trace element content of about 15 $\mu g\ g^{-1}$. For that reasons Norwegian Crystallites AS started seasonal mining of the vein 2 years ago.

However, the multiple overprint of the quartz vein resulted in crystallisation of additional quartz generations mainly at the vein contacts (Fig. 4.23). These younger quartz generations form small volumes but have higher trace element concentrations than the major vein quartz and may thus contaminate the quartz product (Müller unpublished data). Transecting calcite-chlorite-epidote and K-feldspar-quartz veins, macroscopic inclusions (>1 mm) of calcite, chlorite, epidote, K-feldspar, chlorite schist, granitic gneiss, clay, iron oxides and microscopic inclusions of calcite are the challenges for HPQ production from this deposit.

LA-ICP-MS analyses suggest that the hydrothermal quartz had already initially very low concentrations of lattice-bound trace elements. Subsequent, multiple shearing and associated fluid flux resulted in intense recrystallisation and fluid-driven alteration. The latter is evident by the high content of secondary fluid inclusions. CL imaging indicates that about 80–70 vol.% of the primary

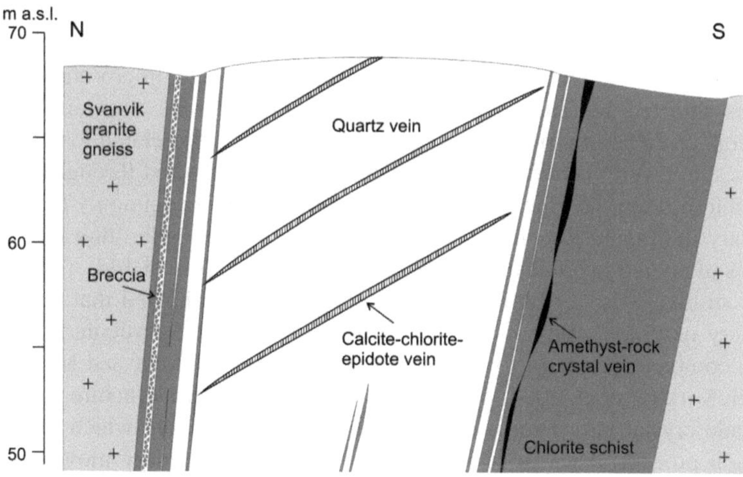


**Fig. 4.23** Geological cross section of the Svanvik quartz vein

hydrothermal quartz was affected and altered by these events. These secondary processes might have contributed to the additional purification of the quartz.

## 4.6 Summary and Outlook

In this study a refined HPQ definition based on concentration limits for the most important detrimental elements is suggested. Increasing demand and quality requirements for HPQ have given rise to a need for such a refined definition. The definition proposed can be applied to natural, untreated quartz for the identification of new deposits, and to processed quartz raw materials as well. The upper concentration limits proposed for Na, K and Ca are relatively high with respect to their average abundance in quartz (Fig. 4.8), because in processed quartz their concentrations are increased by fluid inclusions as well as mica and feldspar micro-inclusions, which are the most common intracrystalline impurities as shown in this study.

LA-ICP-MS is an excellent method for analysing the concentration of lattice-bound trace elements in single quartz crystals in situ. The concentrations determined correspond theoretically to the best achievable quality of the final quartz product if perfect processing is applied, removing all non-lattice-bound impurities. However, trace element concentrations in processed quartz products will rarely attain the concentrations determined by LA-ICP-MS even if the best available processing technology is applied. Some impurities will remain in the product, superimposing the concentrations of lattice-bound trace elements. On the other hand, the experience of the authors showed that some of the interstitially bound $Li^{+}$ and $Fe^{2+}$ could be removed by acid leaching.

This study demonstrates that SEM-CL is an appropriate method for visualising structures within quartz crystals that might be related to HPQ formation. These structures, which are only visible in CL, include healed micro-cracks, quartz domains affected by alteration, newly formed quartz due to grain boundary migration, etc. The first two structure types are presumably related to fluid-driven processes and were formed when fluids penetrated and flushed through the quartz due to micro-scale shearing and brecciation. The fluids are commonly preserved as secondary fluid inclusions in healed micro cracks. The irregular, dull luminescent domains of altered quartz extend from the healed cracks (Fig. 4.14d–f). Studies of similar structures in pegmatite and metamorphic quartz showed that this type of secondary quartz is commonly depleted in trace elements, particularly in Ti, Al and Li, compared to the primary host quartz (Van den Kerkhof and Müller 1999; Van den Kerkhof et al. 2004; Müller et al. 2008a). The third structure is related to retrograde crystal lattice recovery. Grain boundary migration, which is a recrystallisation process, leads to strain reduction by slight boundary movements into neighbouring grains that are more strained. During this process crystal defects and trace elements are expelled from the newly formed quartz and concentrate at the migration front (e.g., Passchier and Trouw 2006). The tongues of newly formed quartz show dull CL, typical for defect-poor quartz, whereas the quartz which becomes replaced has a brighter CL indicating a higher frequency of defects and possibly trace elements (e.g., Müller et al. 2007). Thus, the trace element signature of the quartz has been slightly modified by retrograde recrystallisation. The CL observations indicate that the formation of secondary quartz commonly results in refinement of the quartz quality. However, more work has to be done, particularly experimental work, in order to achieve a better understanding of which type of fluids and what temperature and pressure conditions lead to quartz refinement.

The investigation of HPQ deposits in Norway showed that the processes of HPQ formation differ in the various deposits and that there is no general rule for HPQ formation. The processes revealed include:

- Primary crystallisation of massive vein quartz from hydrothermal fluids mobilised within or in the vicinity of major shear zones at greenschist-facies conditions (Nesodden, Kvalvik, Svanvik).
- Primary crystallisation of quartz from NYF-type pegmatite melts (Nedre Øyvollen).
- Crystallisation (recrystallisation) of quartz-rich rocks at amphibolite-facies conditions resulting in the formation of quartzites (Melkfjell, kyanite quartzites). The temperature–pressure conditions had to be high enough to "delete" the primary trace element signature of the quartz in the protolith and low enough to form quartz with <10 μg $g^{-1}$ Ti. The crystallisation temperature of quartz with 10 μg $g^{-1}$ Ti is 530°C according to the Ti-in-quartz geothermometer (Wark and Watson 2006). However, if the quartz-forming fluid or melt is undersaturated in Ti the crystallisation temperature may be higher.
- Secondary quartz refinement at micro-scale related to post-crystallisation shearing and fluid-driven overprint in greenschist- to lower-amphibolite-facies

conditions (Melkfjell, Norwegian kyanite quartzites, Nesodden, Kvalvik, Svanvik). In this case the primary quartz has to have relatively low contents of lattice-bound trace elements in order to attain HPQ quality on average, because the secondary refinement commonly affected only parts of the quartz.

The list of HPQ-forming processes provided is presumably not complete, because only a few HPQ deposits have been investigated in detail so far and more work has to be done. It is shown that the investigated quartzites have a high content and diversity of micro inclusions and small grain sizes and, thus, it is relative unlikely that HPQ can be produced from these deposits with today's processing technology despite the low content of lattice-bound trace elements. In addition, it is concluded that deposits containing >100,000 tons of HPQ are extremely rare compared to the very high number of exposed and known quartz deposits in Norway and elsewhere.

**Acknowledgments** This study was supported by the Geological Survey of Norway (NGU). We greatly appreciate the language improvement of R. Boyd. We are grateful to the management of the Norwegian Crystallites AS who allowed the publication of data related to the Nedre Øyvollen and Svanvik quartz deposits.

## References

Adachi T, Hokada T, Osanai Y, Toyoshima T, Baba S, Nakano N (2010) Titanium behavior in quartz during retrograde hydration: occurrence of rutile exsolution and implications for metamorphic processes in the Sør Rondane Mountains, East Antarctica. Polar Sci 3:222–234

Åmli R, Lund B (1979) Diamantboringer Nedre Øyvollen kvartsforekomst, Drag i Tysfjord. NGU rapport 1771, 6 pp

Andresen A, Tull JF (1986) Age and tectonic setting of the Tysfjord gneiss granite, Efjord, North Norway. Norsk Geologisk Tidsskrift 66:69–80

Bambauer HU, Brunner GO, Laves F (1962) Wasserstoff-Gehalte in Quarzen aus Zerrklüften der Schweizer Alpen und die Deutung ihrer regionalen Abhängigkeit. Schweizerische Mineralogische und Petrographische Mitteilungen 42:221–236

Bartovic S, Beane R (2007) Analysis of blue color in quartz grains from cushing formation, peaks island, maine. In: Abstracts GSM spring meeting 2007, GSM Newsletter 36/2:4

Beurlen H, Müller A, Silva D, Da Silva MRR (2011) Petrogenetic significance of trace-element data analyzed with LA-ICP-MS in quartz from the Borborema pegmatite province, northeastern Brazil. Mineral Mag 75:2703–2719

Bibikova EV, Ihlen PM, Marker M (2001) Age of the hydrothermal alteration leading to garnetite and kyanite pseudo-quartzite formation in the Khizovaara segment of the late Archean Keret Greenstone Belt, Russian Karelia. EUG XI Strasbourg, 8–12 April 2001. J Conf Abstr 6:277

Björklund L (1989) Geology of the Akkajaure–Tysfjord–Lofoten traverse, N. Scandinavian Caledonides. Ph D thesis, Chalmers Tekniska Högskola och Göteborgs Universitet, publ. A 59, 214 pp

Blankenburg H-J, Götze J, Schulz H (1994) Quarzrohstoffe. Deutscher Verlag für Grundstoffindustrie, Leipzig-Stuttgart, 296 pp

Breiter K, Müller A (2009) Evolution of rare-metal granitic magmas documented by quartz chemistry. Eur J Mineral 21:335–346

Brouard S, Breton J, Girardet G (1995) Small alkali metal clusters on (001) quartz surface: adsorption and diffusion. J Mol Struct (Theochem) 334:145–153

Bruhn F, Bruckschen P, Meijer J, Stephan A, Richter DK, Veizer J (1996) Cathodoluminescence investigations and trace-element analysis of quartz by micro-PIXE: implications for diagenetic and provenance studies in sandstone. Can Mineral 34:1223–1232

Černý P, Ercit TS (2005) The classification of granitic pegmatites. Can Mineral 43:2005–2026

Cherniak DJ, Watson EB, Wark DA (2007) Ti diffusion in quartz. Chem Geol 236:65–74

Dahl Ø (1980) Nasa og Stødi kyanittfelter-resultater fra diamantboring og geologiske undersøkelser sommeren. Aspro rapport 1115. Bergvesenrapport BV 3506, 21 pp

Dennen WH (1966) Stoichiometric substitution in natural quartz. Geochimica et Cosmochimica Acta 30:1235–1241

Flem B, Larsen RB, Grimstvedt A, Mansfeld J (2002) In situ analysis of trace elements in quartz by using laser ablation inductively coupled plasma mass spectrometry. Chem Geol 182:237–247

Flicstein J, Schieber M (1974) Microsegregation of impurities in hydrothermally-grown quartz crystals. J Cryst Growth 24(25):603–609

Fossen H, Hurich CA (2005) The Hardangerfjord Shear Zone in SW Norway and the North Sea: a large-scale low-angle shear zone in the Caledonian crust. J Geol Soc Lond 162:675–687

Frazier AS, Gobel VW (1982) Rutile as cause of blue color of quartz from llanite, Llano County, Texas. In: Abstracts with programs, Geological Society of America 14/3:111

Frezzotti M-L (2001) Silicate-melt inclusions in magmatic rocks: applications to petrology. Lithos 55:273–299

Frondel C (1962) The system of mineralogy: vol III. Silica minerals, Wiley, New York, 602 pp

Fruth M, Blankenburg H-J (1992) Charakterisierung von authigenen idiomorphen Kohle- und Salinarquarzen durch Einschlussuntersuchungen. Neues Jahrbuch für Mineralogie Abhandlungen 165:53–64

Geis HP (1964a) Befaring av kvartsforekomst Nesodden ved Løvfallstrand, Hardanger. NGU Rapport BA 7978, p 5

Geis HP (1964b) Befaring av Nesodden kvartsforekomst, Hardanger. NGU Rapport BA 7981, p 4

Geis HP (1965a) Undersøkelse av kvartsforekomsten Nesodden ved Løvfallstrand. NGU Rapport BA 7979, p 7

Geis HP (1965b) Nesodden kvartsforekomst. NGU Rapport BA 7980, p 10

GeoReM (2011) Geological and environmental reference materials. http://georem.mpch-mainz.gwdg.de. Accessed 2 Feb 2011

Gerler J (1990) Geochemische Untersuchungen an hydrothermalen, metamorphen, granitischen und pegmatitischen Quarzen und deren Flüssigkeitseinschlüssen. Ph.D. thesis, University Göttingen, 169 pp

Gjelle S (1988) Geologisk kart over Norge, berggrunnskart Saltdal, M 1: 250.000. Geological Survey of Norway, Trondheim

Götze J (2009) Chemistry, textures and physical properties of quartz geological interpretation and technical application. Mineral Mag 73:645–671

Götze J (2012) Mineralogy, geochemistry and cathodoluminescence of authigenic quartz from different sedimentary rocks. (this volume)

Götze J, Plötze M (1997) Investigation of trace-element distribution in detrital quartz by Electron Paramagnetic Resonance (EPR). Eur J Mineral 9:529–537

Götze J, Plötze M, Habermann D (2001) Origin, spectral characteristics and practical applications of the cathodoluminescence (CL) of quartz—a review. Mineral Petrol 71:225–250

Götze J, Plötze M, Graupner T, Hallbauer DK, Bray C (2004) Trace element incorporation into quartz: a combined study by ICP-MS, electron spin resonance, cathodoluminescence, capillaryion analysis and gas chromatography. Geochimica et Cosmochimica Acta 68:3741–3759

Gustavson M, Gjelle ST (1991) Berggrunnskart Mo i Rana 1: 250 000. Geological Survey of Norway, Trondheim

Harben PW (2002) The industrial mineral handybook—a guide to markets, specifications and prices, 4th edn. Industrial Mineral Information. Worcester Park, p 412

Haus R (2005) High demands on high purity—processing of high purity quartz and diatomite. Industrial Minerals October 2005, pp 62–67

Hertweck B, Niedermayr G, Beran A (2003) OH zoning ion alpine quartz from Austria. European Geophysical Society (CD-Rom) Vol. 5, EGS-AGU-EUG Joint Assembly, 6th–11th April 2003, Nice, France, 08506

Husdal T (2008) The minerals of the pegmatites within the Tysfjord granite, northern Norway. Bergverksmuseets Skrift 38:5–28

Hyrsl J, Niedermayr G (2003) Magic world: inclusions in quartz—Geheimnisvolle Welt: Einschlüsse im Quarz. Bode Verlag GmbH, Haltern, 240 pp

Ihlen PM (2000) Utilisation of sillimanite minerals, their geology, and potential occurrences in Norway—an overview. NGU Bulletin 436:113–128

Ihlen PM, Müller A (2011) Forekomster av høyren kvarts langs Hardangerfjorden. NGU Rapport 2009.024, Trondheim, 69 pp

Ingdal SE, Torske T, Kvale A (2001) Bergrunnskart Jondal 1315 4, M 1:50000. Geological Survey of Norway, Trondheim

IOTA® (2011) IOTA® high purity quartz. http://www.iotaquartz.com/techiota4data.html Accessed 20 May 2011

Jacamon F, Larsen RB (2009) Trace element evolution of quartz in the charnockitic Kleivan granite, SW Norway: the Ge/Ti ratio of quartz as an index of igneous differentiation. Lithos 107:281–191

Jakobsen BM, Nielsen E (1977) Kyanit kvartsit projektet 1976–1977. Laboratorienrapport. Endogen Laboratorium. Geologisk Institut Aarhus Universitet, Aarhus, Denmark, 19 pp

Jourdan A-L, Vennemann TW, Mullis J, Ramseyer K, Spiers CJ (2009) Evidence of growth and sector zoning in hydrothermal quartz from Alpine veins. Eur J Mineral 21:219–231

Jung L (1992) High purity natural quartz. Part I: High purity natural quartz for industrial use. Part II: High purity natural quartz markets for suppliers and users. Quartz Technology, Liberty Corner , p 657

Korneliussen A, Sawyer EW (1989) The geochemistry of lower proterozoic mafic to felsic igneous rocks, rombak window, North Norway. NGU Bull 415:7–21

Larsen RB, Henderson I, Ihlen PM, Jacamon F (2004) Distribution and petrogenetic behaviour of trace elements in granitic quartz from South Norway. Contributions Mineral Petrol 147:615–628

Larsson D (2001) Transition of granite to quartz-kyanite rock at Hålsjöberg, southern Sweden: consequence of acid leaching and later metamorphism. GFF 123:237–246

Leeder O, Thomas R, Klemm W (1987) Einschlüsse in Mineralen. VEB Deutscher Grundstoffverlag, Leipzig, 180 pp

Levchenkov OA, Levsky LK, Nordgulen Ø, Dobrzhinetskaya, Vetrin VR, Cobbing J, Nilsson LP, Sturt BA (1995) U–Pb zircon ages from Sørvaranger, Norway, and the western part of the Kola Peninsula, Russia. NGU Special Publication 7:29–47

Luckscheiter B, Morteani G (1981) The H contents of quartz from Alpine veins from the penninic rocks of the central and western tauern window (Austria/Italy). Tschermaks Mineralogisch-Petrologische Mitteilungen 28:223–228

Maschmeyer D, Lehmann G (1983) A trapped-hole center causing rose coloration of natural quartz. Zeitschrift für Kristallographie 163:181–196

McLaren AC, Cook RF, Hyde ST, Tobin RC (1983) The mechanism of the formation and growth of water bubbles and associated dislocation loops in synthetic quartz. Phys Chem Miner 9:79–94

Meinhold G (2010) Rutile and its applications in the earth sciences. Earth Sci Rev 102:1–28

Melezhik VA, Sturt BA (1994) A review of the general geology and history of the development of the early Proterozoic Polmalk-Pasvik-Pecheng Imandra/Varzuga-Ust'Ponoy Greenstone Belt. Earth Sci Rev 36:205–241

Miyoshi N, Yamaguchi Y, Makino K (2005) Successive zoning of Al and H in hydrothermal vein quartz. Am Mineral 90:310–315

Monecke T, Kempe U, Götze J (2002) Genetic significance of the trace element content in metamorphic and hydrothermal quartz: A reconnaissance study. Earth Planetary Sci Lett 202:709–724

Müller A, Koch-Müller M (2009) Hydrogen speciation and trace element contents of igneous, hydrothermal and metamorphic quartz from Norway. Mineral Mag 73:569–583

Müller A, Seltmann R, Behr HJ (2000) Application of cathodoluminescence to magmatic quartz in a tin granite—case study from the Schellerhau Granite Complex, Eastern Erzgebirge, Germany. Mineralium Deposita 35:169–189

Müller A, Kronz A, Breiter K (2002a) Trace elements and growth patterns in quartz: a fingerprint of the evolution of the subvolcanic Podlesi Granite System (Krušne Hory, Czech Republic). Bull Czech Geol Surv 77:135–145

Müller A, Lennox P, Trzebski R (2002b) Cathodoluminescence and micro-structural evidence for crystallisation and deformation processes of granites in the Eastern Lachlan Fold Belt (SE Australia). Contributions Mineral Petrol 143:510–524)

Müller A, Wiedenbeck M, van den Kerkhof AM, Kronz A, Simon K (2003a) Trace elements in quartz—a combined electron microprobe, secondary ion mass spectrometry, laser-ablation ICP-MS, and cathodoluminescence study. Eur J Mineral 15:747–763

Müller A, René M, Behr H-J, Kronz A (2003b) Trace elements and cathodoluminescence of igneous quartz in topaz granites from the Hub Stock (Slavkovsky′ Les Mts, Czech Republic). Mineral Petrol 79:167–191

Müller A, Breiter K, Seltmann R, Pécskay Z (2005a) Quartz and feldspar zoning in the Eastern Erzgebirge pluton (Germany, Czech Republic): evidence of multiple magma mixing. Lithos 80:201–227

Müller A, Wanvik JE, Kronz A (2005b) Norwegian kyanite quartzites—potential resources of high purity quartz? NGU Report 2005.039, Trondheim, Norway, 70 pp

Müller A, Williamson BJ, Smith M (2005c) Origin of quartz cores in tourmaline from Roche Rock, SW England. Mineral Mag 69:381–401

Müller A, Ihlen PM, Wanvik JE, Flem B (2007) High-purity quartz mineralisation in kyanite quartzites, Norway. Mineralium Deposita 42:523–535

Müller A, Ihlen PM, Kronz A (2008a) Quartz chemistry in polygeneration Sveconorwegian pegmatites, Froland, Norway. Eur J Mineral 20:447–463

Müller A, Wiedenbeck M, Flem B, Schiellerup H (2008b) Refinement of phosphorus determination in quartz by LA-ICP-MS through defining new reference material values. Geostand Geoanal Res 32(3):361–376

Müller A, Behr H-J, van den Kerkhof AM, Kronz A, Koch-Müller M (2010a) The evolution of late-Hercynian granites and rhyolites documented by quartz—a review. Earth Environ Sci Trans Royal Soc Edinburgh 100:185–204

Müller A, Herrington R, Armstrong R, Seltmann R, Kirwin DJ, Stenina NG, Kronz A (2010b) Trace elements and cathodoluminescence of quartz in stockwork veins of Mongolian porphyry-style deposits. Mineralium Deposita 45:707–727

Neumann H (1952) Feltspat forekomster i Tysfjorddistriktet. NGU Bergarkivrapport nr. 5208

Nordgulen Ø (1999) Geologisk kart over Norge, Berggrunnskart Hamar, M 1: 250.000. Geol Surv Norway, Trondheim

Northrup CJ (1997) Timing structural assembly, metamorphism, and cooling of the Caledonian nappes in the Ofoten-Efjorden area, north Norway: Tectonic insights from U-Pb and 40Ar/39Ar geochronology. J Geol 105:565–582

Norwegian Crystallites AS (2011) http://norcryst.no/. Accessed 21 Jan 2011

Parker RB (1962) Blue quartz from the Wind River Range, Wyoming. Am Mineral 47:1201–1202

Passchier CW, Trouw RAJ (2006) Microtectonics. Springer, Heidelberg 366 pp

Penniston-Dorland SC (2001) Illumination of vein quartz textures in a porphyry copper ore deposits using scanned cathodoluminescence: grasberg igneous complex, Irian Jaya, Indonesia. Am Mineral 86:652–666

Pfenninger H (1961) Diffusion von Kationen und Abscheidung von Metallen in Quarz unter elektrischer Feldeinwirkung. PhD Thesis, University Zürich

Ramseyer K, Mullis J (1990) Factors influencing short-lived blue cathodoluminescence of -quartz. Am Mineral 75:791–800

Richter DK (1971) Fazies- und Diagenesehinweise durch Einschlüsse in authigenen Quarzen. Neues Jahrbuch für Geologie und Paläontologie Monatshefte 10:604–622

Roedder E (1984) Fluid inclusions. Reviews in mineralogy, vol. 12. Mineralogical Society of America, Washington, 644 p

Rusk BG, Lowers HA, Reed MH (2008) Trace elements in hydrothermal quartz: relationships to cathodoluminescence textures and insights into vein formation. Geology 36:547–550

Sawyer E (1986) Metamorphic assemblages and conditions in the Rombak basement window. NGU Rapport 88.116, Trondheim, Norway, 11 pp

Seifert W, Rhede D, Thomas R, Förster H-J, Lucassen F, Dulski P, Wirth R (2011) On the origin of igneous blue quartz: inferences from a multi-analytical study of submicron mineral inclusions. Mineral Mag 75:2519–2534

Shepherd TJ, Rankin AH, Alderton DHM (1985) A practical guide to fluid inclusion studies. Blackie and Sons, Glasgow 239 pp

Siebers FB (1986) InhomogeneVerteilung von Verunreinigungen in gezüchteten und natürlichen Quarzen als Funktion derWachstumsbedingungen und ihr Einfluß auf kristallphysikalische Eigenschaften. PhD Thesis, Ruhr-Universität Bochum, 133 pp

Siedlecka A, Nordgulen Ø (1996) Geologisk kart over Norge, berggrunnskart Kirkenes, M 1:250 000. Geological Survey of Norway, Trondheim, Norway

Sigmond EMO (1998) Geologisk kart over Norge; Berggrunnskart Odda–M 1:250.000. Geological Survey of Norway, Trondheim, Norway

Simon K (2001) Does $\delta D$ from fluid inclusion in quartz reflect the original hydrothermal fluid? Chem Geol 177:483–495

Simpson DR (1977) Aluminum phosphate variants in feldspars. Am Mineral 62:351–355

Solli A, Nordgulen Ø (2006) Bedrock map of Norway and the Caledonides in Sweden and Finland. Scale 1: 2 000 000. Geological Survey of Norway, Trondheim

Stephens MB, Gustavson M, Ramberg IB, Zachrisson E (1985) The Caledonides of central north Scandinavia—a tectonostratigraphic overview. In: Gee DG, Sturt BA (eds) The Caledonide Orogen—Scandinavia and Related Areas. Wiley, New York, pp 135–162

Thomas S-M (2008) Wasserstoff in nominell wasserfreien Mineralen. PhD thesis. TU Berlin, D 83, Berlin, Germany, 134 pp

Thomas R, Webster JD, Davidson P (2006) Understanding pegmatite formation: the melt and fluid inclusion approach. In: Webster JD (ed) Melt inclusions in plutonic rocks. Mineralogical Association of Canada, Short Course Series 36:189–210

Tveten E, Lutro O, Thorsnes T (1998) Geologisk kart over Norge, berggrunnskart Ålesund, 1: 250.000. Geological Survey of Norway, Trondheim, Norway.

Van den Kerkhof AM, Hein UF (2001) Fluid inclusion petrography. Lithos 55:27–47

Van den Kerkhof AM, Müller A (1999) Fluid inclusion re-equilibration and trace element redistribution in quartz: observations by cathodoluminescence microscopy. ECROFI XV 1999 Abstracts and Program, Potsdam, Terra Nostra 99(6):161–162

Van den Kerkhof AM, Kronz A, Simon K, Scherer T (2004) Fluid-controlled quartz recovery in granulite as revealed bycathodoluminescence and trace element analysis (Bamble sector, Norway). Contributions Mineral Petrol 146:637–652

Wanvik JE (1988) Svanvik kvartsforekomst i Pasvik, Sør-Varanger kommune. NGU Rapport 87.081, Trondheim, Norway, 18 pp

Wanvik JE (1989a) Statusrapport 1989 for undersøkelse av Svanvik kvartsforekomst. NGU Rapport 89.078, Trondheim, Norway, 17 pp

Wanvik JE (1989b) Sluttrapport for undersøkelse av Svanvik kvartsforekomst. NGU Rapport 89.165, Trondheim, Norway, 9 pp

Wanvik JE (1998) Kyanite investigations in Tverrådalen, Surnadal. NGU Rapport 98.080, Trondheim, Norway, 24 pp

Wanvik JE (2001) Kvartsressurser i Nordland. NGU Rapport 2001.020, Trondheim, Norway, 103 pp

Wanvik JE (2009) Melkfjell kvartsittforekomst–feltundersøkelser høsten 2008. NGU Rapport 2009.025, Trondheim, Norway, 51 pp

Wark DA, Watson EB (2006) TitaniQ: a titanium-in-quartz geothermometer. Contributions Mineral Petrol 152:743–754

Watt GR, Wright P, Galloway S, McLean C (1997) Cathodoluminescence and trace element zoning in quartz phenocrysts and xenocrysts. Geochimica et Cosmochimica Acta 61: 4337–4348

Webster JD (ed) (2006) Melt inclusions in plutonic rocks. Mineralogical Association of Canada, Short Course Series 36, Montreal, Canada, 237 pp

Weil JA (1984) A review of electron spin spectroscopy and its application to the study of paramagnetic defects in crystalline quartz. Phys Chem Miner 10:149–165

Weil JA (1993) A review of the EPR spectroscopy of the point defects in a-quartz: The decade 1982–1992. In: Helms CR, Deal BE (eds) Physics and Chemistry of $SiO_2$ and the Si-$SiO_2$ interface 2. Plenum Press, New York, pp 131–144

Zolensky ME, Sylvester PJ, Paces JB (1988) Origin and significance of blue coloration in quartz from Llano rhyolite (llanite), north-central Llano County, Texas. Am Mineral 73:313–332

# Chapter 5
# Evaluation of the Potential of the Pegmatitic Quartz Veins of the Sierra de Comechigones (Argentina) as a Source of High Purity Quartz by a Combination of LA-ICP-MS, ICP, Cathodoluminescence, Gas Chromatography, Fluid Inclusion Analysis, Raman and FTIR spectroscopy

**Giulio Morteani, Florian Eichinger, Jens Götze, Alexandre Tarantola and Axel Müller**

**Abstract** Due to the increasing demand for high purity quartz the pegmatitic quartz veins of the Guacha Corral shear zone (Sierra de Comechigones, Sierras Pampeanas, Argentina) get increasing economic interest. The presented combination of accurate field work backed by transmitted light microscopy, cathodoluminescence, LA-ICP-MS spot chemical analyses and analyses of solutes and gases liberated from fluid inclusions was developed in order to produce a robust exploration tool able to select in an early stage of prospection quartz veins that are promising for the production of high purity quartz. One of the important results of this combination of methods is the possibility to define the amount of lattice bound

G. Morteani (✉) · F. Eichinger
Hydroisotop GmbH, Woelkestrasse 9, 85301 Schweitenkirchen, Germany
e-mail: gmorteani@gmx.de

F. Eichinger
e-mail: fe@hydroisotop.de

J. Götze
TU Bergakademie Freiberg, Institut für Mineralogie, Brennhausgasse 14,
09596 Freiberg, Germany

A. Tarantola
G2R-CNRS, Lorraine University, B.P. 70239 54506 Vandœuvre-lès-Nancy, France
e-mail: alexandre.tarantola@univ-lorraine.fr

A. Müller
Geological Survey of Norway, Leiv Eirikssons vei 39, Sluppen 6315,
7491 Trondheim, Norway
e-mail: axel.muller@ngu.no

J. Götze and R. Möckel (eds.), *Quartz: Deposits, Mineralogy and Analytics*, 119
Springer Geology, DOI: 10.1007/978-3-642-22161-3_5,

impurities. Such lattice bound impurities are difficult to eliminate by mineral preparation techniques whereas solid and fluid inclusions are amenable to a mineral preparation. In the present case the combination of methods produced a positive evaluation of the studied quartz veins as an excellent source for the production of high purity quartz.

## 5.1 Introduction

The demand for high purity quartz (HPQ) is increasing worldwide (Moore 2005, Haus 2005, Müller et al. 2007). This increase is driven mainly by the demand as raw material for special applications in the high-tech industry. The main suppliers for high purity quartz are Unimin (USA), the Quartz Corporation (USA) and Norwegian Crystallites (Norway). Unimin and Quartz Corporation mines weakly metamorphosed alaskite, Norwegian Crystallites mines the a zoned Drag pegmatite. In Russia hydrothermal quartz vein deposits that seem to be suitable for the production of high purity quartz are investigated at Ust-Puiva/Saranpaul/Tyumen Oblast (Subpolar Ural) (Burlakov 1995, 1999).

High purity quartz is characterized by less than 50 ppm impurities (e.g. Harben 2002). Lattice-bound trace elements which either substitute $Si^{4+}$ (i.e. $Al^{3+}$, $Fe^{3+}$, $Ti^{4+}$, $Ge^{4+}$, $B^{3+}$, $P^{5+}$) or occur at interstitial channel position such as $Li^{+}$, $K^{+}$, $Na^{+}$, $H^{+}$, $Fe^{2+}$ (e.g. Weil 1984, 1993) are the most common ones. They are the quality determining impurities because they are difficult up to impossible to eliminate by mineral processing (e.g. Jung 1992). Minor solid inclusions, typically rutile, mica and feldspars, that carry additional Ca, Na, K, Mg, Al, Sr, Rb, Sm, Nd and Ti (Rossmann et al. 1987), are amenable to elimination by mineral processing techniques. Fluid inclusions are potential sources for Na, K, Ca, Cl and different gas species such as $CO_2$, $N_2$, $H_2S$, $CH_4$ and higher hydrocarbons. However the components contained by fluid inclusions may be removed by milling and subsequent leaching, thermal treatment and calcination (e.g. Haus 2005). Mineral processing, that may include in addition to grinding, sieving and magnetic separation also thermal and chemical treatment by highly reactive acids such as hydrofluoric acid, can be very expensive. Thus, in the mined raw quartz material the content of intracrystalline impurities should be as low as possible to save treatment costs.

The evaluation of quartz deposits as potential sources of high purity quartz including the design of an optimal mineral processing flow sheet requires a reliable determination of the different impurities that are carried by quartz with a suitable combination of methods in a reasonable time and at tolerable costs.

The studied pegmatitic quartz veins belong to the hundreds of pegmatites dotting the whole Sierra de Comechigones (Argentina) ranging from differentiated and zoned ones to pegmatitic quartz veins with only subordinate K-feldspar and mica contents. The study area is given in Fig. 5.1. The pegmatitic veins of the Sierra de Comechigones belong to the large pegmatite province of the Sierras Pampeanas (Herrera 1968; Morteani et al. 1995). Specifically the pegmatitic quartz

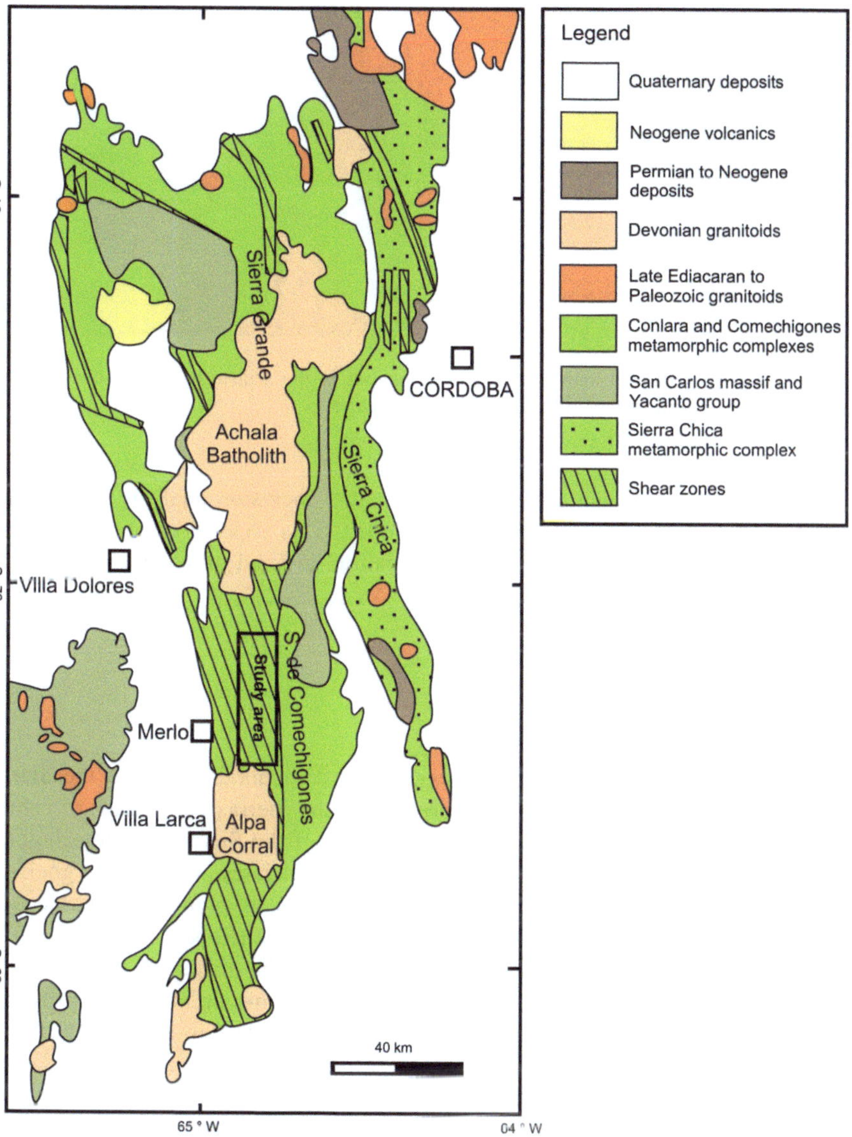


**Fig. 5.1** Geological sketch map of the Sierras Pampeanas according to Martino (2003), Steenken et al. (2006) and Siegesmund et al. (2010)

veins with white fine-grained quartz that are hosted in the Sierra de Comechigones by the mylonites of the Guacha Corral shear zone, get increasing interest as potential suppliers of high purity quartz to the electronic industry.

In order to get a quick and reliable information on the potential of the different quartz veins for the production of high purity quartz a combination of LA-ICP-MS,

**Fig. 5.2** *Left:* boudinaged and faulted quartz vein set in the schistosity of the country rock. *Right:* landscape with subhorizontal pegmatitic quartz veins

cathodoluminescence, gas chromatography, fluid inclusion analysis, Raman and FTIR spectroscopy was successfully developed. This paper presents exemplarily the results of such a multi method investigation on hand of selected quartz vein samples (Fig. 5.1).

## 5.2 Geology

The Sierra de Comechigones is the southernmost mountain range in the Sierras de Cordoba. The Sierras de Cordoba are part of the Sierras Pampeanas of central Argentina. The Sierra de Comechigones consists of the Sierra de Comechigones metamorphic complex, which is a sequence of gneisses, migmatitic gneisses, melt-depleted and melt-enriched diatexites, granulites, amphibolites, marbles and peridotites (Otamendi et al. 1999) with granitoid rocks (Rapela and Shaw 1979; Rapela et al. 1982, 1990). During the Cambrian the rocks of the Sierra de Cordoba underwent Pampean orogeny with a Barrovian-type metamorphism with peak P, T-conditions of 800–900°C and 8.5–9 kbar (Martino et al. 2009, 2010; Steenken et al. 2010, 2011; Otamendi et al. 1998; Martino 2003; Rapela and Shaw 1979). Such P–T conditions are sufficient to generate local partial melts (Sims et al. 1998). The metamorphic peak in the Sierra de Comechigones occurred at 550–540 Ma (Siegesmund et al. 2010). The P, T conditions of the mylonitisation that produced the Guacha Corral shear zone are 540–590°C and 3–6 kbar (Simpson et al. 2003; Whitmeyer and Simpson 2003; Martino 2003). The Guacha Corral shear belt is the most prominent tectonic feature of the Sierra de Comechigones (Fig. 5.1).

The quartz veins crop out as mostly sheared subhorizontal intrafolial irregular folded bodies with thickened fold hinges and flattened, often boudinaged limbs. The size of the bodies ranges from centimetres up to 7 m thickness (Fig. 5.2).

## 5.3 Methods

### *5.3.1 Bulk Chemical Composition by ICP-MS*

Sample aliquots (20–30 g) were at first washed in double rinsed deionised water and subsequently cleaned for three hours in concentrated $HNO_3$ and deionised water at about 100°C. Finally the samples were again rinsed with deionised water and dried in a dust free cabinet desiccator. The cleaned samples were totally dissolved by HF- microwave digestion in a closed vessel and the major and trace element concentration of the solution measured by ICP-MS (Perkin Elmer Elan 6000).

### *5.3.2 Spot Chemical Analysis by LA-ICP-MS*

Concentrations of Li, Be, B, Na, Al, P, K, Ca, Ti, Mn, Fe, and Ge were analysed by laser ablation inductively coupled plasma mass spectrometry (LA-ICP-MS). Point analyses on inclusion-free spots should provide data concerning the lattice-bound trace-element concentrations. The analyses were performed on the double-focusing sector field mass spectrometer (Thermo Instruments ELEMENT XR) which is combined with the New Wave 213-nm laser probe. The laser had a repetition rate of 20 Hz, a speed of 15 μm/s, a spot size of 100 μm, and energy fluence of about 14 mJ/cm$^2$ on the sample surface. Raster ablation was applied in the centre of quartz crystals on an area of approximately 200 × 500 μm. Line ablation was applied to determine trace element concentrations along grain boundaries. The approximate depth of ablation was about 100 μm. The carrier gas for transport of the ablated material to the ICP-MS was He mixed with Ar. External calibration was performed using four silicate glass reference materials (NIST SRM 610, 612, 614 and 616), the NIST SRM 1830 soda-lime float glass, the certified reference material BAM No.1 amorphous $SiO_2$ glass and the Qz-Tu synthetic pure quartz monocrystal. The analytical error ranges within 10% of the absolute concentration of the element. More details of the measurement procedure are provided by Flem et al. (2002).

### *5.3.3 Cathodoluminescence*

Cathodoluminescence (CL) measurements were done on carbon-coated, double-sided polished thick sections with a "hot cathode" CL microscope (HC1-LM, cf., Neuser et al. 1995). The CL system was operated at 14 kV accelerating voltage and a current density of about 10 μA/mm$^2$. Luminescence images were captured "on-line" during CL operations using a Peltier cooled digital video-camera (KAPPA 961-1138 CF 20 DXC). CL spectra in the wavelength range 380–1000 nm were recorded with an Acton Research SP-2356 digital triple-grating spectrograph with a Princeton Spec-10 CCD detector attached to the CL microscope by a silica-glass fiber guide. CL spectra were measured under standardized conditions (wavelength calibration by a Hg-halogen lamp, spot width 30 μm, measuring time 5 s). Spectral measurements in the UV region below 380 nm were not possible because of the UV absorption of the glass optics of the CL system.

### *5.3.4 Fluid Inclusion Investigations*

#### 5.3.4.1 Raman Spectroscopy

The gas phases of individual inclusions were analysed for $CO_2$, $N_2$, $H_2S$, $CH_4$, and $H_2$ at 31°C, just above the critical temperature of pure $CO_2$, using a Jobin–Yvon™ LabRAM HR800 laser-Raman spectrometer coupled with a Linkam™ heating–cooling stage. The excitating radiation, at 514.53 nm, is provided by an ionized argon laser. Raman shifts and scattering cross-sections for each gas species are according to Burke (2001).

#### 5.3.4.2 Fluorescence under UV-Light

The presence of C–C double bonds causes petroleum fluids to fluoresce in the visible range (400–700 nm) under excitation by UV light (Munz 2001). Thick quartz samples (200 μm) were placed under UV light at 365 nm produced by a high-pressure mercury lamp in order to check for the presence of hydrocarbons higher than $CH_4$.

#### 5.3.4.3 Fourier-Transform Infrared Spectroscopy

Two-phase hydrocarbon-bearing fluid inclusions showed high fluorescence under the 514.53 nm laser beam hampering the determination of their chemical composition by Raman spectroscopy. Therefore a qualitative evaluation of their gas composition was obtained using Fourier-Transform Infrared (FTIR) spectroscopy using a polychromatic source in the 1100–2000 nm range (Wopenka et al. 1990).

#### 5.3.4.4 Liberation and Quantification of Entrapped Gases

About 20 g of sample material (3.5–6 g per run) were crushed in an evacuated piston-cylinder device to liberate the gases trapped in the fluid inclusions

(Hämmerli and Diamond 2009; Eichinger et al. 2010). Before crushing, the sample chamber was evacuated, flushed twice with He to avoid any air contamination and filled with He to around 200 mbar. During the crushing process the equipment was heated to 150°C to avoid gas sorption on the freshly crushed quartz surfaces. After crushing the device was directly fitted to a gas chromatograph and the pressure and concentrations of normal gases ($CO_2$, $N_2$, Ar, $O_2$) and hydrocarbons were measured by GC-WLD and GC-FID (Shimadzu GC-17A), respectively. The volume of the extracted gas species per gram of quartz sample were calculated with respect of the weight of the sample, the gas pressure, the total volume of the piston-cylinder device and the temperature. Air contamination was quantified by the oxygen concentration and the results were corrected according to it.

#### 5.3.4.5 Liberation and Quantification of Entrapped Solutes

Aqueous leaching of the solutes entrapped in the fluid inclusions of quartz were conducted following the procedure described by Bottrell et al. (1988). Primarily the samples were cleaned according to the procedure described in Sect. 5.3.1. Subsequently two times 100 g of sample material were milled in a previously very carefully cleaned tungsten-carbide rotary mill for 15 min at 960 revolutions per minute. After comminution the samples were transferred to polycarbonate centrifuge vessels. To guarantee the quantitative transfer to the vessels, the mil was rinsed by extraction solution. Double deionised water was used as extraction solution to determine the concentrations of dissolved anions. To avoid any adsorption of dissolved cations to the quartz surfaces, an extraction solution, consisting of 0.13 mol $HNO_3$ and 200 ppm La (added as $LaCl_3$), was used to determine the concentrations of dissolved cations. The sample to solvent ratio was one. The aqueous extraction was shaken for three hours and subsequently centrifuged for 30 min. The solution was decanted in Teflon vessels. The concentration of major anions and cations of the aqueous extractions was analysed by IC (Dionex ICS 1500), the concentration of trace elements by ICP-MS (Perkin Elmer Elan 6000). The concentration of the leached solute is referred to the weight of the sample and of the leaching solution.

## 5.4 Results

Due to the ongoing exploration for high purity quartz in the Sierra de Comechigones the analytical results are identified only by sample numbers without a reference to specific quartz deposits/veins.

**Fig. 5.3** Transmitted light microphotograph showing coarse quartz grains surrounded by fine grained quartz grains. The texture is a typical mortar texture produced by metamorphic deformation and dynamic recrystallisation

### *5.4.1 Petrography*

The quartz samples of all studied occurrences showed under polarized light a mortar texture caused by a strong deformation with dynamic recrystallisation (Fig. 5.3). The fine-grained intergranular quartz mortar displays equilibrium grain boundaries documenting a post-deformational static recrystallisation. The high amount of fluid inclusions (see Sect. 5.4.5.1), and the absence of solid inclusions identifiable by optical microscopy is remarkable.

### *5.4.2 Cathodoluminescence*

The granular mortar texture of the quartz samples becomes also evident by cathodoluminescence (CL) investigations. All samples exhibit similar luminescence behaviour. Generally quartz shows an initial bluish luminescence, which turns into brownish-violet during electron irradiation (Fig. 5.4). The transient CL behaviour is also visible in the time-dependent CL spectra (Fig. 5.4). The spectrum is dominated initially by a blue emission band around 500 nm, which is typical for pegmatite quartz (Götze et al. 2005). This CL emission band can be related to alkali compensated trace element centres in the quartz structure (Ramseyer and Mullis 1990; Götze et al. 2005).

The large quartz crystals are rimmed by a seam of higher CL luminosity. This rim corresponds to the fine-grained quartz forming the mortar between the coarse quartz grains. In the coarser quartz grains the CL reveals several fluid trails with mostly higher CL intensity. The coarse quartz grains and the quartz forming the mortar showed in spite of different CL luminosity the same spectral characteristics. Therefore, the types of lattice defects are similar, but the higher CL intensity indicates higher defect densities in the mortar quartz.

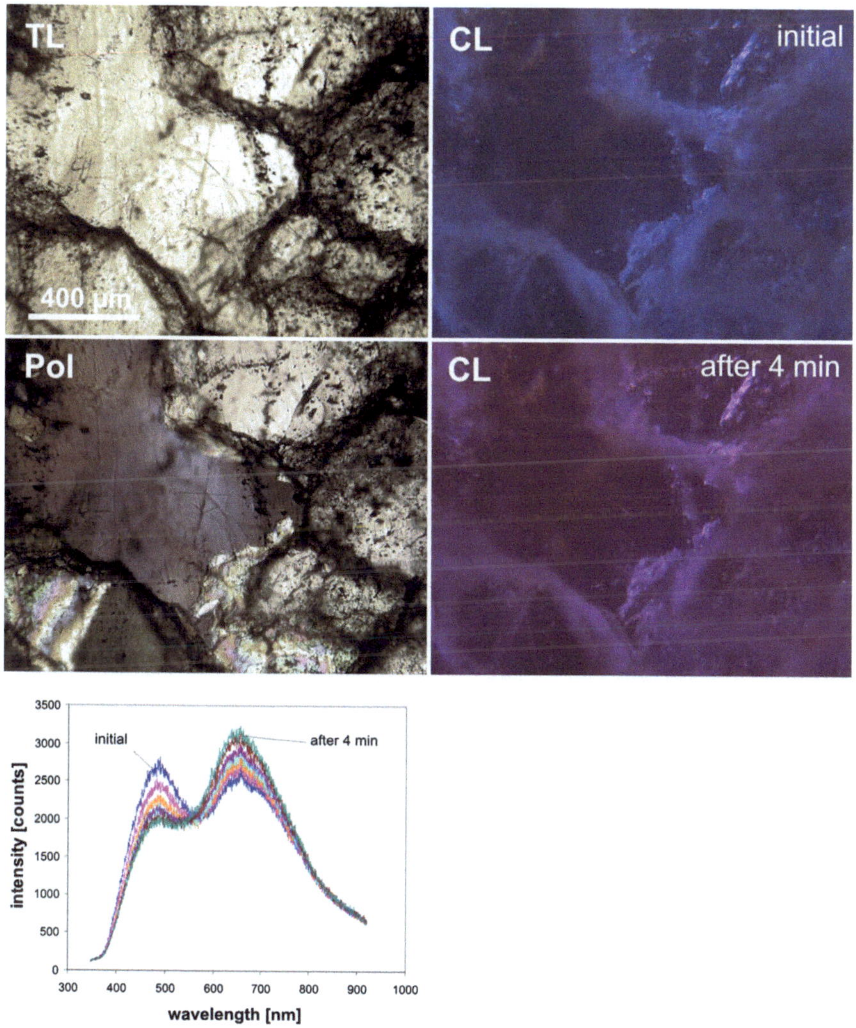


**Fig. 5.4** Microphotographs of the quartz sample 15681 in transmitted light (TL), polarized light (Pol), cathodoluminescence (CL) and time-dependent CL spectrum

## *5.4.3 Bulk Chemical Composition by ICP-MS*

The bulk quartz analyses show that the sum of the analysed elements in the two presented samples is 532 and 883 ppm (Table 5.1). In all analyzed samples the elements Na, K and Ca which are indicative for the presence of feldspar and possibly mica in the analyzed material occur in concentrations up to 200, 100 and 200 ppm respectively (Table 5.1). Magnesium is below the detection limit

**Table 5.1** Trace element concentrations by ICP-MS in two representative quartz samples from two different quartz veins (d.l. = detection limit)

| Element | Unit | Detection limit | 15656 | 15680 |
|---|---|---|---|---|
| Li | ppm | 0.5 | 0.7 | d.l. |
| Na | ppm | 100 | 200 | 200 |
| K | ppm | 100 | 100 | d.l. |
| Rb | ppm | 0.2 | 0.3 | 0.3 |
| Cs | ppm | 0.05 | 0.06 | 0.1 |
| Ca | ppm | 100 | 200 | d.l. |
| Mg | ppm | 0.01 | <0.01 | <0.01 |
| Sr | ppm | 0.2 | 1.4 | 0.3 |
| Ba | ppm | 1 | 0.5 | d.l. |
| Al | ppm | 100 | 300 | 300 |
| Ga | ppm | 0.1 | 0.4 | 0.3 |
| As | ppm | 0.1 | d.l. | d.l. |
| Se | ppm | 0.1 | d.l. | 0.3 |
| Ti | ppm | 5 | 20 | 26 |
| Zr | ppm | 1 | d.l. | 3 |
| Nb | ppm | 0.1 | 0.2 | 0.4 |
| La | ppm | 0.1 | 0.2 | 0.1 |
| Ce | ppm | 0.1 | 0.4 | 0.3 |
| Fe | ppm | 1.6 | d.l. | d.l. |
| SUM | ppm | | 883 | 532 |
| Chemical quality | | | low | low |

indicating the absence of mica inclusions. The concentrations of Al are 300 ppm for both samples and that of Ti are 20 and 26 ppm (Table 5.1). Both Al and Ti contents can be referred to solid inclusions of feldspar and rutile, but alternatively to lattice-bound Al and Ti.

Both samples contain traces of Ga, Rb, Cs, Sr, Nb, La and Ce. In contrast, Li, Ba, Se, and Zr occur as trace elements only in one of the two samples. As and Fe are below the detection limit (Table 5.1). According to Harben (2002) both quartz samples have to be classified as low quality ones.

## 5.4.4 Chemical Composition by LA-ICP-MS Spot Analysis

LA-ICP-MS analysis shows that in general coarse quartz grains have lower concentrations of impurities as compared with the fine-grained quartz grains rimming as mortar the coarser quartz grains (Table 5.2).

The coarse quartz sample 15656, for example, has very low average Li (<1 ppm) and low average Al (14.80 ppm) and Ti (5.5 ppm) (Table 5.2). The aluminium concentrations (14.80 ppm) are moderate. The analyses of the fine

Table 5.2 Chemical composition by LA-ICP-MS spot analysis of two selected quartz samples from two different quartz veins (d.l. = below detection limit)

| Element | Unit | Limit of detection | 15656 | | 15680 | |
|---|---|---|---|---|---|---|
| | | | Rim quartz | Coarse quartz | Rim quartz | Coarse quartz |
| Li | ppm | 0.55 | 0.87 | 0.61 | d.l. | 1.23 |
| Be | ppm | 0.06 | 0.10 | 0.20 | d.l. | 0.15 |
| B | ppm | 1.00 | 1.23 | 2.57 | d.l. | d.l. |
| Na | ppm | 5.5 | d.l. | d.l. | d.l. | d.l. |
| Al | ppm | 6.0 | 26.10 | 14.80 | 62.92 | 23.38 |
| P | ppm | 3.0 | d.l. | d.l. | 4.27 | 9.19 |
| K | ppm | 3.2 | 5.5 | d.l. | 7.49 | d.l. |
| Ca | ppm | 14.0 | 15.4 | d.l. | d.l. | d.l. |
| Ti | ppm | 1.20 | 6.84 | 5.55 | d.l. | d.l. |
| Mn | ppm | 0.50 | d.l. | d.l. | 0.64 | 3.03 |
| Fe | ppm | 1.00 | d.l. | d.l. | 1.35 | d.l. |
| Ge | ppm | 0.17 | 0.48 | 0.42 | 0.43 | 1.12 |
| SUM | ppm | | 56.52 | 24.15 | 77.10 | 38.10 |
| Chemical quality of quartz | | | Medium | High | Medium | High |

grained rim quartz grains show elevated Al, Ca, Ti and K contents compared to the analyses of the coarse grains (Table 5.2).

In the coarse and fine-grained quartz samples 15680 the Ti is below the detection limit of 1.2 ppm. In the fine grained quartz the Al, K and Fe contents in the rim quartz are higher as compared to the coarse quartz. The reverse can be observed for Li, P, Mn and Ge (Table 5.2). According to Harben (2002) and Müller et al. (2007) the total trace element content of less than 50 ppm classifies the coarse quartz grains of both samples given in Table 5.2 as high purity quartz whereas the fine grained quartz has to be classified only as one of medium quality.

### 5.4.5 Fluid Inclusions

#### 5.4.5.1 Petrography

The quartz samples are very dense in fluid inclusion planes. The size of these inclusions is generally smaller than 20 μm. At room temperature three different populations of fluid inclusions are distinguished throughout all the samples:

Population 1 is represented by three-phase inclusions at room temperature (Fig. 5.5a). These inclusions are located in early pseudo-secondary trails within quartz grains showing undulatory extinction. Some trails show low gas/liquid ratio with gas proportions of about 20 Vol.%, some others are characterized by gas proportions up to 80 Vol.%. Variable gas/liquid ratio can also be observed within the same trail.

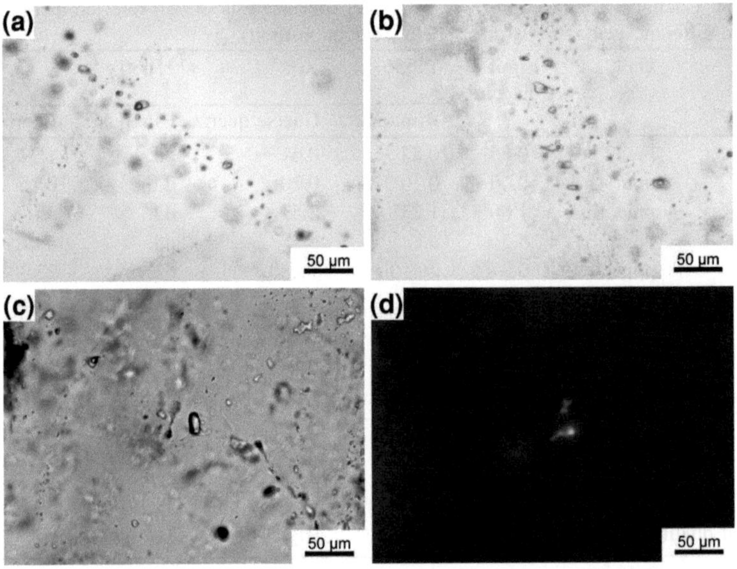


**Fig. 5.5** **a** Three-phase ($L_{aq}$, $L_{car}$, $V_{car}$) inclusions (population 1) located in a pseudosecondary trail; **b** Two-phase inclusions (population 2) located in a secondary plane crosscutting an earlier three-phase inclusions trail; **c,d** Higher hydrocarbon bearing inclusion (sample 15656, population 3) as seen within a 200 μm thick quartz section under transmitted light (**c**) and UV-light (**d**). Only the liquid phase fluoresces and is indicative of the presence of hydrocarbons higher than $C_1$

Population 2 consists of two-phase (gas, liquid) inclusions (Fig. 5.5b). These inclusions are located along late secondary trails crosscutting grain boundaries and recrystallized areas.

Population 3 consists of two-phase, generally flat, inclusions with gas/liquid ratios ranging from 20 to 80 Vol.%. These inclusions show light blue to yellow fluorescence within the liquid phase under UV-light (Fig. 5.5c, d). This feature is indicative of the presence of hydrocarbons (HC) higher than $C_1$ in the liquid phase (e.g. Burruss 2003). These inclusions are rare and seem to be generally located in late secondary cracks. However, many of the HC inclusions are dismembered (Tarantola et al 2010) and some of them appear relatively isolated in the quartz samples.

#### 5.4.5.2 Raman Spectroscopy

The composition of selected fluid inclusions of populations 1 and 2 as given by Raman spectroscopy are given in Table 5.3. The populations 1 and 2 differ in their gas composition:

Population 1: In all analyzed fluid inclusions of the population 1 the composition of the carbonic phase was constant. The composition is dominated by $CO_2$

**Table 5.3** Raman spectroscopy data ranges for selected inclusions of populations 1 (n = 8) and 2 (n = 4)

| Inclusion population | $CO_2$ (mol%) | $N_2$ (mol%) | $H_2S$ (mol%) |
|---|---|---|---|
| 1 | 99.3 | 0.7 | <0.1 |
| 1 | 97.7 | 2.3 | <0.1 |
| 2 | 47.7 | 52.3 | n.d. |
| 2 | 98.9 | 1.1 | n.d. |

The composition in mol% refers to the composition of the gas bubble at 31°C (n.d. = not detected)

(>97.7 mol%) but low amounts of $N_2$ (<2.3 mol%) and traces of $H_2S$ (<0.1 mol%) are always present (Table 5.3).

Population 2: In the fluid inclusions of population 2 the gas bubble has a strong tendency to change position under the Raman laser beam. Raman spectroscopy reveals in the fluid inclusions the presence of a mixture of $CO_2$ and $N_2$. The concentrations of $CO_2$ and $N_2$ vary significantly between 1.1 and 52 mol% for $N_2$ and 47.7 and 98.9 mol% for $CO_2$ (Table 5.3).

#### 5.4.5.3 Fourier-Transform Infrared Spectroscopy

The presence of hydrocarbon molecules in the fluid inclusions of population 3 was confirmed by FTIR spectroscopy. Two different sets of compositions came out. Inclusions with saturated hydrocarbons ($C_2$, $C_3$) only and inclusions with unsaturated hydrocarbons ($C_2$, $C_3$), only. Methane was detected in some inclusions, too.

#### 5.4.5.4 Chemical Composition of Solutes

Aqueous extractions show, that the content in solutes (TDS) varies between 48.0 and 257.0 ppm (Table 5.4). Comparing the amounts of total dissolved solids (TDS) with the total gas volumina (cf. Sect. 5.4.5.5) extracted from the fluid inclusions, it is obvious that the quartz vein samples with the highest TDS have also the highest gas volumina and vice versa. This indicates that the frequency of fluid inclusions is higher in samples with high TDS and gas concentrations than in sample with lower ones. In the analyzed samples the concentrations of Na (11.8, 72.7 ppm), K (2.5, 11.8 ppm) and Cl (19.2, 122.8 ppm) and the molar ratios of (Na + K) to Cl, which are 1.0 and 1.1 indicate that the salt phase dissolved in the fluid inclusions consists predominantly of a Na–K-Cl salt component. The concentrations of Ca and Mg, which are 12.2 and 14.5 ppm for Ca and 0.1 and 6.4 ppm for Mg, respectively, are probably caused by the dissolution of small amounts of Ca–Mg carbonates, which occur in the samples. The concentrations of Al extracted are 1.1 and 10.9 ppm (Table 5.4). The high Al concentrations are probably caused by the dissolution of low amounts of Al-bearing minerals,

**Table 5.4** Chemical composition of the solutes included in fluid inclusions in quartz of two selected samples (ppm = $\mu g/g_{Qtz}$, n.a. = not analysed)

| | | 15656 | 15680 |
|---|---|---|---|
| *Cations* | | | |
| Sodium ($Na^+$) | ppm | 11.8 | 72.7 |
| Potassium ($K^+$) | ppm | 2.5 | 11.8 |
| Calcium ($Ca^{2+}$) | ppm | 12.2 | 14.5 |
| Magnesium ($Mg^{2+}$) | ppm | 0.1 | 6.4 |
| Lithium ($Li^+$) | ppm | 0.02 | 0.09 |
| Strontium ($Sr^{2+}$) | ppm | 0.1 | 0.4 |
| Barium ($Ba^{2+}$) | ppm | 0.04 | 1.00 |
| Aluminum ($Al^{3+}$) | ppm | 1.1 | 10.9 |
| Beryllium ($Be^{3+}$) | ppm | n.a. | <0.01 |
| Germanium ($Ge^{4+}$) | ppm | n.a. | <0.003 |
| Phosphorus (P) | ppm | n.a. | 0.2 |
| Titanium ($Ti^{3+}$) | ppm | n.a. | 0.4 |
| *Anions* | | | |
| Fluoride ($F^-$) | ppm | 0.02 | 1.21 |
| Chloride ($Cl^-$) | ppm | 19.2 | 122.8 |
| Bromide ($Br^-$) | ppm | 0.07 | <0.05 |
| Sulfate ($SO_4^{2-}$) | ppm | 0.8 | 14.5 |
| Total dissolved solids (TDS) | ppm | 48.0 | 257.0 |

like feldspars or micas. Concentration of sulfate is 0.8 and 14.5 ppm (Table 5.4). The detected sulfate is probably formed during the extraction procedure by the oxidation of extracted $H_2S$, which was observed in inclusions of population 1 (cf. Sect. 5.4.5.2). Traces of Li, Sr, Ba, Ti and partly Ge, P, F and Br are also present (Table 5.4).

#### 5.4.5.5 Chemical Composition of Entrapped Gases

The total gas concentrations liberated from the samples are between 58.0 and 79.9 μl/g STP (Table 5.5). In the rock samples, carbon dioxide ($CO_2$) and nitrogen ($N_2$) are the dominant gas species. Their concentrations range between 31.2 and 65.0 μl/g for $CO_2$ and between 14.7 and 26.5 μl/g for $N_2$, respectively. In all samples also argon (Ar) and hydrocarbons (HC) could be detected. Their concentrations are 0.15 and 0.16 μl/g for Ar and 0.03 and 0.13 μl/g for HCs (Table 5.5, Fig. 5.6). The proportions of Ar and HCs of the total gas volume are 0.20 (15680) and 0.26 Vol.% (15656) and 0.04 (15680) and 0.22 Vol.% (15656), respectively.

The different gas concentrations and different proportions of the individual gas species are attributed different distributions of fluid inclusions in the quartz and different abundances of the individual fluid inclusion generations (cf. Sect. 5.4.5.1).

**Table 5.5** Composition of the gases from fluid inclusions (µl/g) in quartz of two selected samples from different occurrences, gas volumina are calculated without water vapour

| Sample | | 15656 | 15680 |
|---|---|---|---|
| *Normal gases* | | | |
| Argon (Ar) | µl/g | 0.15 | 0.16 |
| Nitrogen ($N_2$) | µl/g | 26.5 | 14.7 |
| Carbondioxide ($CO_2$) | µl/g | 31.2 | 65.0 |
| *Saturated hydrocarbons* | | | |
| Methane ($CH_4$) | µl/g | 0.0888 | 0.0238 |
| Ethane ($C_2H_6$) | µl/g | 0.0071 | 0.0017 |
| Propane ($C_3H_8$) | µl/g | 0.0043 | <0.0001 |
| n-Butane (n-$C_4H_{10}$) | µl/g | 0.0019 | 0.0006 |
| i-Butane (i-$C_4H_{10}$) | µl/g | 0.0050 | 0.0006 |
| i-Pentane (i-$C_5H_{12}$) | µl/g | <0.0001 | <0.0001 |
| n-Pentane (n-$C_5H_{12}$) | µl/g | 0.0024 | <0.0001 |
| Hexane ($C_6H_{14}$) | µl/g | 0.0011 | <0.0001 |
| *Unsaturated hydrocarbons* | | | |
| Ethene ($C_2H_4$) | µl/g | 0.0063 | 0.0019 |
| Propene ($C_3H_6$) | µl/g | 0.0094 | <0.0001 |
| Cis(1)-Butene (cis-$C_4H_8$) | µl/g | 0.0041 | <0.0001 |
| Trans-Butene (trans-$C_4H_8$) | µl/g | 0.0001 | <0.0001 |
| Pentene ($C_5H_{10}$) | µl/g | 0.0004 | <0.0001 |
| Hydrocarbons tot | µl/g | 0.131 | 0.03 |
| Gas total | µl/g | 58.0 | 79.9 |

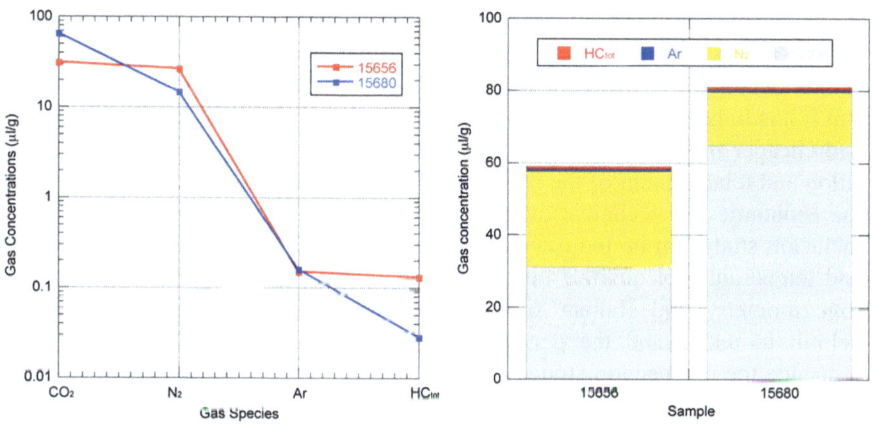


**Fig. 5.6** Concentrations of the individual gas species liberated from pegmatitic quartz veins as function of the gas species (*left*) and as function of the individual samples (*right*); $HC_{tot}$ = total volume of extracted hydrocarbons

The predominant hydrocarbon species, extracted from the rock samples is methane ($CH_4$). In all samples also traces of saturated hydrocarbons (Ethane-Hexane) and unsaturated hydrocarbons (Ethene-Pentene) could be detected (Table 5.5). The abundance of hydrocarbons is related to the occurrence of population three inclusions, which contain methane and higher saturated or unsaturated hydrocarbons (cf. Sect. 5.4.5.3).

## 5.5 Discussion

In the studied samples the LA-ICP-MS analyses reveal that the fine grained quartz found as a mortar between coarse quartz grains is the predominant carrier of impurities (Table 5.2). Already without any mineral preparation the coarser quartz crystals fit with 24.15 and 38.10 ppm of impurities the criteria for high purity quartz. Bulk analysis by total digestion and subsequent ICP-MS analysis only would have missed this point, which is important for the planning of mineral separation tests and flow sheet. Analytical results and microscopy suggest that the impurities in quartz were expelled by the deformation and metamorphic recrystallisation leading to an enrichment of the impurities in the fine grained intragranular quartz mortar.

A comparison of the solution ICP-MS and the LA-ICP-MS data shows that as a whole the content in impurites determined by ICP-MS after total dissolution of the quartz samples is much higher than that determined by LA-ICP-MS spot analysis. This indicates that most of the impurities are bound to solid or fluid inclusions and not to the quartz lattice.

The results obtained from conventional fluid inclusion investigation and the results obtained by aqueous leaching and gas liberation do complement another. But it has to be considered that the extraction methods are comparatively quicker and cheaper methods that can produce reliable information on the bulk composition and total amount of the fluids stored in the inclusions, a key information for the economic and technical evaluation of quartz material. Nevertheless, a fluid inclusion study can be indicated for the estimation of the crystallisation pressure and temperature of quartz from a P,T plot of the isochores obtained after microthermometry and Raman spectroscopy investigations. Such data might be helpful to understand the genesis of the present pegmatitic quartz veins and influence the prospection strategy.

## 5.6 Conclusion

Our multi-technique investigations approach shows that the pegmatitic quartz veins of the Sierra de Comechigones are rightly in the focus of the exploration for high purity quartz. The here given first results show that a combination of careful field work with microscopy in transmitted light, cathodoluminescence, LA-ICP-MS spot analyses and aqueous extraction of the solutes (and gases) liberated from the fluid inclusions by crushing gives within reasonable time and financial effort a robust information about the amount and type of impurities in quartz. The suggested analytical programme allows in an early stage of prospection the selection of individual quartz veins for the production of high purity quartz. A study dealing with the genesis of the pegmatitic quartz veins of the Serra de Comechigones is in preparation.

**Acknowledgments** We thank J. Sfagulla (Cordoba, Argentina), H. J. Heitzig (Pfaffenhofen a. d Glonn, Germany) and G. von Gromann (Cordoba, Argentina) for financial support, help in the field and discussions about geology and economics of quartz. We thank A. M. von der Kerkhof (Göttingen) for a thoughtful review and C. Preinfalk for accurate proof reading.

## References

Bottrell SH, Yardley B, Buckley F (1988) A modified crush-leach method fort he analysis of fluid inclusion electrolytes. Bulletin de Mineralogie 111:279–290

Burke EAJ (2001) Raman microspectrometry of fluid inclusions. Lithos 55:139–158

Burlakov EV (1995) Dodo: Alpine Klüfte im Polar-Ural. Lapis, 20, 13–26

Burlakov EV (1999) The Dodo deposit (subpolar Urals, Russia). Mineral Rec 30:427–442

Burruss RC (2003) Petroleum fluid inclusions, an introduction. In: Samson IM, Anderson AJ, Marshall DD (eds) Fluid inclusions: analysis and interpretation. Mineralogical Association Canada, Short Course Series 32, 159–174

Eichinger F, Meier D, Hämmerli J, Diamond LW (2010) Stable isotope signatures of gases liberated from fluid inclusions in Bedrock at Olkiluoto. Posiva Working Report 2010–88, Posiva Oy, Olkiluoto, Finland. (www.posiva.fi)

Flem B, Larsen RB, Grimstvedt A, Mansfeld J (2002) In situ analysis of trace elements in quartz by using laser ablation inductively coupled plasma mass spectrometry. Chem Geol 182: 237–247

Götze J, Plötze M, Trautmann T (2005) Structure and luminescence characteristics of quartz from pegmatites. Am Mineral 90:13–21

Hämmerli J, Diamond LW (2009) Fluid inclusions in basement rocks at Olkiluoto, Finland, and their implications for a planned nuclear waste repository. In: ECROFI-XX, Granada, Spain, pp 109–110

Harben PW (2002) The industrial mineral handybook—a guide to markets, specifications and prices. Industrial mineral information, Worcester Park, United Kingdom, 4th edn p 412

Haus R (2005) High demands on high purity. Ind Mineral 10:62–67

Herrera AO (1968) Geochemical evolution of zoned pegmatites of Argentina. Econ Geol 63: 13–29

Jung L (1992) High purity natural quartz. Part I: High purity natural quartz for industrial use. Part II: High purity natural quartz markets for suppliers and users. Quartz Technology. Liberty Corner, New Jersey, p 657

Martino R (2003) Las fajas de deformation ductil de la Sierras Pampeanas de Córdoba: una reseña general. Revista de la Association Geologica Argentina 58:549–571

Martino RD, Guereschi AB, Sfragulla JA (2009) Petrology, structure and tectonic significance of the Tuclame banded schists in the Sierras Pampeanas of Còrdoba and its relationships with the metamorphic basement of northwestern Argentina. J South Am Earth Sci 27:280–298

Martino RD, Guereschi AB, Anzil A (2010) Metamorphic and tectonic evolution at 31°36″S across a deep crustal zone from the Sierra Chica of Cordoba, Sierras Pampeanas, Argentina. J South Am Earth Sci 30:12–28

Moore P (2005) High-purity quartz. Ind Minerals 455:53–57

Morteani G, Preinfalk C, Spiegel W, Bonalumi A (1995) The Achala granitic complex and the pegmatites of the Sierras Pampeanas (Northwest Argentina): A study in differentiation. Econ Geol 90:636–647

Müller A, Ihlen PM, Wanvik JE, Flem B (2007) High-purity quartz mineralisation in kyanite quartzites, Norvay. Miner Deposita 42:523–535

Munz IA (2001) Petroleum inclusions in sedimentary basins: systematics, analytical methods and applications. Lithos 55:195–212

Neuser RD, Bruhn F, Götze J, Habermann D, Richter DK (1995) Kathodolumineszenz: Methodik und Anwendung. Zentralblatt für Geologie und Paläontologie Teil I, H 1(2):287–306

Otamendi JE, Nullo FE, Patiño Douce AE, Fagiano M (1998) Geology, mineralogy and geochemistry of syn-anatectic granites from the Achiras complex, Córdoba, Argentina; some petrogenetic and geodynamic implications. J South Am Earth Sci 11:407–423

Otamendi JE, Patino Douce AE, Demichelis AH (1999) Amphibolite to granulite transition in aluminous greywackes from the Sierra de Comechingones, Córdoba, Argentina. J Metamorphic Geol 17:415–434

Ramseyer K, Mullis J (1990) Factors influencing short-lived blue cathodoluminescence of alpha-quartz. Am Mineral 75:791–800

Rapela CW, Shaw DM (1979) Trace and major element models of granitoid genesis in the Pampean Ranges, Argentina. Geochim Cosmochim Acta 43:1117–1129

Rapela CW, Heaman LM, McNutt RH (1982) Rb-Sr Geochronology of granitoid rocks from the Pampean Ranges, Argentina. J Geol 90:574–582

Rapela CW, Toselli A, Heaman L, Saavedra J (1990) Granite plutonism of the Sierras Pampeanas, An inner cordilleran Paleozoic arc in the southern Andes. Geol Soc Am 241:77–90 Special paper

Rossman GR, Weis D, Wasserburg GJ (1987) Rb, Sr, Nd and Sm concentrations in quartz. Geochim Cosmochim Acta 51:2325–2329

Siegesmund S, Steenken A, Martino RD, Wemmer K, Lopez de Luchi M, Frei R, Presnyakov S, Guereschi A (2010) Time constraints on the tectonic evolution of the Eastern Sierras Pampeanas (Central Argentina). Int J Earth Sci 99:1199–1226

Simpson C, Law RD, Gromet LP, Mirò R, Northrup CJ (2003) Paleozoic deformation in the Sierras de Córdoba and Sierra de la Minas, eastern Sierras Pampeanas, Argentina. J South Am Earth Sci 15:749–764

Sims JP, Ireland TR, Camacho A, Lyons P, Pieters PE, Skirrow RG, Stuart-Smith PG, Mirò R (1998) U-Pb, Th-Pb, and Ar–Ar geochronology from the southern Sierras Pampeanas: implications for the Paleozoic tectonic evolution ofr the western Proto-Andean Margin of Gondwana, vol. 142, Geol Soc London Spec Publ 259–281

Steenken A, López de Luchi MG, Martinez Dopico C, Drobe M, Wemmer K, Siegesmund S (2011) The Neoproterozoic-early Paleozoic metamorphic and magmatic evolution of the Eastern Sierras Pampeanas: an overview. Int J Earth Sci 100:465–488

Steenken A, Siegesmund S, López de Luchi MG, Frei R, Wemmer K (2006) Neoproterozoic to Early Palaeozoic events in the Sierra de San Luis: implications fort he Famatinian geodynamics in the Eastern Sierras Pampeanas (Argentina). J Geol Soc 163:965–982

Steenken A, Wemmer K, Martino RD, Lopez de Luchi M, Guareschi A, Siegesmund S (2010) Post-Pampean cooling and uplift of the Sierras Pampeanas in the west of Córdoba (Central Argentina). N Jb Geol Paleont Abh 256:235–255

Tarantola A, Diamond LW, Stünitz H (2010) Modification of fluid inclusions in quartz by deviatoric stress I: Experimentally induced changes in inclusion shapes and microstructures. Contrib Min Pet 160:825–843

Weil JA (1984) A review of electron spin spectroscopy and its application to the study of paramagnetic defects in crystalline quartz. Phys Chem Mineral 10:149–165

Weil JA (1993) A review of the EPR spectroscopy of the point defects in αa-quartz: The decade 1982–1992. In: Helms CR, Deal BE (eds) Physics and chemistry of $SiO_2$ and the Si-$SiO_2$ interface 2. Plenum Press, New York, pp 131–144

Whitmeyer SJ, Simpson C (2003) High strain-rate deformation fabrics characterize a kilometres-thick Paleozoic fault zone in the Eastern Sierras Pampeanas, central Argentina. J Struct Geol 25:909–922

Wopenka B, Pasteris JD, Freeman JJ (1990) Analysis of individual fluid inclusions by Fourier transform infrared and Raman microspectroscopy. Geochim Cosmochim Acta 54:519–533

# Chapter 6
# Brazilian Quartz Deposits with Special Emphasis on Gemstone Quartz and its Color Treatment

**Ricardo Scholz, Mario L. S. C. Chaves, Klaus Krambrock, Maurício V. B. Pinheiro, Sandra B. Barreto and Messias G. de Menezes**

**Abstract** The exploration of Brazilian quartz deposits started in the beginning of the twentieth Century, with intensification of production during the Second World War. Four geological environments are the sources for the different types of quartz in Brazil: (1) Neoproterozoic granitic pegmatites—gemstones and minor industrial quartz (2) Neoproterozoic hydrothermal veins—industrial quartz and minor gemstones, (3) Mesozoic basaltic sheets with amethyst and agate—gemstones and (4) Cenozoic secondary deposits—industrial sands. Industrial quartz occurs in Brazil as lascas and sands and the most important sources are the sedimentary deposits of Botucatu and Piramboia Formations and the hydrothermal veins of the Espinhaço Range. The production is mainly used in the metallurgical industry, in the process of production of ferrosilicon alloys and in the glass industry. Between 1996 and 2005, official data suggest a total production of 1,143,497 tons of lascas and up to 40 Mt of industrial sands. The measured resources of about 2,400 Mt, indicate a potential for growth in the industrial quartz market. The main production is located in the states of São Paulo, Santa Catarina and Minas Gerais.

R. Scholz (✉) · M. G. de Menezes
Departamento de Geologia, Escola de Minas, Universidade Federal de Ouro Preto (UFOP), Ouro Preto, MG 31400-000, Brazil
e-mail: r_scholz_br@yahoo.com

M. L. S. C. Chaves
Departamento de Geologia, IGC, Universidade Federal de Minas Gerais (UFMG), Belo Horizonte, MG 30123-970, Brazil

K. Krambrock · M. V. B. Pinheiro
Departamento de Física, ICEx, Universidade Federal de Minas Gerais (UFMG), CP 702, Belo Horizonte, MG 30123-970, Brazil

S. B. Barreto
Departamento de Geologia, Universidade Federal do Pernambuco (UFPE), Recife, PE 50740-530, Brazil

J. Götze and R. Möckel (eds.), *Quartz: Deposits, Mineralogy and Analytics*, Springer Geology, DOI: 10.1007/978-3-642-22161-3_6,

Transparent single crystals and lascas of quartz from pegmatites, hydrothermal veins and geodes in basalts are the most common gemological material in Brazil. Part of this production is used for color treatment via irradiation and heating, to produce more attractive gemstones, sometimes with colors that will not be found in nature such as the green-gold type.

## 6.1 Introduction

Quartz is an important mineral resource in Brazil. The first ever mentioned finding of quartz crystals was in 1797 in Cristalina, Goiás State, by explorers of gold and emerald. The exploration of quartz intensified at the beginning of the Second World War with strong participation of the United States Geological Survey—USGS (Campbell 1946; Johnston and Butler 1946) due to increasing demand of crystal quartz (Arcoverde and Schobbenhaus 1991). Today Brazil is the leading producer of gemological natural quartz crystals in the world (Drummond 2009), with potential of growth in the production of metallurgical quartz and the glass industry.

With a territory of up to 8,514,876.6 $km^2$ and a complex geological evolution, the environments for formation of quartz mineralizations are manifold. The Brazilian quartz deposits occur in four different geological settings: (1) Neoproterozoic granitic pegmatites, (2) Neoproterozoic hydrothermal veins, (3) Mesozoic basaltic sheets (geodes filled by amethyst and agate in hydrothermal conditions) and (4) Cenozoic secondary deposits. A synthesis of the main Brazilian deposits and their genetic aspects are presented in this study (Fig. 6.1 and Table 6.1).

Brazil has one of the largest resources of industrial quartz in the world, as well as China, Madagascar, South Africa, Canada and Venezuela. The Brazilian resources of quartz lascas were estimated in 17 Mt in 1980 decade, and around 60% of them are to be found in the Minas Gerais State (Alecrim 1982). After this report, only in 2010 official data of estimated and measured resources were published. In 2005, total resources of industrial sand and lascas were measured up to 2,400 Mt. (MME 2010). The world resources of large natural crystals occur almost exclusively in Brazil and, in smaller volumes, in Madagascar (DNPM 2006).

In Brazil the production of quartz is predominant of small mining companies and of informal miners, both in the production of lascas and single crystals. Electronic grade crystals are rare and their production is sporadic. The absence of technological training does not allow the addition of value to the mineral good in the stages of extraction and processing (DNPM 2006). Official production between 1996 and 2005 was measured in 1,143,497 tons of lascas and up to 40 Mt of industrial sands (MME 2010), however the data is questionable.

The most important Brazilian industrial quartz resources are related to hydrothermal veins and secondary deposits. Transparent quartz crystals from Brazilian pegmatites show minor importance for metallurgical and technological industries,

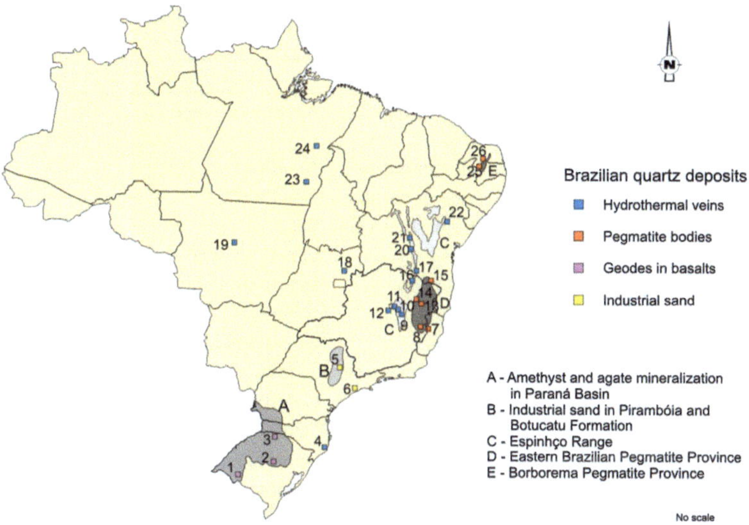


**Fig. 6.1** Location and genetic types of the Brazilian quartz deposits. Localities marked on the map are described in Table 6.1

however in the last 10 years the production is increasing, mainly for the usage as gemstone including color treatment.

## 6.2 Geology of Brazilian Quartz Deposits

### *6.2.1 Brazilian Pegmatite Provinces*

#### 6.2.1.1 Borborema Pegmatite Province

In the northeast region of Brazil the pegmatites are concentrated in the Borborema Pegmatite Province (BPP). The province extends over approximately 75 × 150 km in the eastern-southeastern part of the Seridó Fold Belt in the Northern Domain of the Borborema Tectonic Province, in the States of Paraíba and Rio Grande do Norte (Fig. 6.1).

The Seridó Fold Belt consists of a Paleoproterozoic gneiss-granite-migmatite basement and a Neoproterozoic supracrustal meta-volcano-sedimentary sequence (the Seridó Group) which comprises, from base to top, gneisses and marbles (Jucurutu Formation), quartzites and meta-conglomerates (Equador Formation) and garnet-cordicrite-sillimanite schists (Seridó Formation) (Van Schmus et al. 2003).

In this province more than 1,500 complex pegmatites have been catalogued and over 400 of them have been mined sporadically. The main pegmatite bodies

**Table 6.1** Genetic aspects and type of mineralization of the Brazilian quartz deposits. Localities marked on the map

| | Locality | Type of deposit | Type of quartz mineralization |
|---|---|---|---|
| 1 | Fronteira Sudeste—Quaraí and Livramento | Geodes in basalts | Agate and amethyst |
| 2 | Salto do Jacuí—Soledade | Geodes in basalts | Amethyst |
| 3 | Alto Uruguai | Geodes in basalts | Amethyst |
| 4 | Santa Catarina | Hydrothermal vein | Silex |
| 5 | Descalvado | Sedimentary | Industrial sands |
| 6 | Peruíbe and São Vicente | Sedimentary | Industrial sands |
| 7 | Santa Maria de Jetibá | Pegmatite | Amethyst |
| 8 | Conselheiro Pena District | Pegmatite | Gemological quartz and crystals |
| 9 | Gouveia | Hydrothermal vein | Gemological and industrial quartz |
| 10 | Sopa | Hydrothermal vein | Gemological and industrial quartz |
| 11 | Buenópolis | Hydrothermal vein | Gemological and industrial quartz |
| 12 | Serra do Cabral | Hydrothermal vein | Gemological and industrial quartz |
| 13 | Padre Paraíso | Pegmatite | Gemological quartz and crystals |
| 14 | Araçuaí | Pegmatite | Gemological quartz and crystals |
| 15 | Itambé | Pegmatite | Industrial quartz |
| 16 | Grão Mogol | Hydrothermal vein | Amethyst |
| 17 | Brejinho das Ametistas | Hydrothermal vein | Amethyst |
| 18 | Chapada dos Veadeiros | Hydrothermal vein | Industrial quartz |
| 19 | Chapada dos Guimarães | Hydrothermal vein | Industrial quartz |
| 20 | Novo Horizonte-Ibitira | Hydrothermal vein | Quartz with rutile and hematite inclusions |
| 21 | Oliveira dos Brejinhos | Hydrothermal vein | Gemological quartz |
| 22 | Grota do Coxo | Hydrothermal vein | Amethyst |
| 23 | Pau D'Arco | Sedimentary | Amethyst |
| 24 | Alto Bonito | Hydrothermal vein | Amethyst |
| 25 | Parelhas | Pegmatite | Industrial and rose quartz |
| 26 | Currais Novos | Pegmatite | Industrial and rose quartz |

exploited are located in the municipalities of Picuí, Pedra Lavrada, Nova Palmeira, Carnaúba dos Dantas, São Vincente do Seridó, Juazeirinho, Parelhas, Currais Novos and Equador. More than 750 of them occur as intrusion in garnet-cordierite and/or sillimanite-biotite schists of the Seridó Formation (Da Silva et al. 1995).

The BPP is historically important for its tantalum/tin ore, beryl, and minerals for the ceramics and glass industry (quartz, feldspar and mica). It is also well-known for its small production of gems, including aquamarine, morganite, garnet, euclase and elbaite. Recently, it became famous worldwide for the production of the Cu-bearing elbaite commercially known as "Paraíba Tourmaline".

The pegmatites of BPP were first classified by Johnston and Butler (1946) as being either homogeneous or heterogeneous basing on their internal structure. The homogeneous pegmatites are usually sterile, generally concordant, without internal zoning. Heterogeneous pegmatites are rich in Be–Li-Ta-Sn-minerals and present internal zoning that show variable fabric and mineralogical composition.

**Fig. 6.2** Brazilian quartz deposits: **a** core of rose quartz in the Alto do Feio pegmatite, Pedra Lavrada, Paraíba; **b** quartz vein at Gouveia in Minas Gerais; **c** industrial hydrothermal quartz mine hosted in quartzite, Gouveia, Minas Gerais; **d** amethyst geode in basalt, Soledade, Rio Grande do Sul (Götze photo)

Quartz has been produced predominantly from zoned pegmatites that have a massive quartz core. The quartz is either milky, rose, hyaline or smoked and is extracted to feed, mainly glass, ceramic and the building industries. Secondarily, perfect crystals of quartz, which are more appropriate for electronic components and trade as gems and jewellery, have been found.

In the last 4 years most of the production of rose quartz comes from two complex pegmatites, the Taboa and Alto do Feio pegmatites (Fig. 6.2a). The pegmatites are situated, respectively, in the states of Rio Grande do Norte and Paraíba. Taboa and Alto do Feio zoned pegmatites have a core of rose and milky quartz.

Generally, quartz lascas obtained by the miners are bought by larger companies that refine the selection and sell it with more than 99% silica ($Si0_2$) and less than 0.01% $Fe_20_3$ and 0.1% humidity to industries. The selling price of the lascas have varied greatly in the last 10 years, but its average was around \$ 0.23 per kilo FOB.

Due to the high degree of informality of the mines, no one has, to date, reliable data of production, sale and export of quartz mined from this province, although estimates of such data are available.

### 6.2.1.2 Eastern Brazilian Pegmatite Province

This province is located mainly in the Araçuaí orogen that extends from the eastern edge of the São Francisco craton to the Atlantic cost, occupying a region larger than 400,000 km$^2$ in southeastern Brazil (Pedrosa-Soares et al. 2001, 2008, 2009). The area is covered by Late Neoproterozoic up to Cambro-Ordovician granitic rocks, including the pegmatites, and by Neoproterozoic metassediments (Pedrosa-Soares et al. 2011).

Countless pegmatites of the Eastern Brazilian Pegmatite Province (EBPP) emplaced from ca. 630 to ca. 490 My, i.e., during the Brasiliano event, along the Araçuaí orogen. They comprise two types: anatectic (directly formed from the partial melting of country rocks) or residual pegmatites (fluid-rich silicate melts resulting from the fractional crystallization of parent granites) (Pedrosa-Soares et al. 2009). Mineralizing fluids derived from granite plutons of the G-2, G-4 and G-5 granitic supersuites. The EBPP is divided into eleven districts in Minas Gerais and Espírito Santo States and one in Bahia State (Itambé district) (Pedrosa-Soares et al. 2011).

Important mineralizations of tourmaline, beryl, Nb–Ta minerals, kunzite, hiddenite, rare minerals and industrial minerals such as spodumene and feldspar makes the EBPP the most important pegmatite region of Brazil, and one of the most important worldwide.

Most of the quartz appears as massive, milky type. Crystallized and massive rose quartz occurs in the Conselheiro Pena district, east of Governador Valadares, in the Galiléia county. These pegmatites are famous for the druses and monocrystal specimens of the Alto da Pitorra mine. In the core of Serra do Urucum pegmatite, giant crystals which length up to 3 m along the c-axis were mined. In the same district, in a region named Boa Vista creek, the pegmatites have quartz cores with transparent and colorless quartz which were mined for gemstone and for color enhancement.

During the last 15 years the Padre Paraíso and Araçuaí pegmatite districts, located in the north of Minas Gerais, were the most important sources for pegmatite gemstones in Brazil. Pedrosa-Soares et al. (2009) estimated that more than 300 pegmatites have been mined in the Araçuaí district. Indeed, as well as in the whole province, this activity has been quite intermittent to occasional in most of the pegmatites due to several economic and social factors. Most pegmatites of the Araçuaí district are residual bodies from G-4 intrusive granites, ranging in thickness from a few meters to more than 50 m. Quartz crystals occur in miarolitic cavities, and with minor importance some pegmatites are mined for rose quartz from the core.

Sporadically, the pegmatites of the Espírito Santo district, especially those located in Santa Maria de Jetibá, are producing amethyst and double color smoked/amethyst crystals.

## 6.2.2 Hydrothermal Quartz

### 6.2.2.1 The Espinhaço Range

The Espinhaço Range is a geomorphologic unit located in the southeastern and northeastern part of Brazil (Fig. 6.1). The main present physiography was developed after the Miocene, related to the Pliocene compressional tectonic event, was responsible for the reactivation of Precambrian thrust faults and detachment planes (Saadi 1995). The main geomorphologic unit is divided in four sub-unities: Southern Espinhaço, Central Espinhaço, Northern Espinhaço and Chapada Diamantina.

The quartz from the Espinhaço Range occurs as hyaline, milked, citrine, amethyst and smoky varieties widespread into a large part of the region and is related to hydrothermal quartz veins cutting Proterozoic metasediments of the Rio Paraúna (Paleoproterozoic), Espinhaço (Mesoproterozoic) Supergroups and rhyolites of the Rio dos Remédios Group (Paleoproterozoic). The veins are related to multiple stages of devolatilization during folding metamorphism of clastic sediments during the Neoproterozoic Brasiliano orogenic cycle (640–490 Ma).

In Minas Gerais State three regions are very important sources of hydrothermal quartz. Diamantina is famous for its several small deposits of the hyaline variety (Chaves 1997). Some citrine and smoky quartz occur in that area, and in the vicinities of the Sopa village. Beautiful hyaline druses have been discovered in both areas. Other areas with outstanding importance are Gouveia and Serra do Cabral, where mainly hyaline single crystals and druses of quartz occur. Gouveia deposits are important producers of milky quartz for the metallurgic industry/purpose; veins in this region are larger, sometimes up to 30 m thick (Fig. 6.2b, c occurring in the basal contact zone of Espinhaço Supergroup with schists and granitic rocks of the Basement Complex (Chaves 2007).

Most of the hydrothermal quartz from the Espinhaço Range shows specification to allow the use in different ways by the industry, especially for the production of metallic silicon. Quartz fragments of high purity are used for the growing seeds of synthetic quartz, piezoelectric devices, frequency controllers and electronic-grade silicon. The region is supplying the market with powder quartz of high physical and chemical purities for the production of optical glass to be used in precision instruments, lenses and microscopes.

Quartz amethyst veins are very rare in Minas Gerais and small occurrences are known in Bucnópolis (Southern Espinhaço), and Grão Mogol and Montezuma (Central Espinhaço) counties, revealing generally less than 1 m in thickness (Chaves and Favacho-Silva 2000).

In Bahia state, two main sequences host the hydrothermal veins: acid to intermediary metavolcanic rocks of the Rio dos Remédios Group (Paleoproterozoic), and metasedimentary rocks (mainly phyllite and quartzite) of the Espinhaço Supergroup (Mesoproterozoic). Different types of quartz mineralizations can be found.

The Rio dos Remédios Group outcrops in the southwestern Bahia (Northern Espinhaço). Most prominent quartz deposits occur in the Ibitiara—Novo Horizonte counties (Cassedanne 1991; Couto 2000) where, in the last 20 years, quartz crystals with abundant rutile and/or hematite inclusions (rutilated quartz) have acquired major commercial importance as gems and collection specimens. The main part of the mineralization is hosted in the contact zone of volcanic beds with siliciclastic Espinhaço rocks. Giant crystals measuring several meters in length in the c-axis direction have been discovered in this area.

In the region of Oliveira dos Brejinhos, quartz occurs as clear, transparent and colorless crystals, some with hematite platelets. Quartz extraction occurs directly from the veins, or more rarely in proximal colluvial deposits. Most other deposits within this location are exploited in the same way, including those in the region of northern Espinhaço Range, also belonging to the Espinhaço Supergroup. Prominent deposits, which in addition yield citrine quartz, are located in Brotas de Macaúbas area (Couto 2000).

Amethyst deposits are widespread in the Bahia state and occur mainly in two areas: Brejinho das Ametistas-Serra do Salto (Caetité and Licínio de Almeida counties), and the Grota do Coxo mine (Jacobina). Occurrences in Brejinho das Ametistas are secondary deposits, where amethyst chips and pebbles are erratically scattered in the colluvial surface cover (Couto 2000). Primary mineralization occurs in fractures cutting through quartzites of the Espinhaço Supergroup in the Brejinho das Ametistas and in the Serra do Salto (Salto Mountain, a local name for the Espinhaço Range) in the vicinity of the town Licínio de Almeida. In these areas mineralizations are related to fracture zones that cut fin- to medium- grained sandstone. Amethyst occurs as druses and geodes, in vugs, veins and fractures, or as dissemination of crystals in sandstone. Most amethyst batches examined in this region show good euhedral formation and uniform purple color.

In the Grota do Coxo mine, which is located in the Jacobina county, amethyst crystals usually occur as druses and geodes filling fractures in the quartzites of the Rio do Ouro Formation (Paleoproterozoic, Jacobina Group) in the eastern region of the Espinhaço Range (Cassedanne and Cassedanne 1979; Cassedanne 1991). The crystals are variously colored from reddish violet to velvet violet, and display face edges from 1 to 30 cm in length. Druses of amethyst crystals are sometimes over half a meter wide, and contain brownish, usually opaque crystals (Cassedanne and Cassedanne 1979).

#### 6.2.2.2 The Alto Bonito Amethyst Deposits

The Alto Bonito amethyst deposits are located in the southwest part of Marabá and northeast of the Carajás Range. The exploration started in 1982 by artisanal miners. Two types of amethyst deposits are found in the region. The first type, which is of minor economic importance, comprises alluvial and colluvial deposits with amethyst pebbles. The second, a more important type, is related to a hydrothermal late -metamorphic event (Collyer and Mártires 1991). The hydrothermal veins are developed in metasediments of the Salobo-Pojuca sequence, mainly represented by quartzite. The amethyst crystals occur in veins, dikes and geodes. Large cavities, sometimes connected, can be larger than 150 m.

#### 6.2.2.3 Amethyst and Agate in Basalts of the Rio Grande do Sul

Amethyst and agate are the two most common gemstones in Brazil. The main Brazilian amethyst and agate deposits are located in the north (Alto Uruguai), center (Salto do Jacuí) and southwest (Fronteira Sudoeste) of the Rio Grande do Sul state, in the south of Brazil (Schmitt et al. 1991). The occurrences belong to the Serra Geral Formation, a sequence comprising Cretacic tholeiitic basalts and basaltic andesites locally overlain by rhyolites and rhyodacites. The basaltic flows of southern Brazil and Uruguay are famous worldwide and considered as one of the most important sources of amethyst of the world. The region is responsible for supplying the international market in geodes and crystals for cutting (Fig. 6.2d). The first occurrences were discovered by German immigrants at the beginning of the nineteenth century (Epstein 1988). Official statistic data of the production are doubtful, because the mining has been carried out partially by artisanal methods and the production has not been officially declared. Corrêa et al. (1994) officially reported a production of about 100–200 tons of geodes per month and 20–30 tons of amethyst crystals for gemstone cutting in 1993.

The origin of the Serra Geral Formation comprises continental flood basalts related to the volcanic activity which occurred on the South American platform during the South Atlantic Ocean opening. The lavas of the Paraná Flood Basalt Province show a maximum thickness of about 1.7 km (Bellieni et al. 1984).

Accessory minerals that occur associated with amethyst, filling the geodes, are calcite, pyrite, gypsum, barite and goethite-lepidocrocite (Juchem et al. 1999; Gilg et al. 2003).

The genetic aspects of the origin of geodes and the geode filling are still being debated. The two main genetic proposals are: (1) epigenetic formation at temperatures below 150°C, related to an explosive event with hydraulic basalt fracturing by fluids originated from the Guarany aquifer (Duarte et al. 2005; Morteani et al. 2010), and (2) magmatic gas exsolving from the supersaturated melt with respect to dissolved volatiles, with amethyst crystallization temperature ranging from 204 to 238°C (Scopel et al. 1998; Gilg et al. 2003; Proust and Fontaine 2007). Contradictory data measured in fluid-inclusions suggest temperatures below 50°C (Juchem et al. 1999) and ranging from 370 to 420°C (Thomas and Blankenburg 1981).

### 6.2.3 *Secondary Deposits*

Brazil has favorable weather conditions for the development of secondary deposits. The tropical and sub-tropical climates are responsible for a high degree of weathering and creation of enabling environments for the emergence of sedimentary and colluvial deposits of industrial sands and amethyst to the gemstone market.

#### 6.2.3.1 The Pau D'Árco Amethyst Deposits

The Pau D'Árco Amethyst deposits are located in the southeast region of the Pará state, in the Conceição do Araguaia county, and northwest of Tocantins in the Pau D'Árco county.

The amethyst deposits are related to the weathering of hydrothermal quartz veins. The primary hydrothermal mineralization occurs associated to green schist facies metamorphic rocks of the Couto Magalhães formation, represented by phyllites and quartzites. The secondary deposits are the most important from an economic point of view, corresponding to the best quality material for cutting in Brazil. Secondary deposits of amethyst of Pau D'Árco are represented by Pleistocene polymictic conglomerates of the Araguaia Formation. The conglomerate is matrix-supported and has sand and clay matrix texture with Fe and Mn hydroxides as cement. The size of pebbles, including amethyst, ranges from 1 to 30 cm (Collyer et al. 1991).

#### 6.2.3.2 Quartz Sand Deposits of Pirambóia and Botucatu Formations

Quartz sands are produced from Holocene coastal deposits and Cenozoic sediments related to the Pirambóia and Botucatu formations in the Paraná Basin (Ferreira and Daitx 2003). The measured resources of industrial quartz in 2005 were calculated in 2,439,679,740 tons, and most of them are related to industrial sands (MME 2010).

Industrial quartz, which occurs in nature as sand, sandstone or quartzite, is used in accordance with the specific characteristics and properties of the raw material. Usually, industries that use these products, such as the ferrosilicon and glass industries, have relatively low quality requirements and accept levels of impurities higher than those specified for high technology industry.

The main Brazilian companies that produce different types of sands for the metallurgic and glass industries are located in São Paulo, Minas Gerais, Paraná, Santa Catarina and Rio Grande do Sul states. Located in São Paulo State, the geomorphologic unit named Depressão Periférica Paulista is the main Brazilian producer of industrial sands. The mineralogical, chemical and physical properties are variable and the ore is applied to different types of industries (Ferreira and Daitx 1997).

Important industrial sand deposits from holocenic beachs are located between São Vicente and Peruíbe, in São Paulo State. The production is mainly applied in the glass industry. Measured reserves were calculated up to 1,000 Mt and the indicated and inferred reserves were estimated by about 6,000 Mt (Theodorovicz et al. 1991).

The glass industry accepts $SiO_2$ values of down to 99.0%. The Brazilian glass industry is responsible for 65% of the internal consumption of sands, followed by iron foundry with 30% and other industries such as chemical, blasting and filtration equipments (Luz and Lins 2005).

## 6.3 Color Enhancement of Quartz

### 6.3.1 A Brief Historic

One possible way to increase the commercial value to colorless or pale natural quartz is irradiation in combination with thermal treatment. Although irradiation was only applied to gemstone enhancing in the last century, the color modification of gemstones in general is known since ancient times. As pointed out by Kurt Nassau in his book *Gemstone Enhancement* (Nassau 1984), the first records of treatments to change the color of gems are known from the *Stockholm Papyrus* (Caly 1927) and the works by C. *Pliny Secundus* (23–76 AD). Such treatments ranged from oiling of emerald fractures, dyeing with the help of azurite and malachite, as well as the use of metallic reflecting foils, and even the production of composite stones. Some of these methods are used until today. Among the processes reported, the bleaching of crystals in boiling rice-water and the cleaning (bleaching) of smoky quartz in cauldrons of hot water at public baths are some of the first reports of thermal treatments of quartz done to improve their value. Even though the works of *Pliny* went through the Middle Ages as a reference in mineralogy and gemology, it was *Albertus Magnus* in his work *On Stones* (1260) (Wyckoff 1967) who realised, for the first time, that the color of gemstones is associated to their composition, i.e. to the "different substances of which its parts are composed".

Later, in the Renaissance, the book *The Mirror of Stones* (1502) by Leonardus (1750) discussed the problem of counterfeits in jewellery, like several types of colored glasses and the ways to distinguish synthetic and natural gemstones. One example is the filling of perforated quartz with tincture to simulate the high-valued amethysts. Meanwhile, other works (Biringuccio 1942; Cellini 1967; Agricola 1556; Monton 1734) also dealt with a multitude of ways of either simulating several types of high-valuable gems and/or enhance the value of natural samples by artificially coloring them. Also in the works of *Boetius de Boot* called *Gemmarum et Lapidum Historia* (De Boot 1609), and *John Baptista Porta* in his *Natural Magic* (Porta 1955) decolorizing of several gemstones, including quartz, by thermal treatment was discussed.

However, the first systematic study of gemstone enhancement based on experiments was done by a professional scientist, the chemist *Robert Boyle*, who published his results in 1672. In his essay *About the Origin and Virtue of Gems* (Boyle 1672) the first evidence of color centers and their stability was established in connection with the observation that most gemstones lose their color during heating and, in some cases like agates, the color changes. Boyle (1672) also reported the filling of thermally induced cracks in quartzes with several natural dyes.

In the nineteenth century the first compendium books on gemstones and gemology were published. Among them, one by Kluge (1860) mentioned specifically the heating of several gemstones (amethyst, sapphire, zircon and topaz) being in iron fillings, burned lime or sand as a way to change their colors. In addition, Kluge (1860) provided an extensive discussion of dyeing agates. But it was only at the beginning of the twentieth century that the extensive work of Max Bauer entitled *Edelsteinkunde* (first published in 1896) became a reference on gemstone enhancing. In this work he describes the loss of the smoky color of quartz or its conversion to yellow quartz, and the change of amethyst into citrine after thermal treatments, along with dyeing of cracked and microcrystalline quartz, the removal and restoration of the yellow color in tiger's-eye among several other processes. At that time, the quality of the lapidating industry had also reached a stage where the optimal light reflection in the facets was perfected, as a way to enhance the color of several gemstones and therefore improve their value. With the advent of radioactivity, discovered at the end of the nineteenth and at the beginning of the twentieth centuries, its application to change the gemstone colors was almost immediate. In his book *Die Künstlichen Edelsteine* (1926), Michel reported the irradiation of diamonds, corundum, tourmalines, topazes, kunzites and quartz. Besides the faceting technology, the development of gemstone irradiation became the main process to increase the value of gemstones by enhancing their colors.

Nowadays, with superficial knowledge about the impurities related to the colors and the formation of color centers, as well as the development of color-enhancing processes, several recipes, in particular for quartz, topaz and tourmaline, using a combination of irradiation (mostly gamma, but also highly-energetic electrons and neutrons from linear accelerators and nuclear reactors) and thermal annealing, are globally applied by the gemstone industry despite the fact that most of the recipes for quartz, like those presented in (Nassau 1984), are empiric.

### 6.3.2 *Color-Enhancing of Gemological Crystalline Quartz in Brazil*

Among the exported gemstones, the colored varieties of quartz (smoky, amethyst, citrine, rose quartz and the gamma-irradiated amethyst, green-gold citrine and prasiolite) are an important commodity. Despite its low value compared to other gems like tourmaline, beryls and topaz, the produced volume is by far greater, even when exported without any kind of processing. For example, in 2007 Brazil exported about 22,561 tons of rough/chip quartz (clear crystals, electronic grade) and only 10 tons of processed material (DNPM 2011). For gemological quartz the situation is not very different. Only 12–24 tons of irradiated quartz is officially treated in Brazil per year. Thus, most of the production is exported without any kind of processing. In 2007 the average export price for this rough material (untreated gem quartz) was 217 US$/ton or 0.22 cents/kg. On the other hand, for faceted colored quartz varieties the carat varies from 0.50 cents up to 4 US$ for smoky quartz, 20 US$ for citrine and 30 US$ for amethyst (IBGM 2005). In order to reverse this picture, it has been recognized by the local authorities that it is fundamental to promote the exportation of processed quartz (either natural or irradiated/thermally-treated colored faceted gemstones) with a price gain of three orders of magnitude. After faceting, gamma irradiation and/or thermal-treatment is used to produce the colored quartz varieties in order to increase the value of the primery colorless quartz. Depending on amounts of iron and aluminium, along with other impurities gems such as morion, smoky quartz, green-gold quartz, brown quartz (for Al), amethyst, prasiolite, and citrine (for Fe), (Figs. 6.3 and 6.4) can all be produced artificially. Untreated colorless quartz can cost up to 50 US$/kg, and after being faceted, their prices are comparable to those of natural gemstones, with exception of the high valuable natural deep-colored amethyst.

In Brazil, since the 1970s, gems are gamma-irradiated, either commercially in small scale (or for research), in an informal way in gamma-cells of small irradiators at nuclear research facilities such as in the Centro de Desenvolvimento da Tecnologia Nuclear (CDTN/CNEN) in Belo Horizonte/MG, or in large scale at the industrial facilities of the Companhia Brasileira de Esterlização (CBE-EMBRARAD) in São Paulo. The CBE-EMBRARAD, using panoramic $Co^{60}$ irradiators, e.g. for the sterilization of medical material and reduction of bacterial charge in foodstuff, has also a long tradition in irradiating gemstones. Being the only company of this kind in Brazil, it currently irradiates between 1 and 2 tons of quartz for the gem market every month, even with the present suppressed demand dating back to the global crisis of 2008. Before then, it easily irradiated about 50 tons annually with prices varying from 5 to 80 US$/kg depending mainly on the gamma dose, but also on the origin of quartz, type of final product (color) and type of irradiation process (temperature, continuous or intermittent dose, etc.) (Reichmann 2011).

Apart from the current suppressed demand, several other factors hinder the production of irradiated/thermally treated gemstones in Brazil. The first and perhaps the most important is the artificially highly-valued local currency. With the undervalued US dollar in the last years compared to the Brazilian Real, the high costs of processing the stones in Brazil decreases drastically the export profits. In addition, there are the environmental issues, with mining licenses being more difficult to obtain each day all over the country, and the operational problems like the high-cost of workforce for extraction and preparation of the stones for irradiation/thermal-treatment (which requires mechanic breaking, cutting and pre-forming of the stones for faceting). In Brazil, the faceting, due to its artisanal character and therefore low quality, allied also to the expensive cost of the specialized workforce and automation, is traditionally done overseas (China, India, etc.), for the large scale industry. Despite all these problems, the exportation of color-enhanced (irradiated/thermally-treated) quartz for the jewellery industry is still an important economical sector in the Brazilian mineral economy.

Although recent developments of new and more intense sources of irradiation and the fact that the color enhancement of gemstones became an industrial activity, several detailed receipts and patents for treating a vast range of gemstones, including quartz and its new artificial colored varieties like the green-gold and prasiolite, rely on empirical data produced by prospective irradiations and knowledge of the occurrences, but not much on scientific data on chemical

**Fig. 6.3** Color varieties of Al-rich quartz treated with gamma irradiation and thermal treatment: **a** green-gold quartz; **b** red-brown quartz; **c** brown quartz; and **d** black morion quartz

compositions, structure, formation, and stability of color centers as well as their transition metal solid state chemistry.

## *6.3.3 Quartz Color-Enhancing in Brazil*

The purpose of this section is to review the color-enhancing processes for gemstones of the quartz family, focusing on crystalline varieties of $\alpha$-$SiO_2$ that currently require irradiation and/or thermal treatment to get their value improved. Color-enhancing of microcrystalline and polycrystalline varieties such as, agate, chrysospase, chalcedony, as well as quartz that own their colors due to inclusions like tiger-eye, rutilated quartz, blue quartz and dendritic quartz, and also rose quartz and opal are left out of this discussion because of their smaller economic importance for Brazillian exports, in terms of production volume and price. However, we will also mention two types of quartz that are nowadays appearing with increased frequency in the gemstone industry: the snow-white quartz which can be produced by thermally-induced microfractures in a specific variety of natural $\alpha$-quartz and a combination of thermal annealing, and the pitch-black quartz by gamma irradiation substituting onyx in the jewelry industry.

**Fig. 6.4** Color varieties of Fe-rich quartz treated with gamma irradiation and/or thermal treatment: **a** amethyst; **b** citrine (burned-amethyst); **c** prasiolite; and **d** snow-white quartz

Nassau (1984) distinguished two groups of α-quartz with different color treatment routes based on the primary impurity content: the amethyst group with dominantly Fe impurities and the smoky quartz group with dominantly Al impurities. Based on the Fe and Al impurities, which are responsible for the majority of colors and color centers within each group, we suggest a refined system of color treatment, in accordance with (Rykart 1995). Figure 6.5 shows the typical routes for color enhancing of Brazilian quartz according to our work Pinheiro et al. (1999), and several other authors, e.g. Nassau (1984), Nunes et al. (2009), Nunes and Lameiras (2005), and Drummond et al. (2010). The requirements in terms of impurities, the treatment process, the most accepted color center models from the literature and final colors with ordinary names for the gemstones treated are given in this roadmap.

Very pure colorless quartz does not respond to any irradiation and heating process. All colors produced are directly or indirectly related to impurities. If the quartz is Fe-rich and its concentration is by far higher than that of Al, the main color produced by irradiation is the violet color typical of natural amethysts. The origin of this color has been controversial for a long time. Although the relationship between the violet colors with the Fe-content is known since Goethe (1982), it was explicitly recognized in 1925 by Holden and von Klemm (Holden 1925). Lehmann and Bambauer (1973), as well as Lehmann (1975, 1977, 1978) proposed a color center model in which the impurity $Fe^{3+}$ substituting $Si^{4+}$ loses an electron, after natural irradiation, to an interstitial $Fe^{3+}$ forming $Fe^{4+}$ substitutional and $Fe^{2+}$ interstitial, whereas the violet color should be related to the $Fe^{4+}$ center. They also observed that the depth of the color depends on the right balance of substitutional $Fe^{3+}$ and interstitial $Fe^{3+}$ and that the presence of hydrogen hindered the violet color formation due to its reaction with substitutional $Fe^{3+}$. Later, Nassau (1984) revised this model, suggesting that the substitutional $Fe^{4+}$ is responsible for the violet color and proposed that no interstitial $Fe^{3+}$ was necessary for the formation of $Fe^{4+}$. Hassan (1972) and Cohen (1975) proposed a controversial model in which only the interstitial $Fe^{3+}$ is responsible for the violet color. Later, Cohen (1985) acknowledged the interstitial $Fe^{4+}$ as the violet color center even though he controversially proposed that it could only be produced after recombination of the Al-related smoky quartz color center. Meanwhile Weil (1984, 1993, 1994) and Cox (1976), using electron paramagnetic resonance spectroscopy (EPR), investigated several Fe-related centers in $\alpha$-quartz. In particular, Cox (1976) observed $Fe^{4+}$, and also investigated its optics (Cox 1977), reinforcing the $Fe^{4+}$ model for the amethyst color center. Recent results based on the Mössbauer effect (Dedushenko et al. 2004) and X-ray absorption spectroscopy (XAS) (Benedetto et al. 2010) seem to confirm this model. Therefore, where, in the scheme above, the violet color appeared, we related it to the $Fe^{4+}$ color center.

In samples containing $Al^{3+}$ which substitutes $Si^{4+}$, the trivalent Al is charge compensated by monovalent ions like $H^+$, $Li^+$ and $Na^+$, with concentrations of $[Al] \geq [H] + [Li] + [Na]$ (e.g. Bambauer 1961). Under gamma irradiation, an electron ejects from an oxygen in the vicinity of the Al impurity and is trapped by one of the charge compensating ions, neutralizing them (Nassau 1983). The resulting Al–O bound hole $h^+$, $[Al^{3+}O_4/h^+]^{4-}$, which is isoelectronic to the normal $[SiO_4]^{4-}$ unit, is the color center responsible for the grey-smoky to black colors (Nassau 1984). The $H^0$ center is not stable at room temperature, and thus only Li and Na can stabilize the smoky color center. As a result for the formation of smoky quartz, it is required that $[Li] + [Na] > [H]$ (Rykart 1995). It is well-known that potassium is also a common interstitial charge compensator for Al, however, its role on the stabilization of color centers is by far less understood (Rykart 1995).

Apart from these two color centers much less is known about the other colored varieties of treated quartz. It is known, however, the presence of $Fe^{2+}$ and the high OH-content (in quartz grown in reducing-conditions) (Nassau 1984) for the greenish Prasiolite varieties (Rose and Lietz 1954; Schuman 2007), the presence of $Fe^{3+}$ and $Fe_2O_3$ nanoparticles or a colloidal dispersion of hydrous ferric oxide in citrine (Holden 1923); and the correlation of the yellow colors (Nassau and Prescott 1975) in Al-rich quartz, with the Li-content. For other color varieties such as the sky-blue quartz, the causes are still a matter of discussion (Nunes et al. 2009). Recently it has been discovered that long-term thermal treatment of OH-rich quartz at temperatures above the $\alpha$-/$\beta$-quartz transition (573–800°C) can result in opal-like milky quartz (opaline quartz) with a slight blue tint to completely opaque snow-white quartz (Krambrock et al., 2010, personal communication).

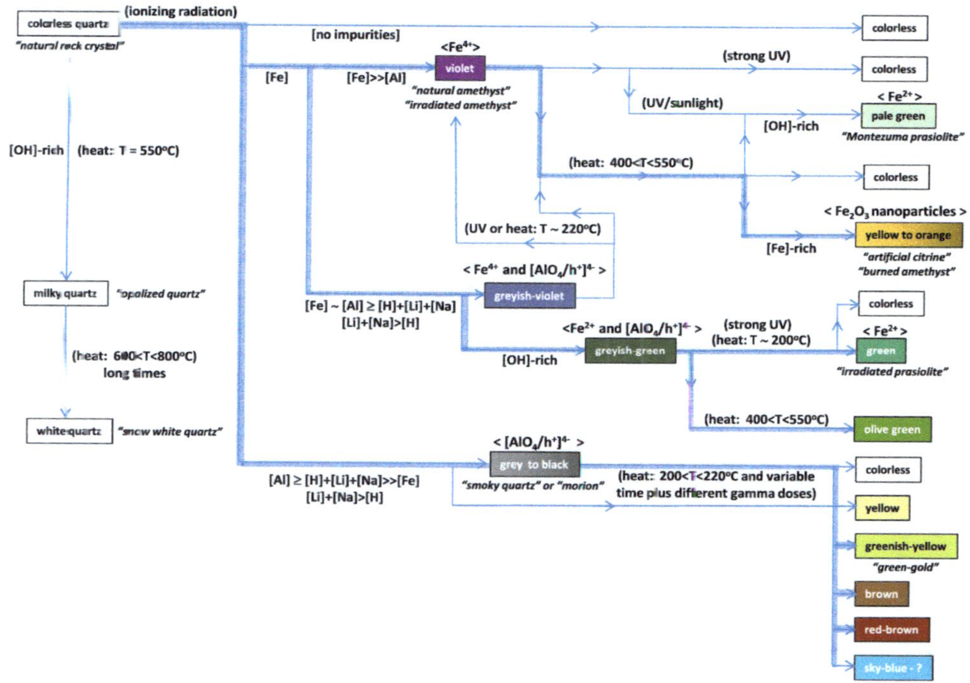


**Fig. 6.5** Summary of all possible color-enhancing routes for Brazilian quartz that involves gamma radiation and heat treatment. See text for details. The *thick lines* represent the treatments done commercially

Despite the lack of knowledge about the relationship of some defects and optical properties, the treatment routes for a large variety of colors are well established, at least empirically, and details such as radiation dose are also known (Pinheiro et al. 1999). On the other hand, the correlation of the treatment result, i.e. color, with particular quartz occurrences and the origin, for example most pegmatite quartz yielding green-gold color (Drummond et al. 2010), are also well known. In order to improve the color-enhancing processes for quartz and to limit losses, systematic research of the spectroscopic properties and atomistic structure of color centers and transition metal impurities, as well as their charge compensation mechanisms and radiation/thermally-induced solid state reactions are necessary.

## 6.4 Concluding Remarks

Currently, Brazil can be considered a minor actor in the quartz market. However, the official data of production is imprecise and questionable, due to an informal market. The Brazilian internal industries responsible for the consumption of industrial sands and lascas are mainly the metallurgical (ferrosilicon alloys) and glass. Resources measured up to 2,400 Mt in 2005 show possibility of growth in the production, and transform Brazil in one of the leaders in the quartz market worldwide.

Brazil is also well-known all over the world to be one of the most important sources of amethyst and agate as geodes and fragments for cutting. In the last 10 years the production of gemological quartz is increasing, specially the transparent and colorless fragments, mined in pegmatites and hydrothermal veins, to be used for color treatment. The installation of private companies to deal with irradiation of gemstones also shows a positive perspective in the gems market.

**Acknowledgments** The authors thank the MCT/CT-MINERAL/VALE/CNPq—Project CORGEMA II grant No. 550319/2010-7 and grants No. 450931/2011-0, 477806/2010-4, and FAPEMIG grants No. APQ-02000-10 and PPM-00222-09. R. Scholz thanks to FAPEMIG grant No. CRA-APQ-03998-10

## References

Agricola G (1556/1955) De Natura Fossilium, Special paper 65. Geogogical Society of America, New York. De Re Metallica, Basel

Alecrim JD (1982) Quartzo—Aspectos Gerais, Mineralogia, Geologia. In: Recursos Minerais do Estado de Minas Gerais. Metamig, Belo Horionte, pp 199–202

Arcoverde WL, Schobbenhaus C (1991) Geologia do quartzo. In: Schobbenhaus C et al (eds) Principais Depósitos Minerais do Brasil. Brasília, Ed. DNPM/CPRM, IV(C), pp 315–324

Bambauer HU (1961) Spurenelementgehalte und $\gamma$-Farbzentren in Quarzen aus Zerrklüften der Schweizer Alpen. Schweiz Mineral Petrogr Mitt 41(2):335–369

Bauer M (1860/1968) Edelsteinkunde; published in English as precious stones, vol 2. Dover Publications, New York

Bellieni G, Comin-Chiaramonti P, Ernesto M, Melfi AJ, Pacca IG, Piccirillo EM (1984) Flood basalt to rhyolite suites in the southern Paraná plateau (Brazil): paleomagnetism, petrogenesis and geodynamic implications. J Petrol 25:579–618

Benedetto F, D'Acapito F, Fornaciai G, Innocenti M, Montegrossi G, Pardi LA, Tesi S, Romanelli MA (2010) Fe K-edge XAS study of amethyst. Phys Chem Mineral 37(5):283–289

Biringuccio V (1942) Pirotechnia. MIT Press, Cambridge

Boyle R (1672) An essay about the origin and virtues of gems. Hafner Pub. Co., New York

Caly ER (1927) The stockholm papyrus. An English translation with brief notes. J Chem Edu 4(8):979–1002

Campbell DF (1946) Quartz crystal deposits in the state of Goiaz, Brazil. Econ Geol 41(8):773–799

Cassedanne JP (1991) Tipologia das jazidas brasileiras de gemas. In: Schobbenhaus C et al (eds) Principais Depósitos Minerais do Brasil. Brasília, Ed. DNPM/CPRM, pp 17–52

Cassedanne JP, Cassedanne JO (1979) La mine d'améthyste de La Grota do Coxo: une mervelle inconnue. Revist Gemmologie AFG 59:2–5

Cellini B (1967) The treatises of Benvenuto Cellini on goldsmithing and sculpure. Dover Publications, New York

Chaves, MLSC, Karfunkel J, Queménéur JJ (1997) Depósitos de quartzo da região de Batatal (Diamantina, Minas Gerais). In: Simpósio de Geologia de Minas Gerais, vol 9. Anais, Ouro Preto, pp 103–104

Chaves MLSC, Favacho-Silva MD (2000) Ocorrências singulares de quartzo gemológico (ametista, fumé e citrino) na Serra do Espinhaço, Minas Gerais. Revista da Escola de Minas 53:181–186

Chaves MLSC (2007) O megaveio de quartzo da Serra da Catinga (Datas/Gouveia, MG). Geociências 26:109–117

Cohen AJ (1975) On the color centers of iron in amethyst and synthetic quartz: a reply. Am Mineral 60:338–339

Cohen AJ (1985) Amethyst color in quartz, the result of radiation protection involving iron. Am Mineral 70:1180–1185

Collyer TA, Mártires RAC (1991) O depósito de ametista do Alto Bonito, município de Marabá, Pará. In: Principais depósitos minerais do Brasil, vol 4A. Gemas e rochas ornamentais. DNPM, Rio de Janeiro, pp 287–293

Collyer TA, Mártires RAC, Machado JIL (1991) O depósito de ametista de Pau D'Árco, município de Conceição do Araguaia, Pará. In: Principais depósitos minerais do Brasil, vol 4A. Gemas e rochas ornamentais. DNPM, Rio de Janeiro, pp 295–302

Corrêa TE, Koppe JC, Costa JFCL, Moraes MAL (1994) Caracterização geológica e critérios de prospecção de depósitos de ametista tipo Alto Uruguai, RS. Anais XXXVIII Congresso Brasileiro de Geologia, Camboriú, Santa Catarina 2:137–138

Couto P (2000) Gemologic map of the state of Bahia—explanatory text. CPRM/SMM/DNPM/SICM Covenor, Salvador, p 76

Cox RT (1976) ESR of an S − 2 center in amethyst quartz and its possible identification as D4 ion $Fe^{4+}$. J Phys C Solid State Phys 9(17):3355–3361

Cox RT (1977) Optical absorption of D4 ion $Fe^{4+}$ in pleochroic amethyst quartz. J Phys C Solid State Phys 10(22):4631–4643

Da Silva MRR, Höll R, Beurlen H (1995) Borborema pegmatitic province: geological and geochemical characteristics. J South Earth Sci 8(3/4):355–364

De Boot AB (1609/1961) Germmarum et Lapidum Historia, cited in J.R. Partington. A history of chemistry, vol 2. Macmillan, London, pp 101–102

Dedushenko SK, Makhina IB, Mar'in AA, Mukhanov VA, Perfiliev YD (2004) What oxidation state of iron determines the amethyst colour? Hyperfine Interact 165(1):417–422

DNPM (2006) Sumário Mineral. Ministério de Minas e Energia—Departamento Nacional de Produção Mineral. p 304

DNPM (2011) Data from the Departamento Nacional de Produção Mineral, Instituto Brasileiro de Gemas e Metais Preciosos (IBGM)

Drummond NF (2009) Ambientes Geológicos e Mudanças de Cor de Quartzo Hialino. Contribuição as Ciências da Terra. Serie M 68(281):195

Drummond NF, Mendes JC, Lameiras FS (2010) Quartz characterization to obtain its gemological varieties by means of gamma irradiation and/or heat treatment. Rem Rev Esc, Minas, Ouro Preto 63(3):449–456

Duarte LDC, Hartmann LA, Vasconcelos MAZ (2005) Epigenetic geode formation in the world-class amethyst deposits of the southern Paraná basaltic province. In: I Simpósio Brasileiro de Metalogenia, Gramado. CD-ROM. Sociedade Brasileira de Geologia, Porto Alegre

Epstein DS (1988) Amethyst mining in Brazil. Gems Gemol 24(1):214–228

Ferreira GC, Daitx EC (1997) Áreas Produtoras de Areia Industrial no Estado de São Paulo. Revista Escola de Minas 50(4):54–60

Ferreira GC, Daitx EC (2003) A mineração de areia industrial na Região Sul do Brasil. Revista Escola de Minas 56(1):59–65

Goethe JW (1982) Theory of colours, trans. Charles. Lock Eastlake, MIT Press, Cambridge, MA

Gilg HA, Morteani G, Kostitsyn Y, Preinfalk C, Gatter I, Strieder AJ (2003) Genesis of amethyst geodes in basaltic rocks of the Serra Geral Formation (Ametista do Sul, Rio Grande do Sul, Brazil): a fluid inclusion, REE, oxygen, carbon, and Sr isotope study of basalt, quartz and calcite. Mineralium Deposita 38:1009–1025

Hassan F (1972) Amethyst. Mineral Rec 3(5):221–225

Holden EF (1923) The color of three varieties of quartz I. Note on the color of ciftine. Am Mineral 8(1):117–118

Holden EF (1925) The Cause of colour in smoky quartz and amethyst. Am Mineral 10(9):203

IBGM (2005) Boletim referencial de preços from the Instituto Brasileiro de Gemas e Metais Preciosos

Johnston WD, Butler RD (1946) Quartz crystal in Brazil. Bull Geol Soc Am 57:601–650

Juchem PL, Fallick AE, Bettencourt JS, Svisero DP (1999) Geoquímica isotopic de oxigênio em geodos mineralizados a ametista da região do Alto Uriguai, RS—um estudo preliminar. 1 Simpósio sobre vulcanismo e ambientes associados, Gramado (Abstract)

Kluge KE (1860) Handbuch der Edelsteinkunde. Brockhaus, Leipzig

Lehmann G (1975) On the colors centers of iron in amethyst and synthetic quartz: a discussion. Am Mineral 60:335–337

Lehmann G (1977) Über die Färbeursachen natüelicher Citrine. Z Dt Gemmol Ges 26(2):53–60

Lehmann G (1978) Farben von Mineralien und ihre Ursache. Fortschr Miner 56:172–252

Lehmann G, Bambauer HU (1973) Quarzkristalle und ihre Farben. Angw Chemie 85:281–289

Leonardus C (1750) The Mirror of stones. J. Freeman, London

Luz AB, Lins FAF (2005) Rochas e Minerais Industriais: Parte II: rochas e minerais industriais: usos e especificações. 5. areia industrial. CETEM/MCT, Rio de Janeiro, p 726

Michel H (1926) Die Künstlichen Edelsteine. W. Diebner, Leipzig

MME – Ministério das Minas e Energias (2010) Areia industrial. Tech rep p 44:52

Monton B (1734) Secretos de Artes Liberales y Mecanicas. En Madrid: En la oficina de Antonio Marin, Madrid

Morteani G, Kostitsyn Y, Preinfalk C, Gilg HA (2010) The genesis of the amethyst geodes at Artigas (Uruguay) and the paleohydrology of the Guaraní aquifer: structural, geochemical, oxygen, carbon, strontium isotope and fluid inclusion study. Int J Earth Sci (Geol Rundschal) 99:927–947

Nassau K (1983) The physics and chemistry of color. Wiley, New York

Nassau K (1984) Gemstone enhancement. Buttler & Tanner, London

Nassau K, Prescott B (1975) A reinterpretation of smoky quartz. Phys Stat Sol (a) 29:659–663

Nunes EH, Melo V, Lameiras F, Liz O, Pinheiro A, Machado G, Vasconcelos W (2009) Determination of the potential for extrinsic color development in natural colorless quartz. Am Mineral 94:935–941

Nunes EHM, Lameiras FS (2005) The optical absorption of gamma irradiated and heat-treated natural quartz. Mater Res 8(3):305–308

Pedrosa-Soares AC, Chaves M, Scholz R (2009) Eastern Brazilian pegmatite province. In: 4th international symposium on granitic pegmaites, field trip guide, p 28

Pedrosa-Soares AC, Noce CM, Wiedemann CM, Pinto CP (2001) The Araçuaí-West Congo orogen in Brazil: an overview of a confined orogen formed during Gondwanland assembly. Precambr Res 110:307–323

Pedrosa-Soares, AC, Alkmim FF, Tack L, Noce CM, Babinski M, Silva LC, Martins-Neto MA (2008) Similarities and differences between the Brazilian and African counterparts of the Neoproterozoic Araçuaí-West Congo orogen. In: Pankhurst RJ, Trouw RAJ., Brito Neves BB, De Wit MJ (eds) West Gondwana: pre-cenozoic correlations across the South Atlantic Region. Geological Society, London, Special Publications 294, pp 153–172

Pedrosa-Soares AC, Campos CP, Noce C, Silva LC, Novo T, Roncato J, Castañeda C, Queiroga G, Dantas E, Dussin I, Alkmim F (2011) Late neoproterozoic—Cambrian granitic magmatism in the Araçuaí orogen (Brazil), the Eastern Brazilian Pegmatite Province and related mineral resources. In: Sial AN et al (eds) Granite-related ore deposits. Geological Society, London, Special Publications 350, pp 25–51

Pinheiro MVB, Lameiras ES, Krambrock K, Karfunkel J, da Silva JB (1999) The effect of the gamma-irradiation dose combined with heat on the color enhancement of colourless quartz. Aust Gemol 20:7

Porta JB (1955) Natural magic. Basic Books, New York

Proust D, Fontaine C (2007) Amethyst-bearing lava flows in the Paraná Basin (Rio Grande do Sul): cooling, vesiculation and formation of the geodic cavitics. Geol Mag 144(1):53–65

Reichmann F (2011) Personal communication from the Operational Director from the CBE-EMBRARAD

Rose H, Lietz J (1954) Ein grün verfarbbarer Amethyst. Naturwissenschaften 41:448

Rykart R (1995) Quarz-Monographie—Die Eigenheiten von Bergkristall, Rauchquarz, Amethyst, Chalcedon, Achat. Opal und anderen Varietäten, Ott Verlag Thun, Schwiz

Saadi A (1995) A geomorfologia como ciência de apoio ao planejamento urbano em Minas gerais. Geonomos 5(2):1–4

Schmitt JCC, Camatti C, Barcellos RC (1991) Depósitos de ametista e ágata no estado do Rio Grande do Sul. In: Principais depósitos minerais do Brasil, vol. 4A. Gemas e rochas ornamentais. DNPM, Rio de Janeiro, pp 271–286

Schuman W (2007) Gemstones of the world, 3 Rev. Exp. edition. Sterling, New York

Scopel RM, Gomes MEBG, Formoso MLL, Proust D (1998) Derrames portadores de ametistas na região de Frederico Westphalen-Irai—Planalto-Ametista do Sul, RS-Brasil. Congresso Uruguaio de Geologia 2, Actas, pp 243–252

Theodorovicz AMG, Francisconi O, Ferrari AP (1991) Depósitos de areia industrial do litoral paulista entre São Vicente e Peruíbe—São Paulo. In: Schobbenhaus C et al (eds) Principais Depósitos Minerais do Brasil. Brasília, Ed. DNPM/CPRM, IV(C), pp 357–363

Thomas R, Blankenburg HJ (1981) Erste Ergebnisse über Einschlussuntersuchungen an Quartzen aus Achatmandeln und Kugeln basischer und sauerer Vulkanite. Z Geol Wiss 9:625–633

Van Schmus WR, Brito Neves BB, Williams IS, Hackspacher PC, Fetter AH, Dantas EL, Babinski M (2003) The Serido Group of NE Brazil, a late Neoproterozoic pre- to syn-collisional basin in West Gondwana: insights from SHRIMP U-Pb detrital zircon ages and Sm-Nd crustal residence (TDM) ages. Precambr Res 127:287–327

Weil JA (1984) A review of electron spin spectroscopy and its application to the study of paramagnetic defects in crystalline quartz. Phys Chem Mineral 10:149–165

Weil JA (1993) A review of the EPR spectroscopy of the point defects in a-quartz. The decade 1982–1992. In: Helms CR, Deal BE (eds) The physics and chemistry of $SiO_2$ and the Si-$SiO_2$ interface 2. Plenum Press, New York, pp 131–144

Weil JA (1994) EPR of iron centers in silicon dioxide. Appl Magn Reson 6:1–16

Wyckoff D (1967) Albertus Magnus: book of minerals. Clarendon Press, Oxford

# Chapter 7
# First-Principles Calculations of the $E'_1$ Center in Quartz: Structural Models, $^{29}$Si Hyperfine Parameters and Association with Al Impurity

**Zucheng Li and Yuanming Pan**

**Abstract** The oxygen-vacancy-related $E'_1$ center in quartz is a prototype defect in $SiO_2$-based materials and has been intensively studied by experimentalists and theoreticians for more than 50 years, but questions remain about its structural model and formation mechanism. We have conducted first-principles calculations for the $E'_1$ center in quartz by using the supercell approach with Gaussian-type all-electron basis sets and hybrid functionals as implemented in the ab initio code CRYSTAL06. Our calculations, featuring both the traditional single-oxygen-vacancy model and a new tri-vacancy model with one silicon and two oxygen vacancies, have improved the reproduction of the experimental $^{29}$Si hyperfine parameters than previous theoretical studies. In particular, our calculated hyperfine coupling constants for the strong and two weak $^{29}$Si hyperfine structures from the new tri-vacancy model are all within 5% of the experimental values, and the calculated directions of the principal hyperfine axes are in agreement with the experimental results as well. Moreover, the tri-vacancy model can account for the common association of the $E'_1$ center with the $[AlO_4]^0$ center and superoxide $O_2^-$ radicals in quartz.

## 7.1 Introduction

Quartz is one of the most abundant minerals in the Earth's crust and is an important material for not only traditional industries such as construction but also high-tech electronic, optical and solar-energy applications. Point defects in quartz,

Z. Li · Y. Pan (✉)
Department of Geological Sciences, University of Saskatchewan,
Saskatoon, SK S7N 5E2, Canada
e-mail: yuanming.pan@usask.ca

J. Götze and R. Möckel (eds.), *Quartz: Deposits, Mineralogy and Analytics*,
Springer Geology, DOI: 10.1007/978-3-642-22161-3_7,

including those associated with silicon or oxygen vacancies, have long been the subject of intense research (Weeks 1956, 1963; Silsbee 1961; Isoya et al. 1981; Jani et al. 1983; Mashkovtsev et al. 2007; Botis et al. 2005, 2008; Perlson and Weil 2008; Weeks et al. 2008; Nilges et al. 2008, 2009; Pan et al. 2008, 2009; Mashkovtsev and Pan 2011), because they are known to exert important (and often adverse) effects on material properties and device performance and have wide applications from radiation dosimetry to geochronology, mineral exploration and quantum computer (e.g., Arends et al. 1963; Ikeya 1993; Stoneham et al. 2003; Pan et al. 2006; Hu et al. 2008). For example, the oxygen-vacancy-related $E'_1$ center, first observed by using electron paramagnetic resonance (EPR) spectroscopy (Weeks 1956), is a fundamental and prototype defect in quartz and other $SiO_2$-based materials (e.g., glasses, thin films and nanoparticles) and has been investigated by numerous experimental and theoretical studies (e.g., Weeks 1956; Silsbee 1961; Feigl et al. 1974; Yip and Fowler 1975; Griscom 1980; Jani et al. 1983; Rudra and Fowler 1987; Edwards et al. 1988; Edwards and Fowler 1990; Boero et al. 1997, 2003, 2004; Pacchioni et al. 1998; Blöchl 2000; Carbonaro et al. 2001; Sulimov et al. 2002; Chadi 2003; Mysovsky et al. 2004; Stesmans et al. 2008; Perlson and Weil 2008; Weeks et al. 2008; Usami et al. 2009; Griscom 2011).

Silsbee (1961) first determined one strong and two weak $^{29}$Si hyperfine matrices $\mathbf{A}(^{29}\mathrm{Si})$ for the $E'_1$ center in neutron-irradiated quartz at room temperature (Table 7.1). Silsbee (1961) also noted another weak $^{29}$Si hyperfine structure with a splitting of 0.07 mT in spectra measured with the magnetic field parallel to the crystal c axis, where the strong and two weak $^{29}$Si hyperfine splittings are 40, 0.9 and 0.8 mT, respectively. Jani et al. (1983), in a combined single-crystal EPR and electron nuclear double resonance (ENDOR) study, clarified the reference coordinate system and re-determined the electronic Zeeman matrix $\mathbf{g}$ and the three $\mathbf{A}(^{29}\mathrm{Si})$ matrices of the $E'_1$ center to higher precisions (Table 7.1), allowing the principal axes of these parameter matrices to be compared with bond directions in the quartz lattice (see also Perlson and Weil 2008).

One of the most salient features of the $E'_1$ center in quartz is its close association with the aluminum-associated oxygen hole center $[AlO_4]^0$ (Jani et al. 1983; Toyoda and Hattori 2000; Usami et al. 2009). For example, Jani et al. (1983) reported a correlation between the growth of the $E'_1$ center and the decay of the $[AlO_4]^0$ center and suggested that the former formed from trapping of the unpaired spin ejected from the latter. Toyoda and Hattori (2000) found that the gamma-ray-induced $E'_1$ center in a natural quartz sample that was annealed at 450°C reaches saturation at the dose of 200 kGy, whereas the $[AlO_4]^0$ center in this sample continues to grow with further irradiation. Usami et al. (2009) confirmed the observation of Toyoda and Hattori (2000) and found that the number of the $E'_1$ center in quartz samples containing small amounts of the $[AlO_4]^0$ center is small and not all oxygen vacancies are converted. These results led Usami et al. (2009) to conclude that the number of electrons ejected from the $[AlO_4]^0$ center plays an important role in the formation of the $E'_1$ center in quartz.

Semi-empirical calculations have been credited for the widely cited single-oxygen-vacancy model for the $E'_1$ center in quartz and amorphous $SiO_2$ (Feigl

**Table 7.1** Comparison of literature data on the $^{29}Si$ hyperfine coupling constants (MHz) and directions of the $E'_1$ center in quartz

| $^{29}Si$ | EPR/ENDOR experiments | | | | | | Theoretical calculations | | | | | | |
|---|---|---|---|---|---|---|---|---|---|---|---|---|---|
| | Silsbee (1961) | | | Jani et al. (1983) | | | Boero et al. (1997) | | | Blöchl (2000) | | | M(2004) |
| | $A_k/h$ | $\theta(°)$ | $\phi(°)$ | $A_k/h$ | $\theta(°)$ | $\phi(°)$ | $A_k/h$ | $\theta(°)$ | $\phi(°)$ | $A_k/h$ | $\theta(°)$ | $\phi(°)$ | $A_k/h$ |
| Strong | 1,271 | 67 | −10 | −1269.72 | 114.1 | 229.7 | 1434.5 | 113.0 | 228.3 | −1272 | 112 | 233 | −1399.3 |
| | 1,091 | ⊥ | ⊥ | −1094.53 | 132.1 | 115.9 | 1248.5 | 59.0 | 70.4 | −1109 | ⊥ | ⊥ | −1221.9 |
| | 1,091 | ⊥ | ⊥ | −1095.02 | 128.3 | 340.4 | 1248.5 | 101.3 | 318.5 | −1109 | ⊥ | ⊥ | −1220.5 |
| Weak1 | 27.43 | 39 | 46 | −27.53 | 140.7 | 284.5 | 61.9 | 141.6 | 281.8 | −39.2 | 149 | 281 | −30.3 |
| | 22.00 | ⊥ | ⊥ | −22.14 | 104.6 | 33.1 | 45.3 | 119.3 | 50.3 | −33.2 | ⊥ | ⊥ | −25.8 |
| | 22.00 | ⊥ | ⊥ | −22.27 | 125.5 | 133.9 | 45.9 | 127.0 | 157.3 | −33.2 | ⊥ | ⊥ | −25.5 |
| Weak2 | 25.81 | 55 | −157 | −26.01 | 58.9 | 260.9 | 67.3 | 50.7 | 252.9 | −42.0 | 56 | 267 | −32.5 |
| | 20.66 | ⊥ | ⊥ | −20.91 | 35.0 | 111.4 | 51.3 | 25.3 | 78.6 | −36.4 | ⊥ | ⊥ | −27.7 |
| | 20.66 | ⊥ | ⊥ | −21.04 | 104.4 | 179.9 | 50.3 | 93.3 | 168.6 | −36.4 | ⊥ | ⊥ | −27.5 |
| Weak3 | 2.0[a] | | | | | | 11.5 | 131.4 | 347.3 | 2.8 | 127 | 13 | 6.2 |
| | | | | | | | 10.2 | 99.6 | 186.6 | 0.0 | ⊥ | ⊥ | 3.6 |
| | | | | | | | 9.6 | 148.7 | 83.3 | 0.0 | ⊥ | ⊥ | 3.9 |

M(2004) is Mysovsky et al. (2004), in which the directions of the principal hyperfine axes were not available

[a] Silsbee (1961) reported a third weak $^{29}Si$ hyperfine splitting of 0.07 mT in spectra measured with the magnetic field parallel to the crystal c axis

et al. 1974; Yip and Fowler 1975; Rudra and Fowler 1987; Edwards et al. 1988; Edwards and Fowler 1990). For example, MINDO/3 calculations by Feigl et al. (1974) and Yip and Fowler (1975), using a $Si_2O_6$ cluster, gave the positively (+1) charged oxygen vacancy $V(O)^+$ model with the unpaired spin largely localized on one of the two asymmetrically relaxed [$SiO_3$] groups. Rudra and Fowler (1987), using larger $Si_8O_7$ and $Si_5O_{16}$ clusters, improved this model and showed that the Si atom containing little unpaired spin relaxes toward or even passes through the basal oxygen plane to form the planar and puckered configurations. The puckered configuration was adopted and supported by several subsequent studies using larger clusters (Snyder and Fowler 1993), including the embedded-cluster treatment (Giordano et al. 2007) and Car-Parrinello molecular dynamics (Boero et al. 1997), both of which explicitly consider the valence electrons only. For example, molecular dynamics calculations by Boero et al. (1997) supported the puckered configuration, but their calculated $^{29}Si$ hyperfine coupling constants differ by >20% for the strong $^{29}Si$ hyperfine structure and as much as 200% for the two weak $^{29}Si$ hyperfine structures (Table 7.1). First-principles calculations by Blöchl (2000) better reproduced the values for the strong $^{29}Si$ hyperfine structure but yielded values >40% too large for the two weak $^{29}Si$ hyperfine structures (Table 7.1).

However, Fiorentini and Carbonaro (1997) suggested that +1 charge state of the oxygen vacancy is not stable and therefore can not be a candidate $E'_1$ center. Fiorentini and Carbonaro (1997) proposed the −3 charge state of the oxygen vacancy as an alternative model but did not calculate $^{29}Si$ hyperfine parameters for comparison with the experimental data. Carbonaro et al. (1997) noted that the ground state of the oxygen vacancy in undoped quartz is the neutral charge state, whereas moderate p- and n-type dopings favor the +3 and −3 states, respectively. Similarly, Chadi (2003) suggested that the +1 charge state of the oxygen vacancy is not stable and proposed $V(SiO_3)^+$ and $V(SiO_4)^+$ as possible models for the $E'_1$ center in quartz. On the other hand, periodic first-principles calculations by Carbonaro et al. (2001) provided compelling evidence for thermodynamic stability of the positively charged oxygen vacancy in quartz. Mysovsky et al. (2004) recalculated the puckered configuration in amorphous $SiO_2$ with the embedded-cluster method and noted that all calculated $^{29}Si$ hyperfine values are overestimated by ~10% (Table 7.1). Obviously, significant questions remain about the structural model for the $E'_1$ center in quartz and amorphous $SiO_2$ (Weeks et al. 2008; Griscom 2011). Also, the close association between the $E'_1$ and $[AlO_4]^0$ centers in both synthetic and natural quartz, as documented by EPR studies (Jani et al. 1983; Toyoda and Hattori 2000; Usami et al. 2009), has not been addressed by any previous theoretical calculations.

Accordingly, we have conducted new periodic first-principles calculations with all-electron Gaussian-type basis sets to further investigate the geometry and electronic structure of the $E'_1$ center in quartz. In particular, our calculations re-evaluate the traditional $V(O)^+$ model but focus on a new tri-vacancy model containing a silicon and two oxygen vacancies. These models are evaluated on the basis of their abilities in reproducing the experimental $^{29}Si$ hyperfine parameters in both principal values and directions. In addition, several of our calculations include a substitutional $Al^{3+}$ ion at a neighboring Si site and provide an opportunity to evaluate the effect of Al impurities on the $E'_1$ center in quartz.

## 7.2 Computation Methodology

Our calculations have been performed by using the supercell approach with all-electron Gaussian-type basis sets and various hybrid functionals as implemented in the ab initio code CRYSTAL06 (Dovesi et al. 2006). Ideally, structural models for a paramagnetic defect such as the $E'_1$ center should be evaluated on the basis of both energy criteria and their success in the reproduction of experimental EPR data. However, the complexity of solid state calculations prevents us from using the former rigorously: i.e., frequency tests for the energetically optimal model as widely used in molecular studies. Moreover, successful reproduction of experimental EPR data appears to be more important for the time being, because a favorable total energy can still fail to locate the correct defect structure (Magagnini et al. 2000). Further more, the boundary conditions used in the supercell approach for charged defects preclude the calculation of a valid total energy, because a uniform background charge is commonly added to maintain the overall electrostatic neutrality (Dovesi et al. 2006). Therefore, we put our trust on the structural optimization process and evaluate the various structural models on the basis of agreement between the calculated and experimental $^{29}Si$ hyperfine constants, rather than the more time-consuming stability tests (e.g., Carbonaro et al. 2001). Whenever possible, however, the formation energy defined below is calculated and included for discussion.

All-electron Gaussian-type basis sets used here are those of the double zeta valence plus polarization (DZVP) quality and are known to be well suitable for periodic calculations. The Si basis set is 8-41G**, the contracted [1s3sp2d] from (20s13p2d) by Pisani et al. (1992), and that for O is 6-31G* with standard contraction of [3s2p1d] used by Gatti et al. (1994). The basis set for Al is 85-11**, the contracted [1s3sp1d] basis by Catti et al. (1994). Pack–Monkhorst and Gilat shrink factors of 4 and 8, the largest possible values, were used to describe electron density and the Fermi surface of the $2 \times 2 \times 2$ supercell of the system, respectively.

The atomic coordinates of $\alpha$-quartz in space group $P3_221$ from Le Page et al. (1980) were used for the optimization of the perfect structure, in which each oxygen atom is connected to two silicon atoms with a short bond (SB) of 1.607 Å and a long bond (LB) of 1.611 Å at room temperature. This difference in the Si–O bond distances, which becomes even smaller at low temperatures (Le Page et al. 1980), has been suggested by Feigl et al. (1974) to be responsible for the asymmetric relaxation that pushes the unpaired spin to localize on the Si0 atom on the SB site of the oxygen vacancy, while leaves Si1 on the LB side almost spin free (Fig. 7.1a).

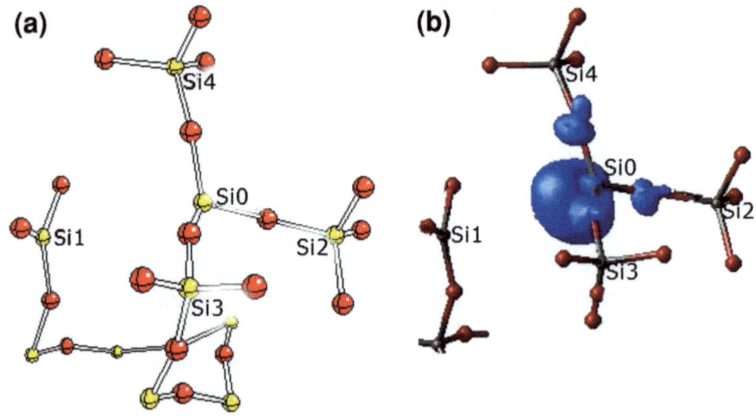


**Fig. 7.1** Local structure (**a**) and electron density (**b**) of the $E'_1$ center in quartz calculated from the single-oxygen-vacancy $V(O)^+$ model. Labels of silicon atoms are similar to those in Jani et al. (1983). Contours are at intervals of 0.005 e/bohr$^3$ and from −0.01 to 0.415 e bohr$^{-3}$

Our calculations started with the optimization of the 2 × 2 × 2 supercell containing 72 atoms for the perfect structure. Subsequently, defects were introduced to the optimized supercell by removing an oxygen atom along with an electron for the $V(O)^+$ model or deleting one silicon and two neighboring oxygen atoms along with an electron for the tri-vacancy model, and the whole structures were then allowed to relax and fully optimized. In addition, calculations have been made with an Al atom replacing a selected Si atom in the defect structures to examine the effects of this impurity. The convergence of defect related properties (i.e. spin density, defect geometry and hyperfine coupling constants or *hfcc*) has been evaluated with respect to unrestricted Hartree–Fock and various DFT methods such as the popular B3LYP and B3PW functionals, but only the results from the best method PBE0 (Adamo and Barone 1999; Li and Pan 2011) are reported herein.

The two parts in *hfcc* for the $^{29}$Si nuclei in the $E'_1$ center, isotropic $a_{iso}$ and anisotropic $T_i$, are calculated as follow (Dovesi et al. 2006):

$$a_{iso} = \frac{(A_1 + A_2 + A_3)}{3} = \frac{2}{3}\mu_0 g_N \beta_N g_e \beta_e \langle \rho^{spin}(\mathbf{r}_A) \rangle \tag{7.1a}$$

$$T_i = \frac{1}{4}\mu_0 g_N \beta_N g_e \beta_e \sum_{\mu\nu} \sum_{w} P^{spin}_{\mu\nu w} \int \varphi_\mu(\mathbf{r}) \left( \frac{r_A^2 \delta_{ij} - 3r_{Ai} r_{Aj}}{r_A^5} \right) \varphi^w_\nu(\mathbf{r}) d\mathbf{r} \Big|_{i=j} \tag{7.1b}$$

Here subscript $i$ takes *xx*, *yy*, *zz*, or simply 1, 2, 3. The EPR experiments usually report $A_i$ defined as $A_i = a_{iso} + T_i$. $\mu_0$ is the permeability of vacuum, $g_N$ is the $^{29}$Si's nuclear g-factor and $g_e$ is the electronic g-factor, $\beta_N$ and $\beta_e$ are the nuclear and Bohr magnetons, respectively, $< \rho^{spin}(\mathbf{r}_A) >$ is the expectation value of the spin density at the nucleus A ($^{29}$Si) at point $\mathbf{r}_A$, $r_A = |\mathbf{r} - \mathbf{r}_A|$, $r_{Ai} = (\mathbf{r} - \mathbf{r}_A)_i$, the

**Table 7.2** Calculated local structure (Å) of the $E'_1$ center in quartz

| Model | Atom pair | PBE0 | Ref[a] | Ref[b] | Ref[c] | Ref[d] |
|---|---|---|---|---|---|---|
| $V(O)^+$ | Si0–Si1 | 4.606 | 4.32 | 4.06 | 4.358 | 4.46 |
| | Si1–$O_B$ | 3.480 | 1.91 | 1.81 | 1.852 | 1.82 |
| | <Si1–O> | 1.587 | / | / | / | / |
| V(O) with Al | Si0–Si1 | 3.986 | / | / | / | / |
| | Al–$O_B$ | 1.988 | / | / | / | / |
| | <Al–O> | 1.692 | / | / | / | / |
| $V(SiO_2)^+$ | Si0–Si1 | 4.543 | / | / | / | / |
| | Si1–$O_B$ | 3.369 | / | / | / | / |
| | <Si1–O> | 1.623 | / | / | / | / |
| $V(SiO_2)$ with $AlO_4$ | Si0–Al | 4.277 | / | / | / | / |
| | Al–$O_B$ | 4.296 | / | / | / | / |
| | <Al–O> | 1.722 | / | / | / | / |
| $V(SiO_2)$ with $AlO_3$ | Si0–Al | 4.929 | / | / | / | / |
| | Al–$O_B$ | 2.742 | / | / | / | / |
| | <Al–O> | 1.670 | / | / | / | / |

The average <Al–O> bond distance excludes the backbonding Al–$O_B$
[a] Data from Mysovsky et al. (2004), where SiS–$O_B$ is probably Si1–$O_B$
[b] Data from Pacchioni et al. (1998)
[c] Data from Blöchl (2000)
[d] Data from Boero et al. (1997)

*i*th component of vector $\mathbf{r}_A$, $\delta_{ij}$ is Kronecker delta, and $P^{spin}_{\mu\nu w}$ is the spin density matrix element between basis functions $\varphi_\mu(\mathbf{r})$ of the reference cell and $\varphi^w_\nu(\mathbf{r})$ of the *w*th cell.

## 7.3 Results and Discussion

### 7.3.1 *Single-Oxygen-Vacancy Models*

Our periodic PBE0 calculation has reproduced most essential features obtained by previous $V(O)^+$ calculations (Feigl et al. 1974; Yip and Fowler 1975; Boero et al. 1997; Blöchl 2000; Mysovsky et al. 2004), but with notable differences. For example, the PBE0 optimized Si0–Si1 distance of 4.606 Å (Table 7.2; Fig. 7.1a) is larger than those in the puckered models of Boero et al. (1997) and Mysovsky et al. (2004). Also, our calculated $Si1O_3$ group is only slightly puckered by 5°, and the Si1–$O_B$ distance of 3.48 Å is significantly larger than those from previous studies as well (Table 7.2). In fact, our calculations show that the puckered configuration of Boero et al. (1997) is not stable (see also Fiorentini and Carbonaro 1997; Chadi 2003) and relaxes to the slightly puckered one (Table 7.2). Also, we did not find the dimer configuration with the Si0–Si1 distance of ~3.0 Å, proposed by Mysovsky et al. (2004).

Figure 7.1b shows that the spin density on Si0 is 0.818 *e* and that small spin densities (0.028, 0.043 and 0.053 *e*) on its three bonded oxygen atoms are also apparent, which are important in accounting for the two weak $^{29}$Si hyperfine tensors (Silsbee 1961; Jani et al. 1983). The calculated spin densities on Si1, Si2, Si3 and Si4 are 0.012, 0.012, 0.010 and 0.009 *e*, respectively. Table 7.3 shows that our calculated $^{29}$Si hyperfine parameters are in better agreement with experimental EPR results than previous theoretical calculations (Boero et al. 1997; Blöchl 2000; Mysovsky et al. 2004). Nevertheless, the two weak $^{29}$Si hyperfine tensors are still not well predicted (Table 7.3). Also, significant discrepancies are apparent between the calculated and experimental directions of the principal hyperfine axes (Table 7.3). Moreover, the calculated principal values for the third weak $^{29}$Si hyperfine structure are overestimated (Table 7.3).

Our calculation for the single-oxygen-vacancy model with an Al impurity at the Si1 site improves the calculated *hfcc* for two of three weak $^{29}$Si hyperfine structures (Table 7.3), suggesting an important role of Al impurity for the $E'_1$ center. Interestingly, the Al atom forms a weak back-bond of 1.988Åwith $O_B$ (Table 7.2), thus making the $AlO_3$ group considerably puckered (an OAlOO torsional angle of −131.5°). However, the calculated directions of the principal hyperfine axes from this model with an immediate Al neighbor are poor in comparison with experimental results (Table 7.3). Therefore, calculations have been made for a more-distant Al impurity (i.e., next to Si1, Si2 or Si3), but they invariably resulted in the delocalization of the unpaired spin from the Si0 atom, inconsistent with the experimental features of the $E'_1$ center (Jani et al. 1983).

### 7.3.2 Tri-Vacancy Models

Other proposed structural models such as the no-oxygen vacancy model of Boero et al. (2003) and the four- or five-vacancy models of Chadi (2003) may be energetically plausible, but do not reproduce the EPR data as well as the single-oxygen-vacancy model. For example, Chadi (2003) suggested that the five-vacancy cluster $V(SiO_4)^+$ is the smallest model to account for the EPR experiment and have the electronic properties close to those predicted by the single-oxygen-vacancy model. However, Chadi (2003) admitted that the formation energy of this five-vacancy cluster is much larger than those of single oxygen vacancies. Herein, we propose a new tri-vacancy model, $V(SiO_2)^+$, which is expected to have a lower formation energy than the $V(SiO_4)^+$ cluster of Chadi (2003). Also, this tri-vacancy defect is neutral and, therefore, represents an ideal candidate for trapping an electron. Another line of evidence for multiple vacancies stems from EPR studies of biradical $E''$ centers (Bossoli et al. 1982; Mashkovtsev et al. 2007; Mashkovtsev and Pan 2011). These $E''$ centers have been interpreted to consist of two interacting $E'$-type defects separated by distances of 5 ~ 6 Å or larger (Mashkovtsev and Pan 2011). For such large distances, the single-oxygen-vacancy models are obviously not plausible.

**Table 7.3** $^{29}$Si hyperfine coupling constants (MHz) and directions of the $E'_1$ center in quartz from periodic PBE0 calculations

| $^{29}$Si | One-oxygen-vacancy model | | | | | | Tri-vacancy model | | | | | | | | |
|---|---|---|---|---|---|---|---|---|---|---|---|---|---|---|---|
| | $V(O)^+$ | | | V(O) with Al | | | $V(SiO_2)^+$ | | | $V(SiO_2)$ with $AlO_4$ | | | $V(SiO_2)$ with $AlO_3$ | | |
| | $A_k/h$ | $\theta(°)$ | $\phi(°)$ | $A_k/h$ | $\theta(°)$ | $\phi(°)$ | $A_k/h$ | $\theta(°)$ | $\phi(°)$ | $A_k/h$ | $\theta(°)$ | $\phi(°)$ | $A_k/h$ | $\theta(°)$ | $\phi(°)$ |
| Strong | −1221.5 | 124.2 | 221.5 | −1230.4 | 109.4 | 216.9 | −1243.3 | 115.5 | 234.5 | −1197.0 | 127.3 | 236.6 | −1282.7 | 117.7 | 234.6 |
| | −1081.0 | 125.0 | 103.1 | −1066.7 | 104.4 | 121.7 | −1076.6 | 145.2 | 101.9 | −1030.0 | 132.7 | 101.9 | −1114.0 | 133.8 | 114.4 |
| | −1081.8 | 126.6 | 341.8 | −1066.8 | 155.5 | 357.4 | −1077.1 | 112.5 | 335.9 | −1030.4 | 124.6 | 347.0 | −1114.5 | 123.5 | 344.9 |
| Weak1 | −34.3 | 118.0 | 302.7 | −26.8 | 106.6 | 271.8 | −29.9 | 149.1 | 277.2 | −28.9 | 148.0 | 265.6 | −32.1 | 144.2 | 279.3 |
| | −28.8 | 143.4 | 18.6 | −21.9 | 100.6 | 118.5 | −24.0 | 91.6 | 27.2 | −23.1 | 113.9 | 40.5 | −26.0 | 103.0 | 27.9 |
| | −28.7 | 111.5 | 140.6 | −21.7 | 160.0 | 127.4 | −24.3 | 118.3 | 123.5 | −23.2 | 110.1 | 139.8 | −26.1 | 122.7 | 126.4 |
| Weak2 | −18.3 | 38.0 | 288.3 | −17.0 | 74.1 | 241.3 | −27.8 | 55.3 | 261.4 | −9.8 | 23.5 | 260.1 | −23.4 | 42.8 | 260.6 |
| | −14.3 | 52.6 | 119.8 | −12.8 | 44.5 | 117.1 | −22.2 | 42.5 | 122.4 | −6.7 | 76.6 | 136.9 | −18.5 | 58.0 | 128.1 |
| | −14.8 | 95.6 | 145.5 | −13.0 | 130.2 | 206.4 | −22.5 | 111.4 | 187.1 | −7.2 | 109.0 | 222.3 | −18.7 | 115.1 | 201.0 |
| Weak3 | −14.1 | 102.5 | 79.8 | −1.4 | 104.6 | 86.6 | 3.3 | 123.7 | 70.9 | 2.4 | 124.9 | 72.5 | −3.5 | 103.6 | 59.8 |
| | −10.9 | 126.9 | 59.3 | −1.1 | 132.9 | 42.6 | 2.6 | 84.1 | 37.0 | 1.2 | 52.2 | 15.3 | −2.7 | 15.5 | 30.5 |
| | −10.5 | 39.6 | 154.2 | −1.1 | 36.5 | 130.9 | 1.6 | 34.3 | 178.2 | 1.2 | 57.0 | 135.5 | −2.5 | 82.7 | 148.0 |

See text for details about the structural models investigated in this study

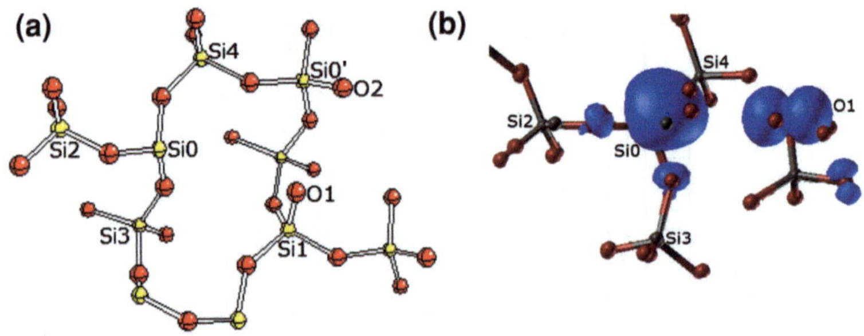


**Fig. 7.2** Local structure (**a**) and electron density (**b**) of the $E'_1$ center in quartz calculated from the tri-vacancy model without an Al impurity. Note that labels of Si0, Si2, Si3 and Si4 are equivalent to those in Fig. 7.1, but Si1 is not. Also note the presence of the unexpected spin on the O1 atom. Contours are at intervals of 0.005 e/bohr$^3$ and from −0.01 to 0.415 e/bohr$^3$

Our tri-vacancy model calculations started with the removal of the two oxygen atoms and the central Si atoms connected by the long bonds, giving the Si0 and Si0′ atoms a dangling short bond each (Fig. 7.2a). The optimized structure turns out to be very interesting, because it does not form the expected two $E'_1$ centers (i.e., biradical or triplet state; Bossoli et al. 1982; Mashkovtsev et al. 2007; Mashkovtsev and Pan 2011). Rather, the Si0′ atom is relaxed to link to a fourth oxygen atom (O2) from a neighboring $SiO_4$ group (i.e., Si0′–$O_2$ = 2.162 Å), and the unpaired spin localizes on the O1 atom of another $SiO_4$ group at the Si1 site (Fig. 7.2b). Nevertheless, the Si0 atom contains 0.846 *e*, hence an $E'_1$ center. In fact, the calculated hyperfine parameters from this model, including the directions of the principal axes, best reproduce the experimental data (Jani et al. 1983), especially for the two difficult weak $^{29}$Si hyperfine structures (Table 3). Interestingly, the relaxed Si0–Si1 distance of 4.543 Å in this tri-vacancy model is only 0.06 Å shorter than the asymmetrically relaxed Si–Si pair in the $V(O)^+$ model (Table 7.2), indicating similar geometry and spin structure between these two models to account for their similar hyperfine coupling constants. However, the Si1 atom from this tri-vacancy model contains an appreciable amount of the unpaired spin (−0.017 *e*), giving rise to significant hyperfine constants $A_k/h$ of 41.1, 37.4 and 39.3 MHz, which have not been observed in EPR experiments (Silsbee 1961; Jani et al. 1983). These positive hyperfine values indicate their origin from spin polarization due to the strong spin on the oxygen atom. We tested the possibility of removing the unexpected spin on the oxygen atom, but the fully optimized structure has the unpaired spin delocalized from Si0, hence inconsistent with the $E'_1$ center.

Further calculations were made for this tri-vacancy model with an Al atom at the Si1 site. This procedure removed the unexpected spin on the O1 atom that forms a weak bond with the Al atom at a distance of 1.905 Å, while the other three Al–O bonds are ~1.722 Å (Fig. 7.3; Table 7.2). This structure gives even better directions of the principal hyperfine axes, but the principal values for one of the weak $^{29}$Si hyperfine structures are not well reproduced (Table 7.3). The geometry

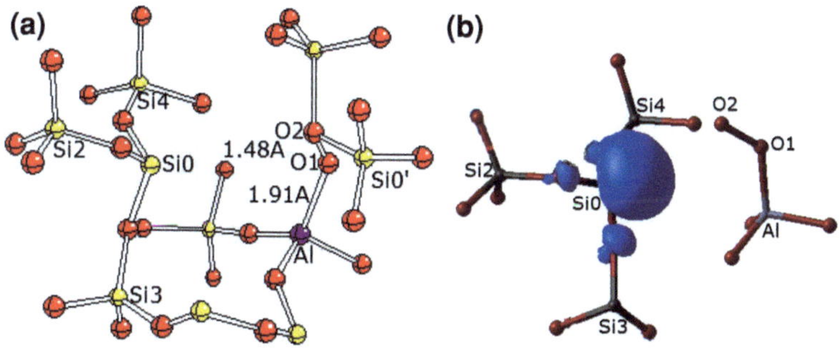


**Fig. 7.3** Local structure (**a**) and electron density (**b**) of the $E'_1$ center in quartz calculated from the tri-vacancy model with an Al atom at the Si1 site. Atom labels are similar to those in Fig. 7.2. Note that the $AlO_4$ group has an elongated Al–O1 bond of 1.906 Å. Contours are at intervals of 0.005 e/bohr$^3$ and from −0.01 to 0.415 e/bohr$^3$

of this $AlO_4$ group, including the elongated Al–O bond, is closely comparable to those of the well-established $[AlO_4]^0$ and $[AlO_4/M^+]^+$ centers in quartz (Nuttall and Weil 1981; Walsby et al. 2003; To et al. 2005; Botis and Pan 2009, 2011). Another interesting feature is that the O1 and O2 atoms are relaxed to form an electron-rich O–O bond of 1.48 Å (Fig. 7.3a), which is close to the peroxy type (Edwards and Fowler 1982; Nilges et al. 2009; Pan et al. 2009). This electron-rich O–O bond represents an ideal precursor for trapping a hole to form superoxide ($O_2^-$) radicals (Nilges et al. 2009; Pan et al. 2009). Such superoxide ($O_2^-$) radicals, which are close analogues of the peroxy radical and the non-bridging oxygen hole center (NBOHC) in amorphous $SiO_2$ (Friebele et al. 1979; Griscom and Friebele 1981; Edwards and Fowler 1982; Uchino et al. 2001), have been found in association with the $E'_1$ center in both artificially irradiated quartz and natural quartz from uranium deposits (Hu et al. 2008; Nilges et al. 2008, 2009; Pan et al. 2008, 2009). Therefore, our calculations of the new tri-vacancy model with an Al impurity not only better reproduces the experimental EPR data for the $E'_1$ center in quartz (Table 7.3) but can now account for its close association with the $[AlO_4]^0$ and $O_2^-$ centers (Jani et al. 1983; Toyoda and Hattori 2000; Nilges et al. 2008, 2009; Pan et al. 2008, 2009; Usami et al. 2009). Further calculations for these pairs of electron–hole paramagnetic centers (i.e., $E'_1$ and $[AlO_4]^0$ vs. $E'_1$ and $O_2^-$) in quartz are currently underway.

Additional calculations were made for this tri-vacancy model by allowing the O1 atom to relax away from the Al atom. The resulting configuration with an $AlO_3$ group (Fig. 7.4a) is 1.061 eV more stable than that with the $AlO_4$ group. This $AlO_3$ group is also slightly puckered (3°) and contains a fourth nearest oxygen atom at a distance of 2.742 Å (Table 7.2), which is considerably shorter than the Si1–$O_B$ distance of 3.480 Åfrom our $V(O)^+$ calculation. This $AlO_3$ group represents a precursor for trapping an electron to form the Al-equivalent $E'_1$ center, which has been reported to occur in irradiated vitreous silica (Brower 1979) but has not yet been found in quartz. Figure 7.4 shows that the peroxy O1–O2 bond of 1.45 Å also occurs in this

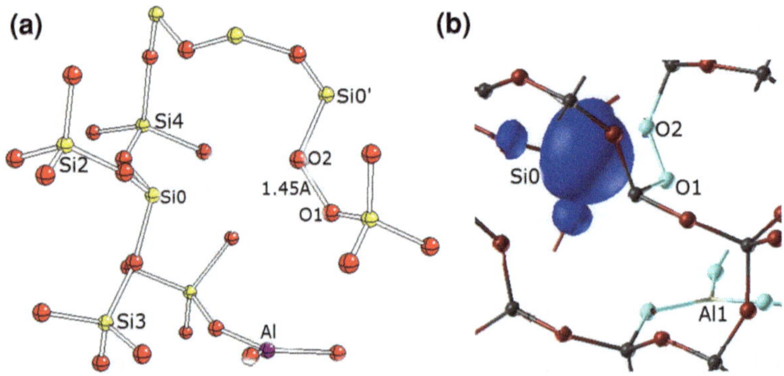


**Fig. 7.4** Local structure (**a**) and electron density (**b**) of the $E'_1$ center in quartz calculated from the tri-vacancy-model with an Al atom at the Si1 site. Atom labels are similar to those in Figs. 7.2 and 7.3. Note that the O1 atom is relaxed away from the $AlO_3$ group. Contours are at intervals of 0.005 e/bohr$^3$ and from −0.01 to 0.415 e/bohr$^3$

configuration and that the spin density on Si0 spreads toward this bond. Moreover, the distance between Si0 and the middle of this O–O bond is only 3.91 Å, which is shorter than the Si0–Al distance (Table 7.2). Therefore, the electrons on the O–O bond may reduce the electron deficiency of Si0 and thus stabilize this configuration. This may explain the stability of the $E'_1$ center in quartz and other $SiO_2$-based materials (Weeks 1956; Silsbee 1961; Griscom 1980; Jani et al. 1983; Weeks et al. 2008; Perlson and Weil 2008). This variant of the tri-vacancy model with an Al impurity also well reproduces the experimental $^{29}$Si hyperfine parameters in both principal values and directions (Table 7.3).

**Acknowledgments** We thank Prof. Jens Götze for invitation to the special session on quartz at BHT2011, Drs. Robert Möckel and Michael Plötze for manuscript review and handling, and Natural Science and Engineering Research Council (NSERC) of Canada for financial support. All calculations in this research have been enabled by the use of Westgrid computing resources, which are funded in part by the Canadian Foundation for Innovation, Alberta Innovation and Science, BC Advanced Education, and the participating research institutions. Westgrid equipment is provided by IBM, Hewlett Packard and SGI.

## References

Adamo C, Barone V (1999) Toward reliable density functional methods without adjustable parameters: the PBE0 model. J Chem Phys 110:6158–6170

Arends J, Dekker AJ, Perdok WG (1963) Color centers in quartz produced by crushing. Phys Stat Sol 3:2275–2279

Blöchl PE (2000) First-principles calculations of defects in oxygen-deficient silica exposed to hydrogen. Phys Rev B 62:6158–6179

Boero M, Pasquarello A, Sarnthein J, Car R (1997) Structure and hyperfine parameters of $E'_1$ centers in α-quartz and in vitreous $SiO_2$. Phys Rev Lett 78:887–890

Boero M, Oshiyama A, Silvestrelli PL (2003) $E'$ centers in a-quartz in the absence of oxygen vacancy: a first-principles molecular-dynamics study. Phys Rev Lett 91:20640-1–20641-4

Boero M, Oshiyama A, Silvestrelli PL (2004) $E'$ centers in silicon dioxide: first-principles molecular dynamics studies. Modern Phys Lett B 18:707–724

Bossoli RB, Jani MG, Halliburton LE (1982) Radiation-induced $E''$ centers in crystalline $SiO_2$. Solid State Commun 44:213–217

Botis SM, Pan Y (2009) Theoretical calculations of $[AlO_4/M^+]^0$ defects in quartz and implications for crystal-chemical control on the uptake of Al. Mineral Mag 73:537–550

Botis SM, Pan Y (2011) Modeling of $[AlO_4/Li^+]^+$ paramagnetic defects in α-quartz. Can J Phys 89:809–816

Botis SM, Nokhrin S, Pan Y, Xu Y, Bonli T, Sopuck V (2005) Natural radiation damages in quartz. I. Cathodoluminescence and correlation with paramagnetic defects. Can Mineral 43:1565–1580

Botis SM, Pan Y, Nokhrin S, Nilges MJ (2008) Natural radiation-induced damage in quartz. III. A new ozonide radical in drusy quartz from the Athabasca basin. Saskatchewan Can Mineral 46:122–135

Brower KL (1979) Electron paramagnetic resonance of Al $E'_1$ centers in vitreous silica. Phys Rev B 20:1799–1811

Carbonaro CM, Fiorentini V, Massidda S (1997) Ab initio study of oxygen vacancies in α-quartz. J Non-Crystal Solids 221:89–96

Carbonaro CM, Fiorentini V, Bernardini F (2001) Proof of the thermodynamic stability of the $E'_1$ center in $SiO_2$. Phys Rev Lett 86:3064–3067

Catti M, Valerio G, Dovesi R, Causa M (1994) Quantum-mechanical calculations of the solid-state equilibrium MgO + alpha-$Al_2O_3$ + $MgAl_2O_4$ (spinel) versus pressure. Phys Rev B 49:14179–14187

Chadi DJ (2003) Negative-U property of the oxygen vacancy defect in $SiO_2$ and its implication for the $E'_1$ center in α-quartz. Appl Phys Lett 83:437–439

Dovesi R, Saunders VR, Roetti C, Orlando R, Zicovich-Wilson CM, Pascale F, Civalleri B, Doll K, Harrison NM, Bush IJ, D'Arco P, Llunnell M (2006) CRYSTAL 2006, User's manual. University of Torino, Torino. http://www.crystal.unito.it

Edwards AH, Fowler WB (1982) Theory of the peroxy-radical defect in a-$SiO_2$. Phys Rev B 26:6649–6660

Edwards AH, Fowler WB (1990) Relationship between hyperfine parameters and the geometry of defects in nonmetallic solids. Phys Rev B 41:10816–10823

Edwards AH, Fowler WB, Feigl FJ (1988) Asymmetrical relaxation of simple $E'$ centers in silicon dioxide isomorphs. Phys Rev B 37:9000–9005

Feigl FJ, Fowler WB, Yip KL (1974) Oxygen vacancy model for the $E'_1$ center in $SiO_2$. Solid State Commun 24:225–229

Fiorentini V, Carbonaro CM (1997) Comment on "structure and hyperfine parameters of $E'$ centers in alpha-quartz and vitreous $SiO_2$". Phys Rev Lett 97:03027

Friebele EJ, Griscom DL, Stapelbroek M, Weeks RA (1979) Fundamental defect centers in glass: the peroxy radical in irradiated, high-purity, fused silica. Phys Rev Let 42:1346–1349

Gatti C, Saunders VR, Roetti C (1994) Crystal-field effects on the topological properties of the electron-density in molecular-crystals. The case of urea. J Chem Phys 101:10686–10696

Giordano L, Sushko PV, Pacchioni G, Shluger AL (2007) Optical and EPR properties of point defects at a crystalline silica surface: *ab initio* embedded-cluster calculations. Phys Rev B 75:024109-1-7

Griscom DL (1980) $E'$ center in glassy $SiO_2$: $^{17}O$, $^1H$, and "very weak" $^{29}Si$ superhyperfine structure. Phys Rev B 22:4192–4202

Griscom DL (2011) Trapped-electron centers in pure and doped glassy silica: a review and synthesis. J Non-Crystal Solids 357:1945–1962

Griscom DL, Friebele EJ (1981) Fundamental defect centers in glass: $^{29}Si$ hyperfine structure of the non-bridging oxygen hole center and the peroxy radical in $SiO_2$. Phys Rev B 24:4896–4898

Hu B, Pan Y, Botis S, Rogers B, Kotzer T, Yeo G (2008) Radiation-induced defects in drusy quartz, Athabasca Basin, Canada: a new aid to exploration of uranium deposits. Econ Geol 103:1571–1580

Ikeya M (1993) New applications of electron paramagnetic resonance: ESR dating, dosimetry, and spectroscopy. World Scientific, Singapore

Isoya J, Weil JA, Halliburton LE (1981) EPR and ab initio SCF–MO studies of the Si·H–Si system in the $E'$ center of $\alpha$-quartz. J Chem Phys 74:5436–5448

Jani MG, Bossoli RB, Halliburton LE (1983) Further characterization of the $E'_1$ center in crystalline $SiO_2$. Phys Rev B 27:2285–2293

Le Page Y, Calvert LD, Gabe EJ (1980) Parameter variation in low-quartz between 94 and 298 K. J Phys Chem Solids 41:721–725

Li Z, Pan Y (2011) First-principles study of boron oxygen hole centers in crystals: electronic structures and nuclear hyperfine and quadrupole parameters. Phys Rev B 84:115112

Magagnini M, Giannozzi P, Corso AD (2000) Microscopic structure of the substitutional Al defect in a quartz. Phys Rev B 61:2621–2625

Mashkovtsev RI, Pan Y (2011) Biradical states of oxygen-vacancy defects in $\alpha$-quartz: centers $E''_2$ and $E''_4$. Phys Chem Minerals 38:647–654

Mashkovtsev RI, Howarth DF, Weil JA (2007) Biradical states of oxygen-vacancy defects in $\alpha$-quartz. Phys Rev B 76:214114

Mysovsky AS, Sushko PV, Mukhopadhyay S, Edwards AH, Shluger AL (2004) Calibration of embedded-cluster method for defect studies in amorphous silica. Phys Rev B 69:085202-/1-/10

Nilges MJ, Pan Y, Mashkovtsev RI (2008) Radiation-induced defects in quartz. I. Single-crystal W-band EPR study of an electron irradiated quartz. Phys Chem Minerals 35:103–115

Nilges MJ, Pan Y, Mashkovtsev RI (2009) Radiation-induced defects in quartz. III. EPR, ENDOR and ESEEM characterization of a peroxy radical. Phys Chem Minerals 36:63–71

Nuttall RHD, Weil JA (1981) The magnetic properties of the oxygen-hole aluminum centres in crystalline $SiO_2$. I. $[AlO_4]^0$. Can J Phys 59:1696–1708

Pacchioni G, Ierano G, Marquez AM (1998) Optical absorption and nonradiative decay mechanism of $E'$ center in silica. Phys Rev Lett 81:377–380

Pan Y, Botis S, Nokhrin S (2006) Applications of natural radiation-induced paramagnetic defects in quartz to exploration in sedimentary basins. J China Univ Geosci 17:258–271

Pan Y, Nilges MJ, Mashkovtsev RI (2008) Radiation-induced defects in quartz. II. W-band single-crystal EPR study of a natural citrine. Phys Chem Minerals 35:387–397

Pan Y, Nilges MJ, Mashkovtsev RI (2009) Multifrequency single-crystal EPR characterization and DFT modeling of new peroxy radicals in quartz. Mineral Mag 73:517–535

Perlson BD, Weil JA (2008) Electron paramagnetic resonance studies of the $E'$ centres in alpha quartz. Can J Phys 86:871–881

Pisani C, Dovesi R, Orlando R (1992) Near-Hartree-Fock wave functions for solids: the case of crystalline silicon. Int J Quant Chem 42:5–33

Rudra JK, Fowler WB (1987) Oxygen vacancy and the E'1 center in crystalline $SiO_2$. Phys Rev B 35:8223–8230

Silsbee RH (1961) Electron spin resonance in neutron-irradiated quartz. J Appl Phys 32:1459–1462

Snyder KC, Fowler BW (1993) Oxygen vacancy in $\alpha$-quartz: a possible bi- and metastable defect. Phys Rev B 48:13238–13243

Stesmans A, Clemer K, Afanas'ev MM (2008) Primary Si-29 hyperfine structure of $E'$ centers in nm-sized silica: probing the microscopic network structure. Phys Rev B 77:094130

Stoneham AM, Fisher AJ, Greenland PT (2003) Optically-driven silicon-based quantum gates with potential for high temperature operation. J Phys Condens Matter 15:L447–L451

Sulimov VB, Sushko PV, Edwards AH, Shluger AL, Stoneham AM (2002) Asymmetry and long-range character of lattice deformation by neutral oxygen vacancy in $\alpha$-quartz. Phys Rev B 66:024108

To J, Sokol AA, French SA, Kaltsoyannis N, Catlow R (2005) Hole localization in $[AlO_4]^0$ defects in silica materials. J Chem Phys 122:144704

Toyoda S, Hattori W (2000) Formation and decay of the $E'_1$ center and its precursor. Appl Rad Isotop 52:1351–1356

Uchino T, Takahashi M, Yoko T (2001) Structure and generation mechanism of the peroxy-radical defect in amorphous silica. Phys Rev Lett 80:4560–4563

Usami T, Toyoda S, Bahadur H, Srivastava AK, Nishido H (2009) Characterization of the $E'_1$ center in quartz: role of aluminum hole centers and oxygen vacancies. Phys B 404:3819–3823

Walsby CJ, Lees NS, Claridge RFC, Weil JA (2003) The magnetic properties of oxygen-hole aluminum centers in crystalline $SiO_2$. VI. A stable $AlO_4$/Li centre. Can J Phys 81:583–598

Weeks RA (1956) Paramagnetic resonance of lattice defects in irradiated quartz. J Appl Phys 27:1376–1381

Weeks RA (1963) Paramagnetic spectra of $E'_2$ centers in crystalline quartz. Phys Rev 130:570–576

Weeks RA, Magruder RH, Stesmans A (2008) Review of some experiments in the 50 year saga of the $E'$ center and suggestions for future research. J Non-Cryst Solids 354:208–216

Yip KL, Fowler WB (1975) Electronic structure of $E'_1$ centers in $SiO_2$. Phys Rev B 11:2327–2338

# Chapter 8
# Gamma-Irradiation Dependency of EPR and TL-Spectra of Quartz

**Michael Plötze, Dieter Wolf and Matthias R. Krbetschek**

**Abstract** Paramagnetic defects formed during experimental $\gamma$-irradiation with various dose (70 to $1.5 \times 10^7$ Gy, $^{60}Co$) in well documented polycrystalline quartz samples from tin-tungsten ore veins, gneiss and granite were investigated by X-band continuous-wave electron paramagnetic resonance spectroscopy at temperatures from 20–295 K. On the same samples spectrally resolved thermoluminescence measurements were carried out (temperature range 50–350 °C, heating rate 2 K/s; wavelength region 200–800 nm) to examine the relationships between the radiation dose, the EPR signal intensity and the thermoluminescence spectra to link the EPR-spectra to the centres responsible for the thermoluminescence peaks in natural quartz. The present study concerns mainly the analysis of the paramagnetic centres $[AlO_4]^0$ and $[TiO_4^-/Li^+]^0$. The investigation of the relationship between the EPR intensity of paramagnetic centres and the $\gamma$-irradiation dose shows an increase of the EPR intensity with values related to the impurity content. Different centres reveal different saturation behaviour but independent from concentration. The $[AlO_4]^0$ centres show saturation at about $1 \times 10^6$ Gy and the $[TiO_4^-/Li^+]^0$ centres already after irradiation with $5 \times 10^3$ Gy with radiogenic annealing at higher irradiation dose. From the irradiation behaviour the $[TiO_4^-/Li^+]^0$ centre is suggested

M. Plötze (✉)
ClayLab, Institute for Geotechnical Engineering,
ETH Zurich, 8093 Zurich, Switzerland
e-mail: michael.ploetze@igt.baug.ethz.ch

D. Wolf
Institute of Mineralogy, TU Bergakademie Freiberg,
Brennhausgasse 14, 09596 Freiberg, Germany

M. R. Krbetschek
Institute for Applied Physics, TU Bergakademie Freiberg,
Leipziger Str 23, 09596 Freiberg, Germany
e-mail: quatmi@mailserver.tu-freiberg.de

J. Götze and R. Möckel (eds.), *Quartz: Deposits, Mineralogy and Analytics*,
Springer Geology, DOI: 10.1007/978-3-642-22161-3_8,

as electron traps for the TL peaks at 150–200 °C/330–340 nm, 200 °C/510 nm and 280 °C/470–510 nm, whereas oxygen-vacancy-centres for the first peak and the $[AlO_4]^0$-centres for the other peaks are working as recombination sites.

## 8.1 Introduction

In the past 50 years a large number of paramagnetic centres in quartz have been detected and crystallographically described using EPR. The results of more than 200 publications about this topic were included in the reviews of Weil 1984 and 1993.

Structural defects are arising from impurities as well as vacancies in the crystalline structure. There are different mechanisms of trace-element input into the quartz. Trace elements in quartz are incorporated into the crystal structure or bound to microinclusions. Due to its structure, quartz is considered to incorporate only small amounts of impurities in its crystal lattice. The most elements are bound to fluid inclusions and to microscopic mineral inclusions (Blankenburg et al. 1994; Götze et al. 1997, 2004). The structural incorporation of Al, Ge, Ti, Ga, Fe or P into the Si tetrahedron is well investigated. The charge deficit is compensated by alkali ions (Li, Na, K) or hydrogen (or Cu, Ag, etc.), which are distributed in structural channels parallel to the c-axis.

Paramagnetic centres can be formed by incorporation of a priori paramagnetic ions such as $Mn^{2+}$, $Fe^{3+}$ or REE in the proper valence state or by ionizing radiation which produces charged electron or electron hole centres on structural defects or atomic vacancies (Table 8.1). Such radiation defects can be used for dating or radiation dose measurements (Ikeya 1993). The spectral analysis of the cathodoluminescence (CL) emission in combination with electron paramagnetic resonance (EPR) measurements and/or trace-element analysis allows to study extrinsic (trace elements) or intrinsic (lattice defects) point defects in the quartz structure. The defects causing the different CL emissions in quartz often reflect the specific physicochemical conditions of crystal growth and therefore, can be used as a signature of genetic conditions (Götze et al. 2001, 2004, 2005; Botis et al. 2005).

The Al centre is caused by substitution of $Al^{3+}$ for $Si^{4+}$ with an electron hole trapped by a nonbonding 2p orbital at one of the four nearest $O^{2-}$ ions, forming an $O^-$ type centre. This was the paramagnetic centre discovered first in quartz by Griffiths et al. (1954). The precursor state for this centre is the diamagnetic $[AlO_4/M^+]^0$ associated with an adjacent charge compensating cation $M^+$ ($H^+$, $Li^+$). During irradiation at 295 K the $M^+$-ion may diffuse away yielding the paramagnetic $[AlO_4]^0$ (Nuttall and Weil 1981). The EPR of the centre is observable at temperatures lower than 150 K and was measured at 70 K, at which the line intensity is maximum. For polycrystalline samples the EPR spectrum of $[AlO_4]^0$ exhibits a multiline spectrum of superimposed 6-line hyperfine patterns (Fig. 8.1). The $[TiO_4^-/M^+]^0$ centres are produced by irradiation of the diamagnetic precursor $[TiO_4]^0$ ($Ti^{3+}$, i.e., electron centre at $Ti^{4+}$) which is formed by substitution of $Ti^{4+}$ for $Si^{4+}$ at the Si position, where charge compensation is achieved by a proton or

**Table 8.1** Classification of selected paramagnetic centres in natural quartz (notation after Weil 1984)

| | | Impurities | | Vacancies | |
|---|---|---|---|---|---|
| | | Substitutional for $Si^{4+}$ | Interstitial | $O^{2-}$ vacancy | $Si^{4+}$ vacancy |
| Metastable | Electron centres ($+e^-$) | $[TiO_4]^-$ ] $[TiO_4/M^+]^0$ $[GeO_4]^-$ $[GeO_4/M^+]$ | $M^+ = H^+, Li^+$ | E′-centres: $[SiO_3]^{3-}$ | |
| | Electron hole centres ($-e^-$) | $[AlO_4]^0$ $[FeO_4]^0$ $[H_4O_4]^+$ $[H_3O_4]^0$ | | | $O_2^-$, $O_3^-$ |
| Stable (paramagnetic without charge trapping) | | $[FeO_4/M^+]^0$ (*S*) $[FeO_4]^-$ | $M^+ = H^+$, $Li^+$, $Na^+$ | | |

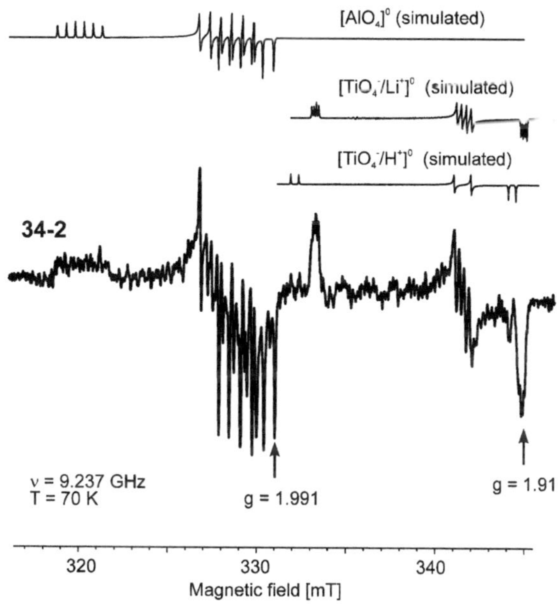


**Fig. 8.1** EPR powder spectrum of natural granite quartz with simulated spectra of the $[AlO_4]^0$ and $[TiO_4^-/M^+]^0$ centres (adapted Nuttal and Weil 1981 and Bershov 1970 respectively). The line at $g = 1.991$ was used for determination of the $[AlO_4]^0$ centre concentration and the line at $g = 1.911$ for $[TiO_4^-/Li^+]^0$

$Li^+$ ion at a channel position nearby (Wright et al. 1963; Bershov 1970; Rinneberg and Weil 1972). These centres can be observed between 5 and 120 K. The spectrum of the polycrystalline sample exhibits a characteristic pattern that consists of three well-separated doublets for the $[TiO_4^-/H^+]^0$ and quartets for $[TiO_4^-/Li^+]^0$ (Fig. 8.1). For both centres two types can be distinguished (Okada et al. 1971; Rinneberg and Weil 1972): One stable at temperatures below 100 K and the other one stable at room temperature. Only the type which is stable at room temperature is observable after $\gamma$-irradiation at 295 K.

The Fe-related centres has been shown to be the uncompensated $[FeO_4]^-$ centre (Mombourquette et al. 1986), and the compensated so called S centres $[FeO_4/M^+]^0$, where $M = H^+$, $Li^+$ and $Na^+$ (Stegger and Lehmann 1989a, b; Mombourquette et al. 1989).

A multiproton-centre is formed by four $H^+$ replacing a single $Si^{4+}$. During irradiation, one oxygen ion is trapping an electron hole to produce a $[H_4O_4]^+$ centre or one of the protons is exchanged with an electron hole to form the $[H_3O_4]^0$ centre (Weil 1993; Cordier et al. 1994).

The wide family of electron and electron hole centres on structural vacancies involves oxygen-vacancy related centres, the so called E centres (Weeks 1956; Jani et al. 1983; Orlenov 1984; Mashkovtsev et al. 2007; Usami et al. 2009) and centres at oxygens near a silicon-vacancy. The $O_2^{3-}$ and $O_2^{3-}$-$M^+$ centres, proposed by Marfunin and others (Marfunin 1979; Mashkovtsev et al. 1978; Orlenov 1984), have now been shown to be superoxide ($O_2^-$) radicals (Maschmeyer and Lehmann 1983; Nilges et al. 2008, 2009; Pan et al. 2008, 2009). Another silicon-vacancy hole centre is the ozonide ($O_3^-$) radical (Botis et al. 2008). The vacancy-related centres are stable up to temperatures of 570–670 K and the diamagnetic precursor even to at least 1100 K (Pan and Hu 2009).

The simplest view on the relation of the concentration of radiation-induced paramagnetic centres to the irradiation dose is the product of dose rate and time. However, radiation-induced centres are metastable and fade out at any temperature. The annealing rate increases with temperature, which is the base of the thermoluminescence where such radiation-induced centres act e.g. as recombination sites. The thermal stability and therefore the TL-properties depend from different factors such as the defect structure and the radiation history but also the sample treatment and measurement procedure (Kuhn et al. 2000; Vyatkin et al. 2007; Furetta 2010). TL measurements of minerals were carried out to investigate the defect structure, for example for genetic studies. TL-spectra provide important information on the nature of both intrinsic and extrinsic metastable defects. Thermoluminescence dosimetry with natural minerals has found wide application in Earth Sciences e.g., in TL-dating. Much less is known about the TL-properties of quartz, such as emission spectra, impurities and related (paramagnetic) centres and the radiation dependence of these centres than in the case of the more intensively studied alkali halides. There are a few publications dealing with TL-emission spectra from quartz above room temperature (Table 8.2). Emission wavelengths are reported in the bands of about 330, 380, 470 and 620 nm. The emission spectra vary from samples of different origin and are related to different impurities and centres. An important part for the TL emission play Al, Ti, and Ge substitutions for $Si^{4+}$ and the charge compensating alkali ions $H^+$, $Na^+$ and $Li^+$. The $[TiO_4/M^+]^0$ and $[GeO_4/M^+]^0$-centres act as electron traps and the $[AlO_4]^0$-, the $[H_3O_4]^0$- and the vacancy-centres as recombination sites. The recombination at the $[AlO_4]^0$ centre yields emission near 470 nm, whereas the recombination at $[H_3O_4]^0$ and at oxygen-vacancy centres yields emissions near 380 and 330–365 nm respectively (Serebrennikov et al. 1982; McKeever 1984; McKeever 1991; Rink et al. 1993).

Our object was to examine the relationships between the radiation dose, the EPR signal intensity and the TL-spectra to link the EPR-spectra to the centres responsible for the TL-peaks in natural quartz.

**Table 8.2** Selected TL peaks above room temperature

| Peak T (°C) | Wavelength (nm) | Defect/mechanism | Authors |
|---|---|---|---|
| >20 | 365 | Recombination from E′ centres at hole centres ($O^-$type defects) | Serebrennikov et al. (1982) |
| >20 | 420–490 | $[AlO_4]^0$ as recombination centre | Lysakov et al. (1969) McKeever (1984) |
| 100 | 380 | Recombination from $[GeO_4]^-$ at $[H_3O_4]^0$ | McKeever (1991) |
| 100 | 470 | Recombination from $[GeO_4]^-$ at $[AlO_4]^0$ | McKeever (1991) |
| 100 | 380, 470 | $[AlO_4]^0$, $[AlO_4/M^+]^0$ | Martini et al. (1995) |
| 110 | | $E_1' + Li^+$, $Na^+$ | Martini et al. (1985) |
| 180 | | $[GeO_4/M^+]^0$ | Arnold (1976) |
| >200 | 470 | $[AlO_4]^0$ as recombination centre | Jani et al. (1983) |
| 200–220 | 380 | $[AlO_4]^0$ | Marfunin (1979) |
| 230 | 280, 330 | Recombination at oxygen vacancies (E =) (?) | Rink et al. (1993) |
| 230 | 470 | Recombination from $[GeO_4/M^+]^0$ at $[AlO_4]^0$ | Rink et al. (1993) |
| 280–300 | 440–480 | $Ti^{3+}$, $Ge^{3+}$ | Marfunin (1979) |
| 300 | | $Ti^{4+}$ | Medlin (1963) |
| 330 | | $[AlO_4]^0$ | Arnold (1976) |
| 330–350 | | $Li^+$, $Na^+$ | Batrak (1958) |
| 340 | 450 | ? | Mejdahl (1986) |
| 375 | 470 | $[AlO_4]^0$ | Hashimoto et al. (1987) |
| 380 | 620 | ? | Hashimoto et al. (1987); Ichikawa (1967) |

## 8.2 Experimental

Spectrally resolved TL and EPR measurements have been carried out on quartz from the tin tungsten deposit Ehrenfriedersdorf (Germany), located in the exo-contact of a Hercynian Li–F granite intrusion. The samples are quartz from veins (Q-2, Q-51) and from the wall rock gneiss (Q-101). The second sample set consists of quartz from a drilling profile in the Eldzhurtinskiy Granite (Northern Caucasus, Russia) (Lyakhovitch and Gurbanov 1992).

The quartz grains were carefully crushed to a grain size <200 μm and sieved in the fraction of 100–200 μm. This fraction was treated with HCl to remove carbonates and with HF to remove feldspars, cleaned with distilled water, dried at 25 °C, and hand-picked under a binocular microscope in order to get pure quartz samples.

The separated quartz samples were analysed for trace-element content by atomic emission and absorption spectroscopy.

Before irradiation, the samples were heated at 400 °C for 5 h to anneal the paramagnetic centres formed by natural irradiation. After thermal treatment the samples were divided in aliquots and $\gamma$-irradiated with a $^{60}Co$ source at room temperature in separate doses from 70 Gy to $1.5 \times 10^7$ Gy and a dose rate of appr. $1.5 \times 10^2$ Gy/h. For doses higher than $1 \times 10^3$ Gy the dose rate was $1 \times 10^4$ Gy/h. The sample sets were stored at 250 K after the irradiation for 2 weeks.

The paramagnetic centres of polycrystalline quartz samples were investigated by EPR at X-band frequencies (9.2 GHz) at 20, 70, and 295 K. The spectra were recorded by a Varian E-line spectrometer. The sample weight was 20 mg. The influence of technical parameters such as modulation amplitude, microwave power, temperature, scan time etc. on the spectra was checked for the optimal settings for recording the spectra. The sample temperature was controlled with a low temperature unit based on a helium gas flow device (Oxford ESR 900A). These settings (modulation field $H_M = 1$ G, microwave power $p = 7$ mW, temperature $T = 70$ K for $[AlO_4]^0$ and $[TiO_4^-/Li^+]^0$ centres) were kept constant throughout all the measurements to allow correct comparison between the signal intensities of different spectra. The intensity of $[TiO_4^-/M^+]^0$ centres was measured using the same conditions as for Al centres. The concentration of the paramagnetic centres was determined as peak to peak or peak to base intensity at the analytical lines (Moiseev 1985). The specific peak positions of the paramagnetic centres were drawn from simulated spectra with the data from literature (Nettar and Villafranca 1985). The variation of intensity detected by repeated measurements of selected analytical lines was up to 10%. The concentration of Al centres was quantified using a reference sample with known $[AlO_4]^0$ concentration (Moiseev 1985). All other centres were calculated in relative amounts.

Spectrally resolved TL measurements were carried out in the temperature range from 50 to 350 °C (heating rate 2 K/s) with a laboratory made CCD-camera based high sensitivity TL/OSL-spectrometer (Rieser et al. 1994). With a holographic concave grating the spectrum range of 200–800 nm is dispersed onto a $N_2$(liq) cooled Astromed CCD-chip. To obtain the whole spectral data to make a temperature/wavelength 3-D-plot 40 short exposures were made. Further data processing on a PC puts these slices with the corresponding temperature and displays the data as 3-D-plots or contour maps.

## 8.3 Results and Discussion

*Chemical analysis.* Selected trace-elements contents are shown in Table 8.3. The vein quartzes are relatively similar in trace-element concentrations. The high content of Na and Li in the quartz Q-51 is probably caused by fluid inclusions, which also cause the milky appearance of this quartz. The quartz from the wall rock gneiss has the lowest content of Al, Ti, Na and Li. The granite quartz shows the highest concentrations in Al and Ti.

*EPR spectroscopy.* In the samples the paramagnetic centres $[AlO_4]^0$, $[TiO_4/Li^+]^0$, $[TiO_4/H^+]^0$, $[TiO_4]^-$, $[GeO_4/Li^+]^0$, electron hole varieties and $Fe^{3+}$

**Table 8.3** Content of selected trace elements (ppm) in the quartz samples

| Sample | Al | Ti | Ge | Fe | Na | Li |
|---|---|---|---|---|---|---|
| Vein quartz (Q-2) | 176 | 10.8 | 0.9 | 20 | 146 | 5 |
| Vein quartz (Q-51) | 157 | 2.4 | 0.7 | 30 | 750 | 47 |
| Wall rock quartz (Q-101) | 103 | <2.4 | nd | 24 | 66 | <1 |
| Granite quartz (34-2/90) | 407 | 47.0 | nd | 46 | 300 | 29 |

were detected (Figs. 8.1 and 8.4). The $[AlO_4]^0$ and $[TiO_4^-/Li^+]^0$ were the most intense detected centres whereas the $[GeO_4^-/M^+]^0$ centres show only very weak intensities. The EPR-spectra of the various vein quartzes (Q-2, Q-51) are relatively similar. The quartz from the wall rock (Q-101) with the lowest impurity content shows lower intensity of $[AlO_4]^0$ and no $[TiO_4^-/Li^+]^0$ centres. Consequentially, the quartz with the highest content of impurities (granite quartz 34-2/90) shows also the highest EPR intensities. The differences in $[AlO_4]^0$ centre concentrations and the occurrence of $[TiO_4^-/Li^+]^0$ between the samples indicate their different genesis. The higher amounts of the centres in irradiated samples from veins can be correlated with the higher amounts of Al and Ti substituted for Si. The difference may be indicative for the specific H and Li concentrations in the respective hydrothermal solutions.

The investigation of the relationship between the EPR intensity of paramagnetic centres and the $\gamma$-irradiation dose reveal different behaviour for different paramagnetic centres. Generally, $\gamma$-irradiation caused a transformation of diamagnetic precursor centres into these paramagnetic centres resulting in an increase of the EPR intensity. The intensity of EPR-spectra of the $[AlO_4]^0$ centre increases with the $\gamma$-dose and reaches saturation at a dose of about $1 \times 10^6$ Gy (Fig. 8.2). This irradiation behaviour corresponds with the time dependent decreasing intensity of the 380 nm cathodoluminescence emission (Götze et al. 2001). The related diamagnetic $Al/Li^+$ defect is transformed into the paramagnetic Al-defect during the $e^-$ irradiation. The $[TiO_4^-/Li^+]^0$ spectrum shows the highest intensity already after irradiation with $5 \times 10^3$ Gy (Fig. 8.3). Further irradiation leads to a decrease of the centre concentration (Plötze and Wolf 1996). This behaviour was described as "radiogenic annealing" (Moiseev 1985). Another explanation could by the diffusion of charge compensating $M^+$ ions. In this case the intensity of the uncompensated Ti-centre should increase with decreasing intensity of the compensated centres. However, this could not be observed. Only after a strong $\gamma$-irradiation ($>1 \times 10^6$ Gy) some varieties of electron hole centres were detected in EPR at 295 K (Figs. 8.4, 8.5). This irradiation behaviour corresponds with the time instable behaviour of the 620 nm cathodoluminescence emission (Götze et al. 2001).

*Thermoluminescence spectra.* The TL-spectra for the different vein quartz samples are, like in EPR, very similar. Because of the spectrometer sensitivity no thermoluminescence is observable until a $\gamma$-dose of 700 Gy. Three main peaks are detected in the spectra after an irradiation with $1 \times 10^4$ Gy: 150–200 °C/330–340 nm, 200 °C/510 nm and 280 °C/470–510 nm (Fig. 8.6). The quartz from the wall rock (Q–101) shows in the EPR no $[TiO_4^-/Li^+]^0$ centres and in TL only

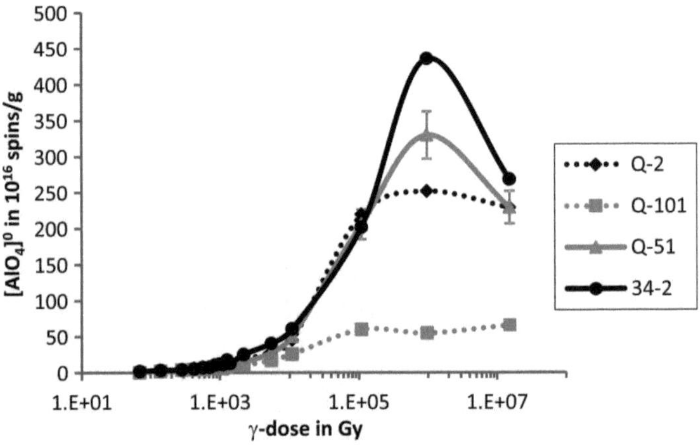


**Fig. 8.2** Radiation dependence of $[AlO_4]^0$ centres

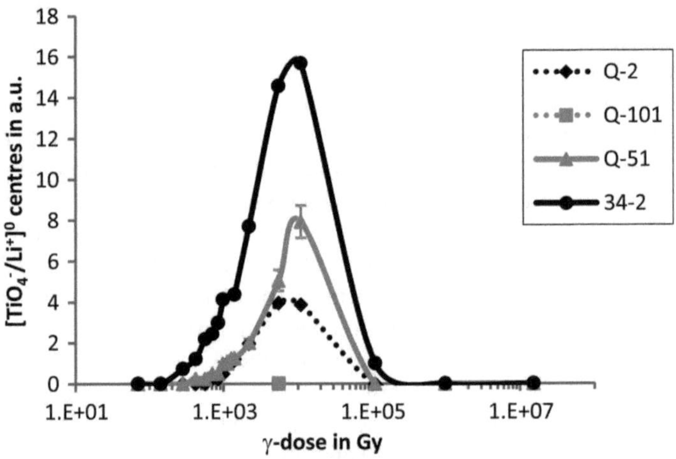


**Fig. 8.3** Radiation dependence of $[TiO_4^-/Li^+]^0$ centres

the peak at 280 °C (Fig. 8.7). Furthermore, an intensive TL-signal is visible at 450–550 nm at temperatures higher than 330 °C in all samples. This peak is probably related to electron hole centres and $[AlO_4]^0$-centres because of the same $\gamma$-dose response (Figs. 8.2, 8.5). The intensity of this signal increases by higher irradiation ($1 \times 10^6$ Gy) whereas the peaks at 150–200 and 200 °C disappear (Fig. 8.8).

Similar TL-spectra were published by Rink et al. (1993) for quartz samples from pegmatites. The authors claimed that the peak at 230 °C/470 nm (near our peak at 200 °C/510 nm) is related to $[GeO_4/M^+]^0$ centres because of the same $\gamma$-dose response. However, they did not study EPR at low temperatures so that the

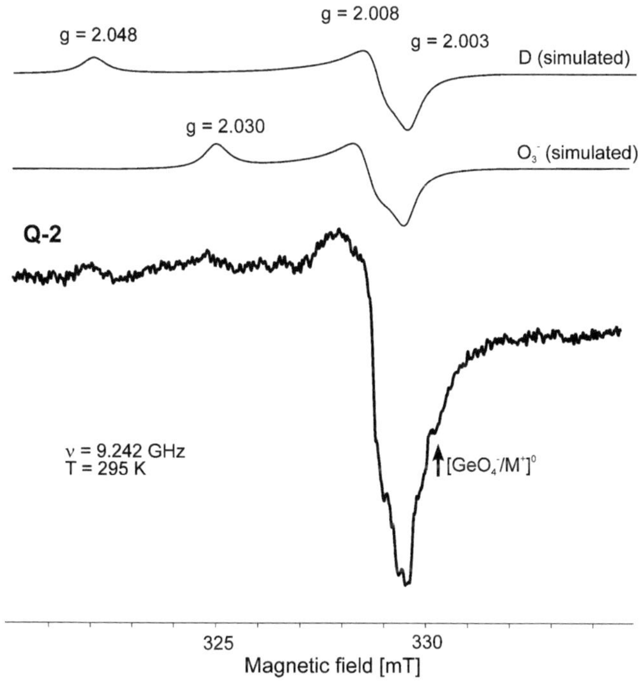


**Fig. 8.4** EPR powder spectrum of vein quartz after $10^7$ Gy $\gamma$-irradiation with simulated spectra of varieties of Si vacancy related electron hole centres (centre D: Maschmeyer and Lehmann 1983, centre $O_3^-$: adapted Nilges et al. 2009)

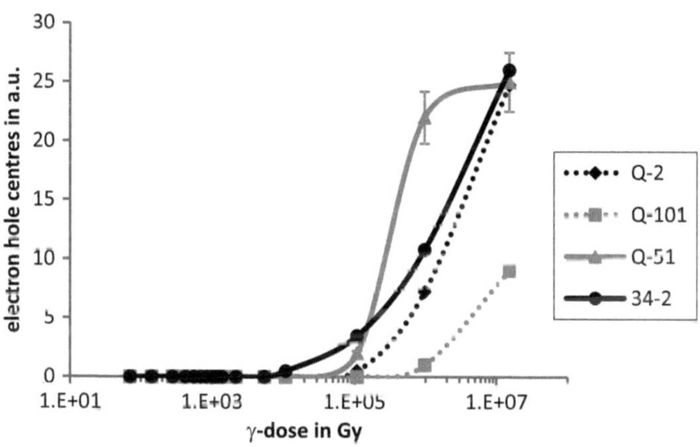


**Fig. 8.5** Radiation dependence of Si vacancy related electron hole centres (sum)

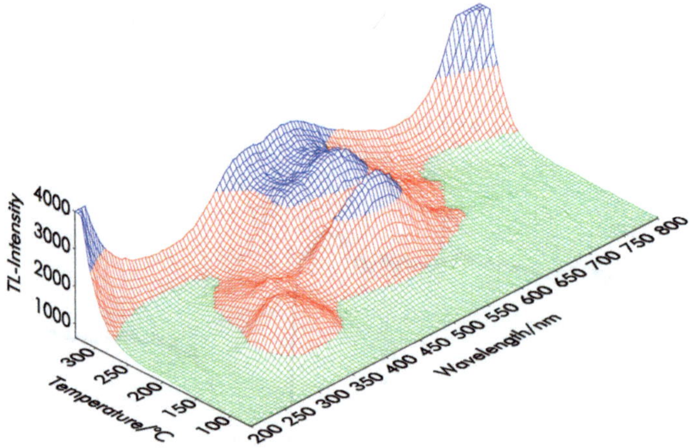


**Fig. 8.6** TL-spectrum of vein quartz (Q-2, $\gamma$-dose $^{60}Co$ 1 × $10^4$ Gy)

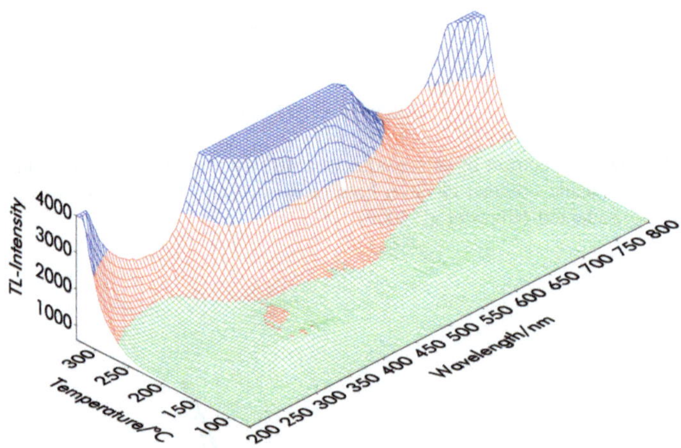


**Fig. 8.7** TL-spectrum of quartz of the wall rock (Q-101, $\gamma$-dose $^{60}Co$ 1 × $10^4$ Gy)

Ti-centres could not be detected. Furthermore, the total Ge content and the concentration of paramagnetic $[GeO_4^-/M^+]^0$ centres in our samples is very low. Therefore, the $[TiO_4^-/Li^+]^0$ centres can be suggested as possible electron traps for the TL-peaks at 150–200 °C/330–340 nm, 200 °C/510 nm and 280 °C/470–510 nm. This suggestion for the latter peak is in good agreement with results from Marfunin (1979) who related this peak also to Ti- and Ge-centres. For the TL-peak at 150–200 °C/330–340 nm oxygen-vacancy centres are working as recombination sites (Serebrennikov et al. 1982, Rink et al. 1993) and for the TL-peaks at 200 °C/

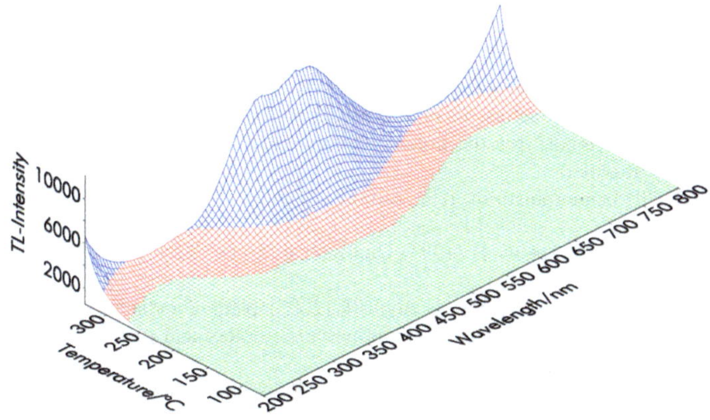


**Fig. 8.8** TL-spectrum of vein quartz (Q-2, $\gamma$-dose $^{60}Co$ 1 × $10^6$ Gy)

510 nm and 280 °C/470–510 nm the $[AlO_4]^0$ centres (Jani et al. 1983, McKeever 1984, 1991, Rink et al. 1993).

## 8.4 Conclusion

The most intense detected paramagnetic centres are $[AlO_4]^0$, $[TiO_4/Li^+]^0$, and after heavy irradiation different electron hole varieties. The EPR intensity of different centres is related to the impurity content and reflects different formation conditions. The investigation of the relationship between the EPR intensity of paramagnetic centres and the $\gamma$-irradiation dose shows an increase of the EPR intensity, but reveal different saturation behaviour. The $[AlO_4]^0$ centre reaches saturation at a dose of about $1 \times 10^6$ Gy. The $[TiO_4^-/Li^+]^0$ spectrum, however, shows the highest intensity already after irradiation with $5 \times 10^3$ Gy and a decrease after higher irradiation dose. The saturation intensity in EPR is related to the chemical impurity concentration. Because of the same $\gamma$-dose response in TL and the very low $[GeO_4^-/M^+]^0$ centre concentration, the $[TiO_4^-/Li^+]^0$ centres are suggested as possible electron traps for the TL-peaks at 150–200 °C/330–340 nm, 200 °C/510 nm and 280 °C/470–510 nm.

**Acknowledgments** The presented studies were carried out with the support of the Deutsche Forschungs gemeinschaft DFG (grant Wo 489/1). We acknowledge S.S. Hafner (University Marburg) for permission to carry out the EPR measurements. We thank S.M. Sukharjevski (University St. Petersburg) for stimulating discussions and Y. Pan (University of Saskatchewan) for helpful suggestions.

# References


Arnold GW (1976) Thermoluminescence of ion-implanted $SiO_2$. In: Chernaw F (ed) Ion implantation in semiconductors. Plenum Press, New York

Batrak EN (1958) A model for the colour and emission centres in quartz. Kristallografiya 3:626–627 (in Russian)

Bershov LV (1970) Isomorphism of Ti in natural minerals. Izv AN SSSR Ser Geol n12:47–54 (in Russian)

Blankenburg HJ, Götze J, Schulz H (1994) Quarzrohstoffe. Dt. Verl. f. Grundstoffind, Leipzig-Stuttgart

Botis SM, Nokhrin SM, Pan Y, Xu Y, Bonli T (2005) Natural radiation-induced damage in quartz. I. Correlations between cathodoluminescence colors and paramagnetic defects. Can Mineral 43:1565–1580

Botis SM, Pan Y, Nokhrin S, Nilges MJ (2008) Natural radiation-induced damage in quartz III. A new ozonide radical in drusy quartz from the Athabasca basin, Saskatchewan. Can Mineral 46:125–138

Cordier P, Weil JA, Howarth DF, Doukhan JC (1994) Influence of the $(4H)_{Si}$ defect on dislocation motion in crystalline quartz. Eur J Mineral 6:17–22

Furetta C (2010) Handbook of thermoluminescence. World Scientific Publishing, Singapore

Götze J, Plötze M, Trautmann T (2005) Structure and luminescence characteristics of quartz from pegmatites. Am Mineral 90:13–21

Götze J, Plötze M, Graupner T, Hallbauer DK, Bray CJ (2004) Trace element distribution in pegmatite quartz: a combined study by ICP-MS, Electron Spin Resonance (ESR), Capillary Ion Analysis (CIA) and Gas Chromatography (GC). Geochim Cosmochim Ac 68:3741–3759

Götze J, Plötze M, Habermann D (2001) Origin, spectral characteristics and practical applications of the cathodoluminescence (CL) of quartz—a review. Miner Petrol 71:225–250

Götze J, Plötze M (1997) Investigation of trace-element distribution in detrital quartz by electron paramagnetic resonance (EPR). Eur J Mineral 9:529–537

Griffiths JHE, Owen J, Ward IM (1954) Paramagnetic resonance in neutron-irradiated diamond and smoky quartz. Nature 173:439–442

Hashimoto T, Yokosaka K, Habuki H (1987) Emission properties of TL from natural quartz. Nucl Tracks Radiat Meas 13:57–66

Ichikawa Y (1967) Thermoluminescence of natural quartz irradiated by X-rays. Jap J Appl Phys 7:220–226

Ikeya M (1993) New applications of electron spin resonance: dating, dosimetry and microscopy. World Scientific Publishing, Singapore

Jani MG, Halliburton LE, Kohnke EE (1983) Point defects in crystalline $SiO_2$: Thermally stimulated luminescence above room temperature. J Appl Phys 54:6321–6328

Kuhn R, Trautmann T, Singhvi AK, Krbetschek MR, Wagner GA, Stolz W (2000) A study of thermoluminescence emission spectra and optical stimulation spectra of quartz from different provenances. Radiat Meas 32:653–657

Lyakhovich VV, Gurbanov AG (1992) Geochemistry and conditions of formation of Eljurtinsky massif (North Caucasus). Geokhimiya 6:800-812 (in Russian)

Lysakov VS, Serebrennikov AI, Solntsev VP (1969) Sources and spectra of the thermoluminescence for natural quartz crystals. Zh Prikl Spektroskopii 11:757–760 (in Russian)

Marfunin AS (1979) Spectroscopy luminescence and radiation centers in minerals. Springer, New York

Martini M, Sibilia E, Spinolo G, Vedda A (1985) Pre-dose, TSL and a.c. conductance interrelation in quartz. Nucl Tracks Radiat Meas 10:497–502

Martini M, Paleari A, Spinolo G, Vedda A (1995) Role of $[AlO_4]^0$ centers in the 380 nm thermoluminescence of quartz. Phys Rev B 52:138–142

Maschmeyer D, Lehmann G (1983) New hole centers in natural quartz. Phys Chem Miner 10:84–88

Mashkovtsev RI, Shcherbakova MYA, Solntsev VP (1978) EPR of radiation hole centres in α-quartz. In: Sobolev VS (ed) Rentgenografiya i spectroscopiya mineralov. Nauka, Novosibirsk pp 78–86

McKeever SWS (1984) Thermoluminescence in quartz and silica. Radiat Prot Dosim 8:81–98

Mashkovtsev RI, Howarth DF, Weil JA (2007) Biradical states of oxygen-vacancy defects in alpha-quartz. Phys Rev B 76:214114

McKeever SWS (1991) Mechanisms of thermoluminescence production: some problems and a few answers? Nucl Tracks Radiat Meas 18:5–12

Medlin WL (1963) Thermoluminescence in quartz. J Chem Phys 38:1132–1143

Mejdahl V (1986) Thermoluminescence dating of sediments. Radiat Prot Dosim 17:219–227

Moiseev BM (1985) Natural radiation processes in minerals. Nedra, Moskau (in Russian)

Mombourquette MJ, Tennant WC, Weil JA (1986) EPR study of $Fe^{3+}$ in alpha-quartz: a reexamination of the so-called I-center. J Chem Phys 85:68–79

Mombourquette MJ, Minge J, Hantehzadeh MR, Weil JA, Halliburton LE (1989) Electron-paramagnetic resonance study of $Fe^{3+}$ in alpha-quartz: hydrogen-compensated center. Phys Rev B 39:4004–4008

Nettar D, Villafranca JJ (1985) A program for electron-paramagnetic-res powder spectrum simulation. J Magn Res 64:61–65

Nilges MJ, Pan Y, Mashkovtsev RI (2008) Radiation-damage-induced defects in quartz. I. Single-crystal W-band EPR study of hole centers in an electron-irradiated quartz. Phys Chem Miner 35:103–115

Nilges MJ, Pan Y, Mashkovtsev RI (2009) Radiation-induced defects in quartz. III. Single crystal EPR, ENDOR and ESEEM study of a peroxy radical. Phys Chem Miner 36:61–73

Nuttall RHD, Weil JA (1981) The magnetic properties of the oxygen-hole aluminium centers in cristalline $SiO_2$. I. $[AlO_4]^0$. Can J Phys 59:1696–1707

Okada M, Rinneberg H, Weil JA, Wright PM (1971) EPR of $Ti^{3+}$ centers in alpha-quartz. Chem Phys Lett 11:275–276

Orlenov PO (1984) Stable paramagnetic centres in natural quartz: method of concentration measurement in powder. Mineral Zh 6:17–24 (in Russian)

Pan Y, Nilges MJ, Mashkovtsev RI (2008) Radiation-induced defects in quartz. II. Single-crystal W-band EPR study of a natural citrine quartz. Phys Chem Miner 35:387–397

Pan Y, Nilges MJ, Mashkovtsev RI (2009) Radiation-induced defects in quartz: a multifrequency EPR study and DFT modelling of new peroxy radicals. Mineral Mag 73:519–535

Pan Y, Hu B (2009) Radiation-induced defects in quartz. IV. Thermal properties and implications. Phys Chem Miner 36:421–430

Plötze M, Wolf D (1996) EPR and TL-spectra of quartz: radiation dependency of the $[TiO_4^-/Li^+]^0$-centre. Eur J Mineral 8(suppl.1):227 (in German)

Rieser U, Krbetschek MR, Stolz W (1994) CCD-camera based high sensitivity TL/OSL-spectrometer. Radiat Meas 23:523–528

Rink WJ, Rendell H, Marseglia EA, Luff BJ, Townsend PD (1993) Thermoluminescence spectra of igneous quartz and hydrothermal vein quartz. Phys Chem Miner 20:353–361

Rinneberg H, Weil JA (1972) EPR studies of $Ti^{3+}$-$H^+$ centers in X-irradiated alpha-quartz. J Chem Phys 56:2019–2028

Serebrennikov AI, Valter AA, Mashkovtsev RI, Shcherbakova MYA (1982) The investigation of defects in shock-metamorphosed quartz. Phys Chem Miner 8:153–157

Stegger P, Lehmann G (1989a) The structures of three centers of trivalent iron in alpha-quartz. Phys Chem Miner 16:401–407

Stegger P, Lehmann G (1989b) Dynamic effects in a new substitutional center of trivalent iron in quartz. Phys Stat Sol B151:55–59

Usami T, Toyoda S, Bahadur H, Srivastava AK, Nishido H (2009) Characterization of the E1' center in quartz: role of aluminium hole centers and oxygen vacancies. Phys B 404:3819–3823

Vyatkin SV, Koshchug DG, Makhotin SS (2007) Various recombination kinetics of Al centers in quartz from the Elbrus volcano and the Eldzhurtinsky granite rocks. Appl Magn Reson 32:333–344

Weeks RA (1956) Paramagnetic resonance of lattice defects in irradiated quartz. J Appl Phys 27:1376–1381

Weil JA (1984) A review of electron spin spectroscopy and its application to the study of paramagnetic defects in crystalline quartz. Phys Chem Miner 10:149–165

Weil JA (1993) A review of the EPR spectroscopy of the point defects in alpha-quartz: the decade 1982–1992. In: Helms CR, Deal BE (eds) The physics and chemistry of $SiO_2$ and the Si-$SiO_2$ interface 2. Plenum Press, New York, pp 131–144

Wright PM, Weil JA, Buch T, Anderson JH (1963) Titanium colour centres in rose quartz. Nature 197:246–248

# Chapter 9
# Analysis of Low Element Concentrations in Quartz by Electron Microprobe

**Andreas Kronz, Alfons M. Van den Kerkhof and Axel Müller**

**Abstract** Although quartz is one of the most abundant minerals in many rock types, it has not been the focus of in situ quantitative chemical analysis by electron microprobe for a long time. This was simply due to its high purity. Since cathodoluminescence observations reveal a great variety of complex structures within quartz, in situ chemical analysis methods like laser ablation inductively coupled plasma mass spectrometry (LA-ICPMS), secondary ion mass spectrometry (SIMS), and electron microprobe (EMP) applied to quartz have received increasing interest from geoscientists. Although the concentrations of many trace elements in quartz are far below the detection limits of an electron microprobe, Al, K, Ti, and Fe are, amongst others, suitable candidates for quantification. The advantage of EMP analysis over other methods is its high spatial resolution combined with high accuracy. Monte Carlo simulations of the elements listed above in a quartz matrix indicate sampling depths of <2.7 μm for 99% of the acquired X-ray photons. Sampling volumes range from 75 to 250 $\mu m^3$, and depend on excitation energy and defocusing of the electron beam. Unfortunately, beam-induced damage of the quartz lattice limits the use of high beam currents and focused beams. Irradiation induced damage strongly influences the low energy X-ray lines like Al-K$\alpha$. The beam sensitivity of the various quartz samples needs to be frequently tested and the analysis protocol must be adapted according to this signal behaviour. To minimise the effect, we propose dividing the measurement of Al into several subsets. Furthermore, exact investigation of the curvature of the background signal is required to avoid systematic errors. Secondary fluorescence

A. Kronz (✉) · A. M. Van den Kerkhof
Geowissenschaftliches Zentrum, Universität Göttingen, Göttingen, Germany
e-mail: akronz@gwdg.de

A. Müller
Geological Survey of Norway, Leiv Eirikssons vei 39, Sluppen 6315, 7491 Trondheim, Norway

J. Götze and R. Möckel (eds.), *Quartz: Deposits, Mineralogy and Analytics*, Springer Geology, DOI: 10.1007/978-3-642-22161-3_9,

of adjacent minerals is an often-neglected problem of trace element analysis by EMP. Joined crystals of pure quartz connected to $TiO_2$, ilmenite ($FeTiO_3$) and sanidine ($KAlSi_3O_8$) were used to quantify the effect of secondary fluorescence in quartz. Depending on the location of the disturbing phase, along the "line of sight" of the spectrometer or perpendicular to it, measurable effects above the detection limits can be recognized at distances up to 40 μm for Al, 60 μm for K, 200 μm for Ti and 220 μm for Fe. Depending on the position of the spectrometer relative to an adjacent phase, secondary fluorescence effects vary for Ti and Fe even at larger distances, which has to be taken into account when very low concentrations need to be detected. This effect complicates the application of empirical corrections for secondary fluorescence near phase boundaries. Setting of specific elements on multiple different spectrometers will increase the statistical certainty and can point to secondary fluorescence effects. Using our analysis protocol, detection limits of <10–15 $\mu g\ g^{-1}$ for the elements Al, K, Ti and Fe in quartz can be achieved.

## 9.1 Introduction

Quartz occurs in many rock types. It is very stable during weathering and thus dominates the mineralogical composition of most sedimentary rocks. The genesis of quartz covers a large temperature range: from magmatic conditions to moderate-T hydrothermal conditions to low-T conditions during authigenetic formation in sedimentary rocks. Hence it is an ideal candidate to record the petrological conditions of its formation as well as indicate its source for provenance studies. Unfortunately, quartz is a very pure mineral. This is why it has not been the focus micron-scale analysis by electron microprobe (EMP) in the early "microprobing decades".

The electron microprobe is not suitable for analysis of trace concentrations down to a sub-ppm level (given here as mass ppm, $\mu g\ g^{-1}$), as possible when using secondary ion mass spectrometry (SIMS) or laser-ablation inductively coupled plasma mass spectrometry (LA-ICPMS). Besides, light elements with an atomic number below that of Na are difficult to analyse by EMP with correspondingly poor detection limits. Only a few traces in quartz exceed a concentration of 1 $\mu g\ g^{-1}$ in quartz (Götze 2009, and references therein). This limits the number of measureable elements to Al, K, Ti and Fe. Na, which is frequently present in quantities above the detection limits of an electron microprobe, is difficult to analyse for reasons that will be explained below. Other elements only exceptionally reach a concentration which allows them to be analysed by microprobe. Those that are listed above the 10 $\mu g\ g^{-1}$ level in the literature either occur only in rare cases in higher concentrations in quartz (P, Ge) or appear to be mostly concentrated in micro-inclusions (Na, Cl, K, Ca, Rb, Sr, Ba, Götze 2009 and references therein). High purity quartz is of increasing importance in industry. The high purity requirements for quartz—used for fibre optics, optical glass, or as raw material for solar panels—precludes the mining of many quartz deposits. From this point of view, it is important to know how certain trace elements are distributed in the raw material (e.g. Müller et al. 2005). Quartz in which traces are concentrated in fluid or mineral inclusions can be more easily purified than quartz where those same trace elements are bound in the crystal lattice. Thus, analytical methods of high spatial resolution used in combination with bulk trace-analysis play a crucial role in deciphering the distribution of trace elements in quartz. One of the most prominent applications for the use of quartz as a petrological indicator is the single-phase thermometer "TitaniQ" (developed by Wark and Watson 2006, refined by Thomas et al. 2010 and again revised by Huang and Audétat 2011). Beside other newly developed single-phase thermometers like Zr in rutile (Zack et al. 2004; Watson et al. 2006; Tomkins et al. 2007), Ti in zircon (Watson et al. 2006), and Zr in sphene (Hayden et al. 2008), it extends the capabilities to evaluate formation temperatures from quartz in magmatic, metamorphic and hydrothermal systems.

The analysis of trace elements in quartz, eventually combined with cathodoluminescence techniques, offers not only new insights in metamorphic studies or the reconstruction of the genesis of deposits (e.g. Takahashi et al. 2008; Müller et al. 2010a) but also promises to be a useful tool for provenance studies, especially for the reconstruction of orogenic cycles (Owen 1991; Bernet and Bassett 2005).

One of the outstanding characteristics of electron microprobe analysis is its nearly non-destructive behaviour (aside from the sample consumed because of demanding requirements on sample preparation). In contrast to SIMS and LA-ICPMS, no material is consumed during the measurement process itself. In theory, this would allow analyses that are not limited by sample loss during prolonged analysis. But, in reality, beam induced damage of certain phases place practical restrictions on this concept. Other disadvantages of EMP analysis are generally low signal strength and poor signal/noise ratio compared to SIMS or LA-ICPMS. This contribution exclusively reports methodological details of the analysis of quartz by electron microprobe and will not discuss results of different case studies. For those we refer to the literature (Müller et al. 2002, 2003a, b, 2005, 2006, 2008a, b, 2010a, b; Takahashi et al. 2008; van den Kerkhof 2004a, b). All analyses were conducted on a JEOL JXA 8900 RL instrument at the Department of Geochemistry, Geowissenschaftliches Zentrum of the University Göttingen. The electron probe is equipped with 5 wavelength dispersive spectrometers (WDS), an energy dispersive (EDS)-system, a panchromatic cathodoluminescence detector detecting the wavelength range from 200 to 900 nm, as well as secondary and backscattered electron detectors. One spectrometer is a so-called H-type equipped with LIF and PET analysing crystals ("LIFH", "PETH") of Johansson-type geometry (Johansson 1933) and a smaller Rowland-circle (R = 100 mm). This leads to an approx. 5–7 times higher performance, but at the expense of a somewhat poorer spectral resolution compared to the "normal" spectrometers equipped with LIF/PET (Rowland circle, R = 140 mm).

## 9.2 Excited Volume: Spatial Resolution

The high spatial resolution is one of the benefits of EMP. Monte-Carlo simulations (Drouin et al. 2007) provide a straightforward method of estimating sampling

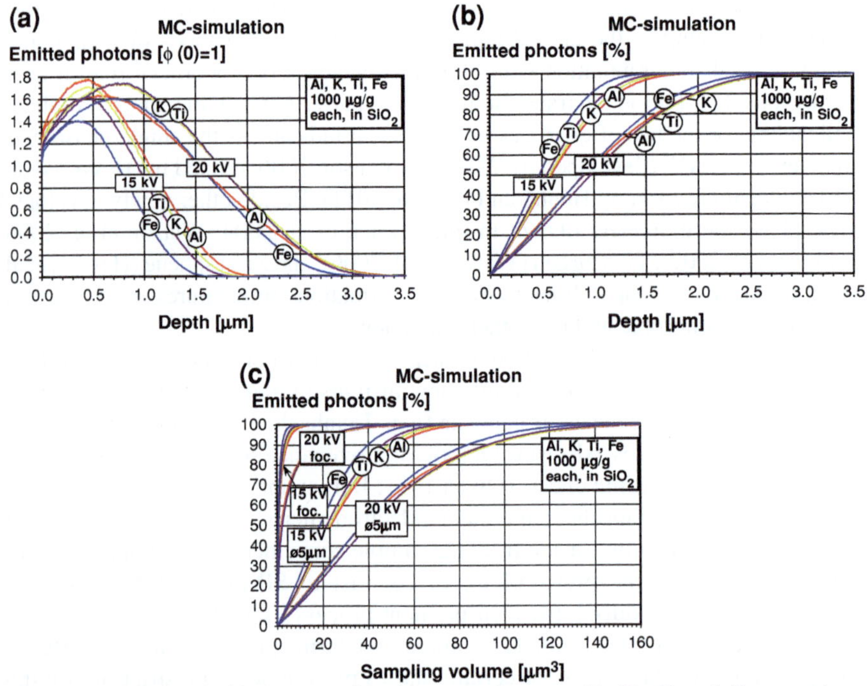


**Fig. 9.1** **a** Calculated $\phi(\rho z)$-curves for the trace elements Al, K, Ti and Fe assuming a concentration of 1000 µg $g^{-1}$ each in pure $SiO_2$ (Drouin et al. 2007). For the calculations accelerating voltages of 15 and 20 kV, a take-off angle of 40°, and a density of 2.65 g/cm$^3$ was chosen. $2 \times 10^5$ electron trajectories were calculated. **b** Cumulative emission depth of Al, K, Ti and Fe from quartz (1000 µg $g^{-1}$ each). **c** Cumulative interaction volume in quartz simplified to a nearly hemispheric excitation, calculated for maximally focused beam and a beam diameter of 5 µm respectively. (modified after Müller et al. 2003b)

depths and excited volumes for different elements in a certain phase. We have simulated the distribution of the K$\alpha$-X-ray signals of Al, K, Ti and Fe, each with a concentration of 1,000 µg $g^{-1}$ in a pure quartz matrix at accelerating voltages of 15, 20, and 25 kV, respectively, for maximally focused and defocused electron beam conditions (Fig. 9.1, Table 9.1, see also Müller et al. 2003b). For analysing Fe-K$\alpha$, a minimum excitation energy of 15 kV is required. Although high accelerating voltages lead to strong signals, energies >20 keV are not recommended for the analysis of trace elements in matrices, such as quartz, that contain predominantly light elements. Here the excited volume expands to several hundred µm$^3$. This counteracts the advantage using an electron microprobe: its good spatial resolution. Furthermore, signal drift arising from beam-induced damage also increases with beam energy. Using an accelerating voltage of 20 kV, 99% of the K$\alpha$ X-rays are emitted from a sample depth in the range of 2.5 (Fe-K$\alpha$) to 2.75 (Al-K$\alpha$) µm (Fig. 9.1a, b). Calculating the volume from which 99% of the X-ray photons are released for different beam diameters, we obtain 32–44 µm$^3$ for a maximum focused beam and 128–151 µm$^3$ for a beam set to a diameter of 5 µm

**Table 9.1** Results of the Monte-Carlo simulation (Drouin et al. 2007) calculated for 1000 μg $g^{-1}$ Al, K, Ti and Fe in pure $SiO_2$-Matrix respectively (density of 2.65 $g/cm^3$)

| 99% Cumulative emission from: | | 15 kV | | 20 kV | | 25 kV | |
|---|---|---|---|---|---|---|---|
| | | Max. focus | ø 5 μm | Max. focus | ø 5 μm | Max. focus | ø 5 μm |
| Al-Kα | Depth (μm) | 1.68 | | 2.74 | | 3.98 | |
| | Volume ($μm^3$) | 11 | 76 | 44 | 151 | 131 | 286 |
| K-Kα | Depth (μm) | 1.60 | | 2.71 | | 4.07 | |
| | Volume ($μm^3$) | 9.1 | 71 | 42 | 148 | 141 | 300 |
| Ti-Kα | Depth (μm) | 1.51 | | 2.64 | | 4.04 | |
| | Volume ($μm^3$) | 7.7 | 66 | 39 | 142 | 138 | 294 |
| Fe-Kα | Depth (μm) | 1.32 | | 2.46 | | 3.87 | |
| | Volume ($μm^3$) | 4.5 | 53 | 32 | 128 | 122 | 274 |

Calculations were performed for 15, 20 and 25 kV using either a maximum focused beam or a beam diameter of 5 μm. The values show the depths [μm] and the sampling volumes ($μm^3$ ) when 99% of the emitted X-rays are reached

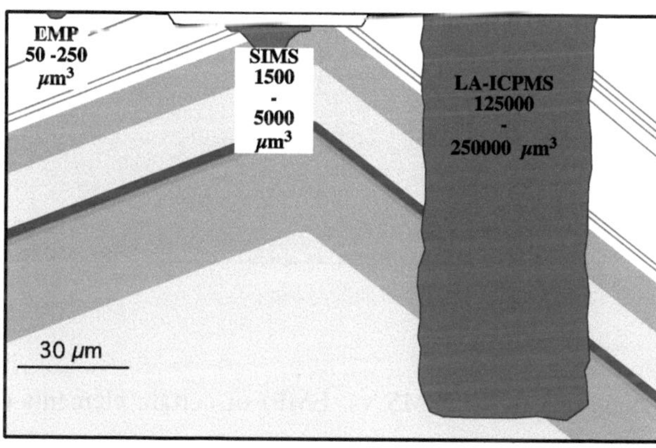


**Fig. 9.2** Schematic illustration of typical sampling volumes of EMP, SIMS and LA-ICPMS in quartz (modified after Müller et al. 2003b)

(Fig. 9.1c). Comparing this to the typical sampling volumes of SIMS and LA-ICPMS, the EMP samples a volume at least one order of magnitude smaller than SIMS and three orders of magnitude smaller than LA-ICPMS (see Müller et al. 2003b, Fig. 9.2), although this depends on the ablation conditions used for LA-ICPMS. However, when the laser is focused down to less than 20 μm diameter, analytical sensitivity decreases accordingly. Figure 9.3 illustrates an electron beam impact of approximately 7 μm in diameter, in comparison to the typical ablation of SIMS and LA-ICPMS (Müller et al. 2003b). We recommend using a multi-analytical approach, combining LA-ICPMS or SIMS with EMP, if possible. Especially small-scale micro-inclusions lead to characteristic offsets in the binary

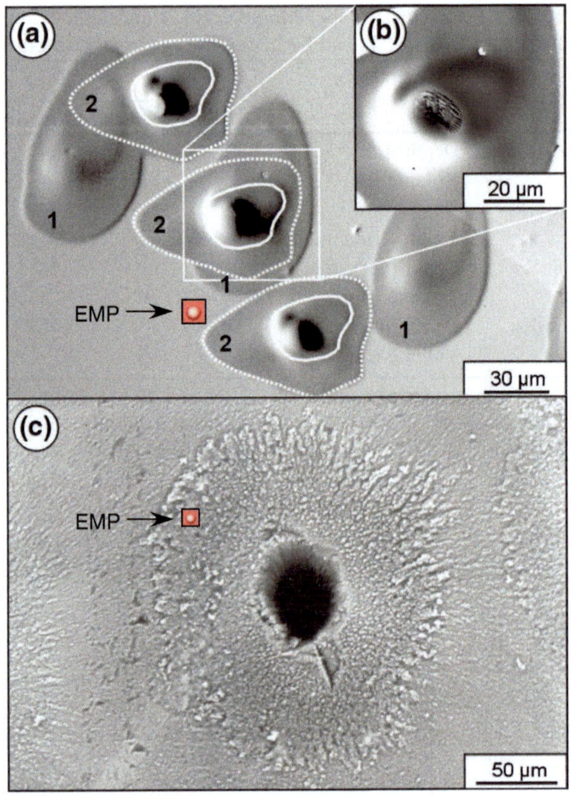


**Fig. 9.3** Backscattered electron images showing ablation craters of SIMS (**a**, **b**) and LA-ICPMS (**c**) Insets visualize electron-beam impacts of 7 μm diameter, respectively (modified after Müller et al. 2003b)

concentration diagrams (LA-ICPMS vs. EMP) of certain elements (Müller et al. 2003b: Fig. 9.8).

## 9.3 Pitfalls in Quartz Analysis

### *9.3.1 Influence of Bremsstrahlung*

A crucial drawback of analysing traces by EMP in general is the high background signal caused by the bremsstrahlung. It leads to a low signal–noise ratio. For example at the 100 μg $g^{-1}$ level, the background signal for Al analysed on TAP and Ti analysed on PET crystal is about 15 times higher than the net signal. Thus, trace element analysis by EMP always requires a careful observation of the background signals for each element in each combination of elements in a certain phase. Systematic errors arise from curvature of the background signal. Tailings of an adjacent major-element signal on a trace element line leading to curvatures in the background signal are especially important. Absorption edges, either generated within the sample itself or detector-specific, can cause huge systematic offsets. Since most of the built-in software used by microprobes only applies a linear fit model, and background curvature tends to be concave, trace element concentrations will be underestimated (Fialin et al. 1999; Donovan et al. 2011). To quantify these effects, long-duration qualitative scans were performed on the K$\alpha$-signals of Al, K, Ti and Fe respectively (Fig. 9.4, Van den Kerkhof et al. 2004b). The spectra background of K-K$\alpha$, Ti-K$\alpha$ and Fe-K$\alpha$ behaves linearly in quartz within the range of background signal measurement positions used for all these elements (Fig. 9.4a). At the lower wavelength side of the Ti-K$\alpha$ position, the spectrum displays a pronounced step at $\lambda = 2.592$ Å. It is related to the Xe-$L_3$ absorption edge of the gas proportional counter. No such effect is observable when using an Ar(90)$CH_4$(10)-gas proportional counter. Contrary to these elements, a significant curvature of the background signal was found adjacent to Al-K$\alpha$ in all Si bearing phases (pure Si, $Si_3N_4$, quartz and $SiO_2$-glass), analysed by a TAP-crystal (Thallium acid phtalate, $C_8H_5O_4Tl$) (Figs. 9.4b, c). The curvature of the background signal cannot be ascribed to the direct influence of the highly intensive Si-K$\alpha$ line as stated by Donovan et al. (2011). In Fig. 9.4b blurred signals occur in the left part of the spectra. The "peaks" can be observed in all phases, independent from applying an energy-filter or not. The origin of these "peaks" is yet unclear, but neither can be assigned to any high order reflection nor directly to any other Si-emission X-ray line. It appears that it is related to another TAP-crystal-specific diffraction lattice plane of certain Si-emission lines.

We have tested the effect of background curvature on a "blank"-quartz and found a difference of 54 μg $g^{-1}$ for which Al would be underestimated (Fig. 9.4c; see also Van den Kerkhof et al. 2004b; Müller et al. 2003b) in accordance with literature (Donovan et al. 2011). Regardless of origin, this value is dependent on how far from the peak signal the background is measured (Fig. 9.4c). Since the software in most electron probes (including our instrument) is only capable of applying a linear background interpolation model, the raw data need to be corrected off-line. Donovan et al. (2011) propose two models to correct for the background curvature for the Al-K$\alpha$-signal. In practice it would also be possible to set the background measurement positions somewhat closer towards the peak signal position, which minimizes the effect. This requires verification using "zero concentration" (blank-) quartz samples.

### 9.3.2 Beam Induced Damage

The hardness and structure of quartz imply stability, but under the electron beam quartz behaves as a very unstable mineral. Fortunately, the lattice breakdown appears not to affect the strongly-bonded trace elements (Al, Ti and Fe)

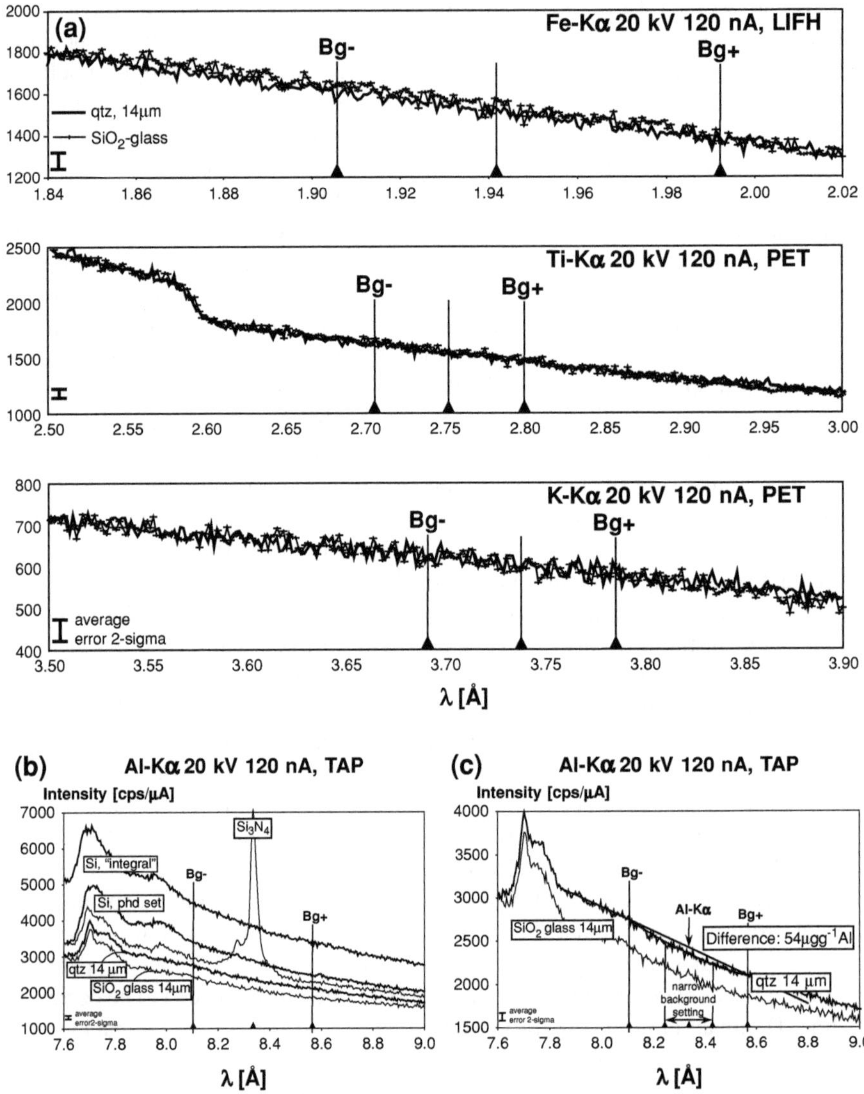


significantly, unlike feldspars where a strong beam-induced alkali drift limits counting time and/or energy. For Na we have found strong fluctuations of the raw signals during measurement. This element is not suitable to be analysed in quartz by EMP. Electron-beam induced damage effects in quartz are described in detail by Stevens-Kalceff (2009), Stevens-Kalceff and Phillips (1995) and Stevens-

◀ **Fig. 9.4** **a** Scans of the continuum around the K-Kα, Ti-Kα and Fe-Kα X-ray lines on quartz and $SiO_2$-glass respectively. **b** Long-duration scan in the vicinity of the Al-Kα emission line on different Si-bearing phases (pure Si, $Si_3N_4$–0.15 mass% Al-, quartz and $SiO_2$-glass). Note the blurred signals in the left part of the spectra, which occur in all species. For explanation see text. **c** Detail of Fig. 9.4b: curvature of the background signal around the Al-Kα emission line leads to a systematic underestimation of concentration by 54 μg $g^{-1}$ Al, if background positions are set to ±0.23 Å (for JEOL spectrometer, R = 140 mm, TAP-crystal: ±2.5 mm). Although a narrower background setting of ±0.092 Å (±1 mm) does not eliminate the effect, the systematic error is reduced to insignificance

**Table 9.2** LA-ICPMS analysis of quartz reference materials carried out at the Norwegian Geological Survey, Trondheim by A. Müller. Values are given in (μg $g^{-1}$)

| Sample<br>el/mass | **LOD**<br>3–σ | **QzGA2**<br>n = 3 | **QzGA1**<br>n = 3 | **QzTU**<br>n = 3 | **Qz_MGB**<br>n = 3 | **QzGCh**<br>n = 3 |
|---|---|---|---|---|---|---|
| Li7 | 0.467 | 1.33 | 3.124 | 1.832 | 3.713 | 2.64 |
| Be9 | 0.25 | 3.612 | 3.124 | <0.25 | <0.25 | 2.846 |
| B11 | 0.304 | 1.574 | 0.693 | 0.382 | 2.758 | 0.897 |
| **Al27** | **2.12** | **5.96** | **24.55** | **8.03** | **24.31** | **11.31** |
| P31 | 4.37 | <4.37 | <4.37 | <4.37 | <4.37 | <4.37 |
| **K39** | **1.921** | **<1.92** | **4.79** | **3.61** | **2.367** | **3.00** |
| **Ti47** | **0.1** | **<0.1** | **<0.1** | **<0.1** | **<0.1** | **<0.1** |
| Mn55 | 0.615 | 2.577 | <0.62 | 2.544 | 1.088 | <0.62 |
| **Fe56** | **3.974** | **<3.97** | **<3.97** | **<3.97** | **<3.97** | **<3.97** |
| Ge74 | 0.544 | <0.54 | 1.264 | <0.54 | 0.746 | <0.54 |
| Rb85 | 0.176 | <0.18 | <0.18 | 0.224 | <0.18 | <0.18 |
| Sr88 | 0.201 | <0.20 | <0.20 | <0.20 | <0.20 | 0.198 |
| Ba138 | 0.003 | 0.02 | 0.024 | 0.022 | 0.01 | 0.008 |
| Pb208 | 0.003 | 0.02 | 0.079 | 0.016 | 0.018 | 0.014 |
| U238 | 0.002 | <0.002 | <0.002 | <0.002 | <0.002 | <0.002 |

The samples are from:
QzGA2: Quartz, Gamsberg, Namibia
QzGA1: Quartz, Gamsberg, Namibia
QzTU: Quartz, synthetic, Mineral collection Univ. Tübingen
Qz_MGB: Quartz, Minas Gerais, Brazil
QzGCh: Quartz, Gotthard, Switzerland
**EMP relevant elements in bold**

Kalceff et al. (2000). To estimate the amount of the beam induced damage for trace analytics, a series of exposure experiments on different quartz crystals and $SiO_2$-glass were conducted using an accelerating voltage of 20 kV and a beam current (Faraday) of 80 nA: The beam diameter was varied from maximum focussed to 3.5, 7 and 14 μm, respectively. A carbon coating of approximately 20 nm was applied. $SiO_2$-glass, synthetic quartz ("QzTu", Collection of the Fachbereich Geowissenschaften, University of Tübingen) that had been cut parallel and perpendicular to its c-axis, and two natural samples (QzGA2: Gamsberg, Namibia and QzGCh: Gotthard, Switzerland) were irradiated. All of these samples have relatively low trace element contents: Al: 6–12 μg $g^{-1}$, K: <4 μg $g^{-1}$ Ti:

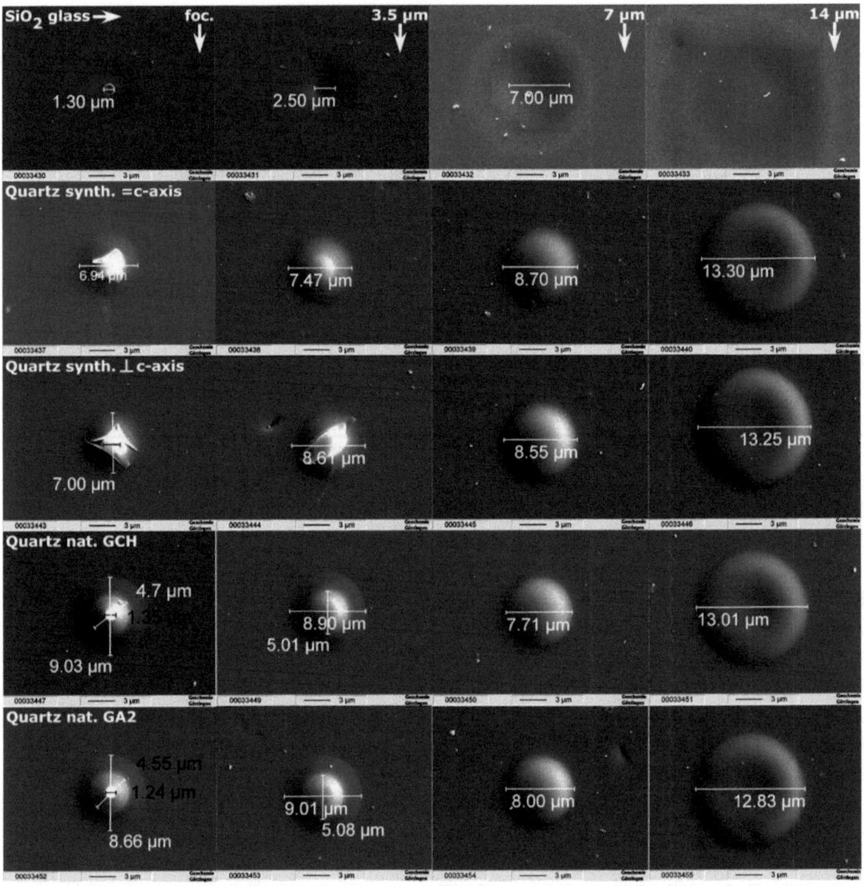


**Fig. 9.5** Surface morphology of $SiO_2$-glass and various quartz samples after 12 min irradiation. An accelerating voltage of 20 kV was used and the beam current was set to 80 nA. "foc.": maximum beam focusing. For the used beam conditions the beam diameter is approx. 0.4 μm, according to a diagram provided by JEOL company for the JXA 8900RL EMP), "quartz synth.": hydrothermal synthetic quartz sample "QzTu", "= c-axis": beam parallel to c-axis. "Quartz nat. GCH": natural quartz from Gotthard/Switzerland, beam parallel to the c-axis "Quartz nat. GA2": Gamsberg/Namibia, beam parallel to the c-axis

$<0.1$ μg $g^{-1}$ and Fe: $<4$ μg $g^{-1}$. Compositions analysed by LA-ICPMS are listed in Table 9.2. Figure 9.5 shows the spots resulting from the measurements, visualized using secondary electron images. To avoid any disturbances resulting from charging effects, which might occur when irradiated by high beam energies, the samples were coated again after the experiment by a layer of 10 nm Au–Pd (60/40), a procedure which additionally allows a superior secondary-electron image quality. All of the quartz crystals, synthetic and natural, display comparable visual effects after irradiation, independent from the crystallographic orientation, however $SiO_2$-glass behaves differently. All of the quartz crystals exhibit a volume expansion observable as a significant outgrowth. The $SiO_2$-glass in contrast displays a small but visible volume decrease after irradiation. Both effects were also described by Stevens-Kalceff (2000). It is noteworthy that the diameter of the outgrowths for the 7 and 14 μm diameter electron exposure is in the range of the real beam diameter, whereas the smaller beam diameter (maximum focused and 3.5 μm ø) produce a nearly constant outgrowth region of 7–9 μm in diameter. This might imply that even for beam diameters smaller than 8 μm, the beam induced destruction zone is larger than the distance of electron penetration through the lattice.

Monitoring the X-ray signals at the positions of K-K$\alpha$, Ti-K$\alpha$ and Fe-K$\alpha$ during irradiation show no significant change in this region of higher X-ray energies between 3.3 and 6.4 keV (Fig. 9.6a). For Si-K$\alpha$, a pronounced increase of the initial signal occurs in quartz when the beam is focused, whereas $SiO_2$-glass behaves in an opposite way showing a decreasing signal with time (Fig. 9.6b). Both effects attenuate with increasing beam diameter. For 14 μm beam diameter, nearly constant conditions are reached for both quartz and $SiO_2$-glass. No significant differences due to the crystallographic orientation of the quartz crystals are observable.

The X-ray signal at the Al-K$\alpha$ position behaves like Si, displaying a strong initial increase in quartz within the first 2 min for a focused beam followed by relatively constant signal behaviour (Fig. 9.6c, see also Müller et al. 2003b; van den Kerkhof et al. 2004b). It has to be emphasised that this signal drift is due solely to a change in the intensity of the background continuum since the samples do not contain significant amounts of Al ($<6$ μg $g^{-1}$ Al). Again, for a beam-diameter of 14 μm a stable X-ray signal is achieved. However, if the blurred peaks in the higher energy-vicinity of the Al-K$\alpha$ X-ray line (Fig. 9.4b, c) can be attributed directly to a Si-X-ray signal, their increase might be related to the "outgrowth" of the material during irradiation. Monitoring the absorbed (specimen) current during irradiation generates more complex results (Fig. 9.6d). For the used accelerating voltages the largest amount of the total primary beam energy contributes to the production of backscattered electrons and absorbed current, whereas the secondary electrons and the produced X-ray have only a minor energetic contribution. Hence, the amount of absorbed current can be regarded nearly opposite to the backscattered signal. For all quartz samples a decrease of the absorbed current occurs during irradiation. In $SiO_2$-glass the absorbed current decreases only when a focused beam is used. At 3.5 μm or larger beam diameters the signal intensity is constant. Again, the smallest effects can be observed at large beam diameters in quartz. However, in contrast to the generated X-ray signals, significant differences for the absorbed currents occur when synthetic quartz is compared to natural samples. Furthermore, the crystallographic orientation does have an influence on the attenuation of the absorbed currents. Thus, the observation of the absorbed current appears to be the most sensitive tool for an evidence of beam induced damage effects. According to these results, the backscattered electron (BSE) intensity increases with surface power density. As shown in Fig. 9.7, decreasing spot increases BSE signal for a constant irradiation time. In other studies it was shown that a change in chemical

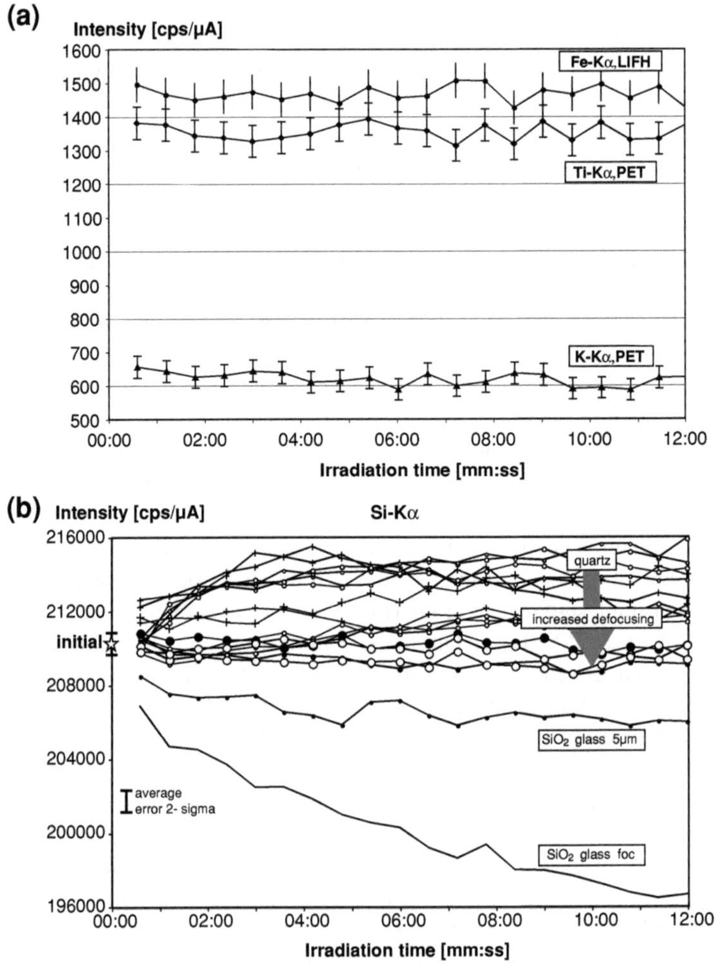


composition cannot alone lead to a BSE increase, because an amorphous state of matter produces more backscattering at the same average atomic number of a certain substance compared to a crystal. This can also be demonstrated when pure quartz is compared with pure $SiO_2$-glass, which produces a generally higher BSE signal (Nasdala et al. 2007). The same is true for zircons, where zones of lattice breakdown due to radioactive decay of U and Th display a higher BSE signal compared with intact zones of an equal composition (Nasdala et al. 2006). Because a change in chemical composition is not observable in the large beam-diameter (14 µm) spots, structural damage must contribute notably to the BSE-increase during electron irradiation. Progressive amorphisation during irradiation inhibits electron channelling and hence increases the BSE-signal and decreases

◀ **Fig. 9.6** Signal variations in $SiO_2$-glass and various quartz samples during 12 min irradiation. Accelerating voltage was set to 20 kV and a beam current of 80 nA was used. None of the tested samples contain any trace elements above 10 μg $g^{-1}$. **a** At the K-Kα (3.312 keV, $\lambda = 3.742$ Å, PET-crystal), Ti-Kα (4.508 keV, $\lambda = 2.749$ Å, PET) and Fe-Kα (6.398 keV, $\lambda = 1.937$ Å, LIFH) spectrometer-positions no significant signal drift can be observed during irradiation in quartz. (Note: the sample does not contain any amount of these elements in concentrations larger than 5 μg $g^{-1}$, hence the values represent pure background continuum.) **b** Si-Kα-X-ray line (1.739 keV, $\lambda = 7.13$ Å), PET-crystal. During irradiation the Si-Kα-signal decreases in $SiO_2$-glass, whereas quartz shows the opposite behaviour. Using a focused beam, the signal increases within the first 3 min. For a beam diameter of 14 μm, nearly constant signals could be obtained both for $SiO_2$-glass and quartz. **c** Al-Kα-X-ray line (1.486 keV, $\lambda = 8.34$ Å), TAP-crystal. The X-ray signal at the Al-Kα position in $SiO_2$-glass decreases with time. The effect is stronger when a focused beam is used. Quartz behaves in an opposite way: a strong increase within the first two minutes of irradiation is observable. Use of an increasingly defocused beam minimizes the effect. At 14 μm beam diameter no measurable shift occurs. The synthetic quartz (Qz-Tu) was mounted both parallel and perpendicular to its c-axis. No orientation effects were observable at any beam diameter. Also no differences between synthetic and natural quartz could be detected. **d** The absorbed current decreases in all quartz samples with time, when irradiated. Defocusing the beam minimizes the effect, but even at 14 μm beam-diameter no constant conditions are obtained. $SiO_2$-glass behaves stably for beam diameters larger than 3.5 μm. The effect is coupled to the increasing amorphisation of the sample, which produces also an increased backscatter electron signal under irradiation

the absorbed current and is dependant on the crystallographic orientation of the crystal.

The reasons for the extreme sensitivity of quartz due to electron irradiation are vague. Although, for example, in feldspars drift processes of the alkalis already occur even at low excitation energies, most other silicates are much more stable. Also, quartz is unique in that no such volume expansion after beam exposure is observable in other silicates.

Temperature rise might play a role, but is moderate for the beam energies used. Based on a calculation of temperature change using the empirical formula of Castaing (1960), amounts of heating that result from beam currents around 100 nA, even for a maximum focused beam (Fig. 9.8) are still well below decomposition temperatures. Furthermore, the temperature increase is comparable to the other silicates, owing to similar thermal conductivity. Thus, the fundamental cause for the quartz instability appears to be more related to its extremely low electrical conductivity leading to a bond breakup and rapid amorphisation.

The following results may be summarized from the irradiation experiments:

1. Electron bombardment even at moderate energies induces notable defects in quartz, expressed in a large drift of the background signal near the Al-Kα line and for the Si-Kα X-ray line.
2. The behaviour of quartz is different from $SiO_2$-glass. Whereas $SiO_2$-glass decreases slightly in volume, quartz always increases in volume during irradiation and produces a bulge on its surface (Fig. 9.5).

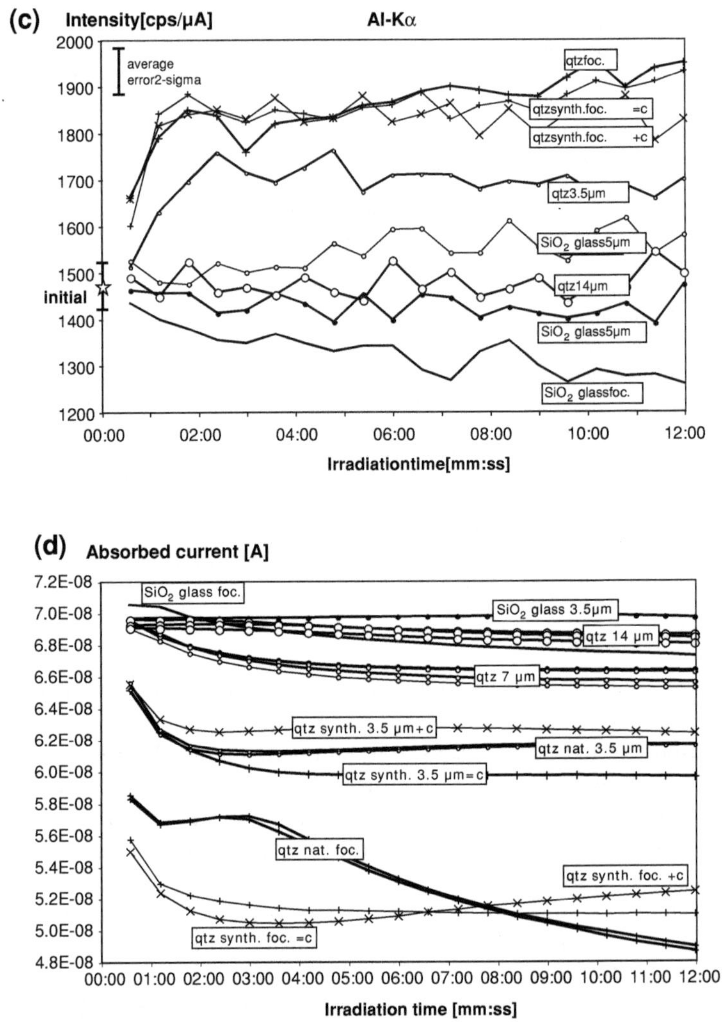


**Fig. 9.6** (continued)

3. No significant differences are observable for the analysed X-ray signals by varying crystallographic orientation of the quartz.
4. The absorbed current and the intensity of the BSE signal point to a progressive amorphisation within the electron interaction volume during irradiation.

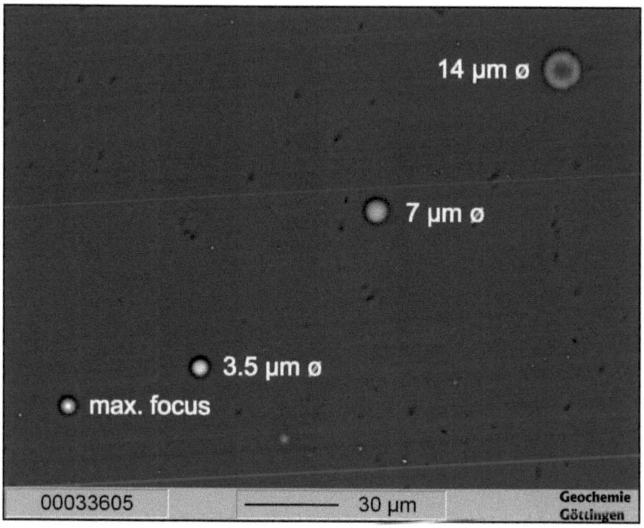


**Fig. 9.7** Backscattered electron (BSE) image in "composition mode" of the 12 min irradiated (20 kV, 80 nA) spots of different size in quartz. Note the intensity increase with decrease in beam diameter

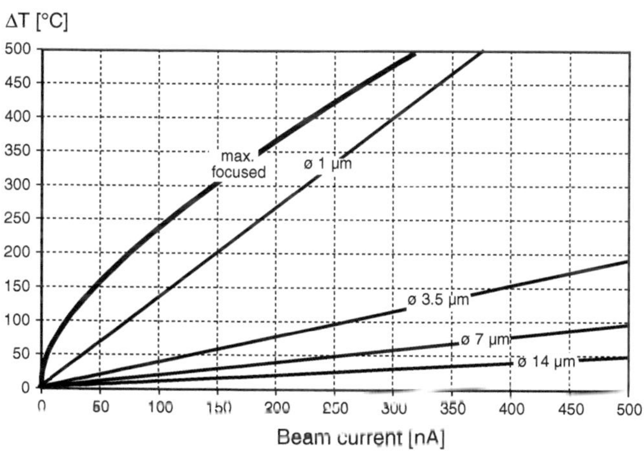


**Fig. 9.8** Temperature increase in quartz calculated after the empirical formula given by Castaing (1960) for different beam diameters at 20 kV accelerating voltage and 80 nA beam current. Heat conductivity of quartz is perpendicular to the c-axis: 7.25 (W/m/K), parallel to the c-axis (not shown): 13.2 (W/m/K) (Kleber 1985)

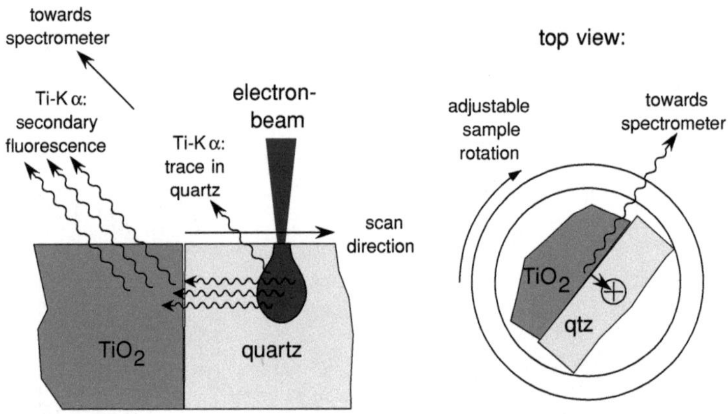


**Fig. 9.9** Schematic illustration of a test sample for evaluating the effect of secondary fluorescence (SF)

## *9.3.3 Secondary Fluorescence Near Phase Boundaries*

Secondary fluorescence (SF) defines the excitation of an element in an adjacent phase by generated characteristic X-rays or continuous X-rays (bremsstrahlung) some distance far from the measurement spot. Since X-rays of energies typically generated in an electron microprobe are capable of penetrating matter over distances of several tens of μm, an excited major element in an adjacent phase could produce an apparent signal that contributes to the trace element intensity within the analysed phase. E.g., a neighbouring rutile close to quartz or an inclusion will produce an apparent Ti-signal within the quartz, which increases towards the rutile and—so much the worse- imitates diffusion behaviour: The attenuation of the X-rays follows the Lambert–Beer law, and thus the apparent concentration profile may be indistinguishable from a typical concentration profile generated by diffusion of the fluorescent element.

The Na–K$\alpha$ (1.041 keV, $\lambda = 11.910$ Å) and Al-K$\alpha$ (1.486 keV, $\lambda = 8.340$ Å) emission lines can be directly generated by excitation from the characteristic Si-K$\alpha$ (1.739 keV, $\lambda = 7.126$ Å) and Si-K$\beta$ (1.836 keV, $\lambda = 6.753$ Å) lines. Although in a $SiO_2$-matrix neither the oxygen nor the silicon is capable of exciting the K$\alpha$-lines of the other considered elements (K, Ti, and Fe), excitation by the continuous X-rays is significant at concentrations below 100 μg $g^{-1}$.

Furthermore, the magnitude of the effect depends on the respective direction of the WD-spectrometer (Dalton and Lane 1996). If the adjacent disturbing phase is along the "line of sight" of the spectrometer, the emitted radiation appears to be out of focus at a distance larger than approximately 50–100 μm (depending on the type of spectrometer and analysing crystal; see Fig. 9.10), whereas a disturbing phase located along a line perpendicular to the viewing direction can have the same effect even over longer distances.

**Table 9.3** Composition of Minerals used for secondary fluorescence tests near phase boundaries

| Mass% | Sanidine Sd_X7Ef Eifel Germany | Ilmenite Il-MO4 Monastery South Africa | Magnetite Mt-PAR Parinacota Chile |
|---|---|---|---|
| $SiO_2$ | 64.3 | <0.09 | <0.08 |
| $TiO_2$ | NA | 49.6 | 3.78 |
| $Al_2O_3$ | **18.1** | 0.95 | 0.85 |
| FeO | 0.14 | **37.6** | **87.5** |
| MnO | NA | 0.19 | 1.01 |
| MgO | NA | 10.54 | 0.5 |
| CaO | <0.03 | <0.03 | <0.05 |
| $K_2O$ | **14.33** | <0.02 | <0.04 |
| $Na_2O$ | 1.5 | <0.03 | <0.04 |
| SrO | 0.11 | NA | NA |
| $Rb_2O$ | 0.016 | NA | NA |
| BaO | 0.89 | NA | NA |

Sd_X7Ef: Sanidine Volkesfeld, Eifel, Germany (courtesy of Gerhard Wörner, Geowissenschaftliches Zentrum, Universität Göttingen)
Il-MO4: Ilmenite Monastery South Africa (courtesy of Thomas Zack, University Mainz, Germany)
Mt-PAR: Magnetite, Parincota volcano, Chile, courtesy of John Hora, Geowissenschaftliches Zentrum, Universität Göttingen
*NA* not analysed
**Elements used for fluroscence tests highlighted in bold**

A theoretical procedure for calculating the effects by a distance function, without taking focusing into account, was first described by Bastin et al. (1983, 1984). However, it is not applicable to complex compounds and does not correct for the influence of the continuum spectrum. Recently, software based on Monte Carlo simulations has been used with success (Fournelle 2007). Also, fast numerical correction calculations now exist (Escuder et al. 2010). Several case studies exist in the literature dealing with the problem of secondary fluorescence (e.g. Llovet and Galan 2003; Fournelle et al. 2005; Wark and Watson 2006; Jercinovich et al. 2008). To evaluate the influence of SF for the trace elements in quartz analysed here, several experiments were carried out. Three couples of two minerals each (quartz-rutile, quartz-ilmenite and quartz sanidine) were contacted with polished surfaces together. The joined couples were mounted with epoxy into brass tubes, such that the contact layer was oriented perpendicular to the surface. After curing, the samples were ground and polished. The mounts could be rotated and two quantitative line profiles, each of 200 micron length were conducted. Figure 9.9 schematically illustrates one example. One profile was set along the "line of sight" of a certain spectrometer, the other perpendicular to it. The chemical compositions of the test samples are listed in Table 9.3. The synthetic Rutile used is very pure with respect to EMP analysis (For composition see: Luvizotto et al. 2008). It has to be emphasised that the quartz used (QzGA2) does not contain concentrations of Al, K, Ti or Fe higher than 6 $\mu g\ g^{-1}$ each (Table 9.2).

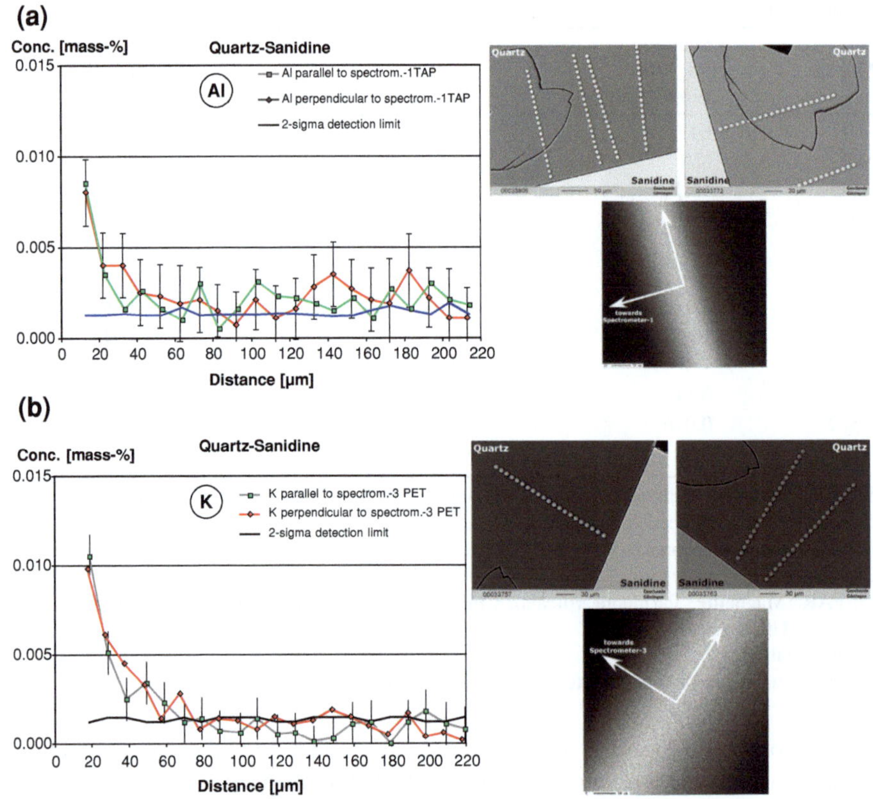


**Fig. 9.10** Results of the experiments that test the effect of secondary fluorescence. On the left concentrations of two scans respectively, along the "line of sight" of the spectrometer and perpendicular to it versus distance to the phase boundary are shown. Error bars are 2-sigma values of the error by counting statistics. Detection limits are 2-sigma values of the background noise. On the *upper right* the direction of the profiles for each mineral-couple are visualized as BSE images. On the *lower right* an element map of the respective element +WD-spectrometer combination is shown in beamscan-mode on the pure phase producing a certain SF ($TiO_2$, sanidine, ilmenite). Hence the focusing behaviour of each element-spectrometer-crystal combination is obvious relative to the marked scan profiles. **a** Quartz-sanidine couple: Al-K$\alpha$ on TAP, spectrometer-1. **b** Quartz-sanidine couple: K-K$\alpha$ on PET, spectrometer-3 **c** Quartz-$TiO_2$ (snyth.) couple: Ti-K$\alpha$ on PET, spectrometer-4 **d** Quartz-magnetite couple: Fe-K$\alpha$ on LIFH, spetrometer-5

Figure 9.10 a–d summarizes the results of the SF tests. For all measured K$\alpha$-X-ray lines, a significant disturbing influence of the respective adjacent major-element-phase can be demonstrated. Although Al-K$\alpha$ (1.486 keV) might be excited by the characteristic Si-K$\alpha$-line (1.739 keV) the effect is not stronger than the other higher energy X-ray lines. For Al and K excited in an adjacent sanidine, the analysable influence of SF ends at a distance of approx. 40 µm (Al-K$\alpha$) and 60 µm (K-K$\alpha$), respectively (Fig. 9.10a, b) where the signal reaches the detection

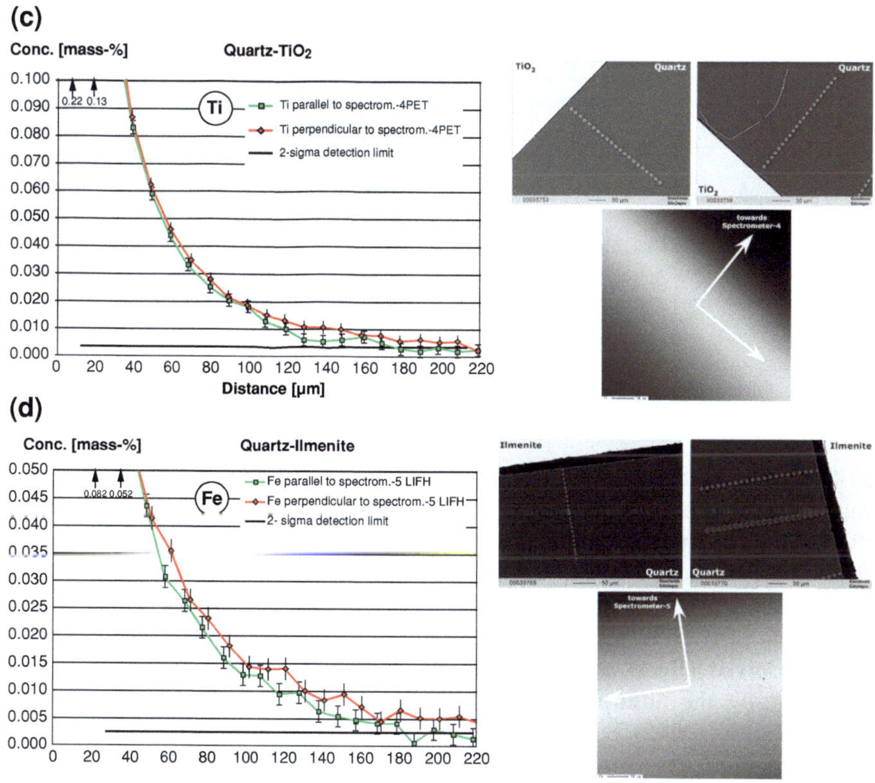


**Fig. 9.10** (continued)

limit. No systematic difference according to the orientation of the line of spots is apparent. For Ti-K$\alpha$ in the rutile (synthethic)—quartz couple, an elevated signal is present up to a distance of 160–200 µm. This result corroborates the results of Wark and Watson (2006), although they give neither error-bars nor detection limits in their graph (Wark and Watson 2006: Fig. 9.3).

A slight systematic difference between the orientation of the scans (either along or perpendicular to spectrometer line-of-sight) could be distinguished (Fig. 9.10c). Because only 2-sigma confidence is given for the detection limits and errors, this difference is not very significant for a certain measurement point, but obviously all spots at distances larger than 100 µm give higher concentrations for the scan perpendicular to the spectrometer direction, which is in fact within the focusing of the spectrometer. Note that the detection limit is only about 33 µg $g^{-1}$ $TiO_2$ (2-sigma confidence) for the used analysing conditions. For longer counting times, where detection limits of <10 µg $g^{-1}$ $TiO_2$ are obtained, the effect of SF dependant on the spectrometer direction might be apparent at distances of 200 µm or even larger.

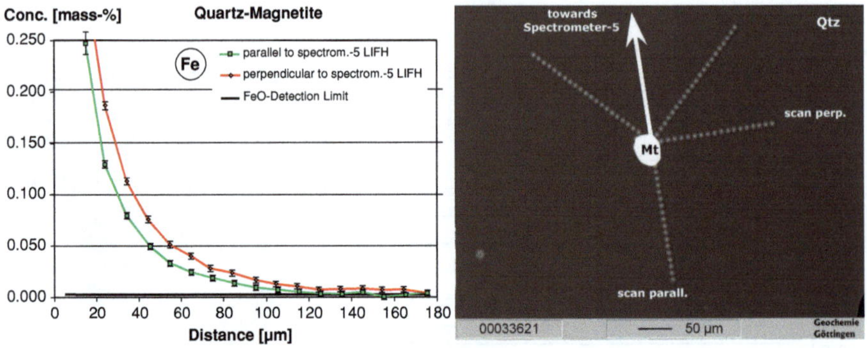


**Fig. 9.11** Observation of SF in a natural quartz (Parinacota dacite, Chile) near a magnetite inclusion: Fe-Kα on LIFH, spectrometer-5: along and perpendicular to its "line of sight"

Similarly, in the case of Fe in the quartz-ilmenite couple (Fig. 9.10d), the measurement spots that remain in the focal plane of the spectrometer (perpendicular to its "line of sight"), show higher apparent concentrations within the quartz. These values are significantly above the detection limits even at a distance of 200 µm. For the scan towards the spectrometer direction the apparent concentrations after 160–180 µm reach the detection limit. It has to be mentioned that these observed effects at larger distances from the phase boundary cannot be explained by the absorption-dependant effects alone, which are described by Fournelle et al. (2005). Such effects occur at smaller distances when the generated SF-X-rays either pass through the initial phase (here: quartz) or the excited phase (here: $TiO_2$, ilmenite or sanidine), where different absorption occurs. On the other hand, defocusing of the WD-spectrometer does not play a role at small distances (<50 µm, see the beam-scanned element maps in Fig. 9.10). The model of a flat phase boundary perpendicular to the surface and of "infinite" dimension (in respect to the sphere of X-ray interaction) is not necessarily applicable in natural samples. Insofar it is difficult to apply correction procedures in order to subtract the amount of SF close to phase boundaries. The simplest strategy is a careful observation and the avoidance of neighbouring Ti- and Fe-bearing phases. An example is given in Fig. 9.11. Here a small magnetite of approx. 60 µm diameter occurs in volcanic quartz (Parinacota volcano, Chile, courtesy of John Hora, Göttingen). Also for this nearly spherical inclusion, a severe influence is visible and the difference of the two scan directions is significant.

## 9.4 Detection Limits, Precision and Accuracy

Limits of detection (LOD) and errors can be calculated from the single counting-statistical errors of each counting event (peak-counting and background counting on each side of an emission X-ray line) following the Gaussian law of error propagation:

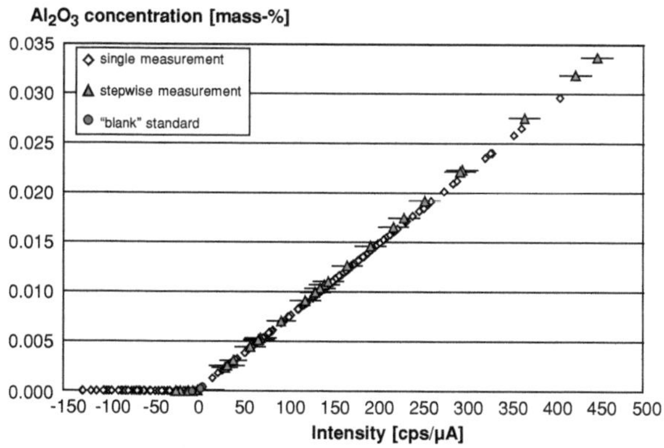


**Fig. 9.12** Net intensity versus calculated concentration of $Al_2O_3$ in various quartz samples (see also Müller et al. 2003a). For the uniform quartz matrix the ZAF–factor remains constant, leading to an ideal linear relationship between net intensities and calculated concentrations. Negative net-counts are produced by subtraction of too large background values and were set by the software to "zero-concentration". For a first set of measurements negative net countrates of −130 cps/µA equals to an underestimation of approx. 100 µg g$^{-1}$ $Al_2O_3$, although the background curvature was already corrected. This was due to the large signal shift during measurement: when measuring in one single step (300 s on peak and thereafter 150 s. on each background) the continuous signal increase subtracts a too large background value. Subdividing the Al-measurement into 5 distinctive steps of 60 s on peak and 30 s. on each background respectively minimizes the systematic error ("stepwise measurement"). A nearly Al-free quartz (GA2, blank-reference; 6 µg g$^{-1}$ Al) was used to confirm the "zero"-concentration

The simple method for calculating LOD, when a single measurement is estimated, is given by:

$$N_{LOD} = \sigma \cdot \frac{1}{2}\sqrt{N_{bg^-} + N_{bg+}}$$

where:

$N_{LOD}$ is the error in total counts

$\sigma$ is the sigma factor depending on the applied confidence value (usually 3)

$N_{bg}-$ an $N_{bg}+$ are the total accumulated counts on each background signal.

After transforming the error of total counts into a count-rate ($I_{LOD}$) this value can simply be calculated into a concentration ($C_{LOD}$) by applying the certain matrix correction procedure. Because of the strong linear behaviour of net-counts of a certain trace element in an overall uniform matrix (quartz), a simple linear equation can be applied, after deducing the calibration factor from multiple measurements (Fig. 9.12):

$$C_{LOD}[\mu g\ g^{-1}] = \text{factor} * I_{LOD}$$

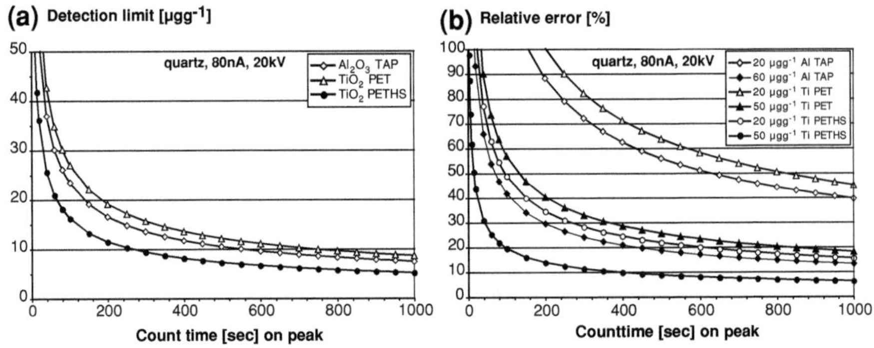


**Fig. 9.13** **a** Calculated detection limits by counting statistics (2 sigma of the background noise) for Al and Ti as a function of counting time. **b** Calculated relative errors by counting statistics (2-sigma) for Al and Ti as a function of counting time at different concentration levels. For Ti the values are calculated for a "normal" spectrometer equipped with a PET analyser crystal and for a high sensitive spectrometer ("PETH")

We recommended comparing this method of "LOD by single measurement" with the counting statistics of the background noise obtained by multiple measurements on a "blank"-reference material.

Limits of detection (LOD) are calculated from n background measurements with 95% confidence utilising the Student's *t*-distribution according to the following equation

$$I_{LOD} = \mathrm{t_z}\,(\mathrm{P};\mathrm{f})\,\sigma_{\mathrm{BG}}$$

where $I_{LOD}$ is the intensity of the LOD [cts]; $\sigma_{\mathrm{BG}}$ = standard deviation of the average background; $\mathrm{t_z}$ (P;f) = the Student's $t$ for binomial limitation determined by the confidence P and the degrees of freedom f = n – 2, where n is the number of background measurements. The Student's $t$ with 95% confidence amounts 2.0 for 50 measurements and 2.3 for 10 measurements.

Both methods give comparable results, but we prefer the random sampling using Student's t-test (Müller et al. 2002). The latter is more sensitive to systematic errors and the occurrence of drift-phenomena.

If a certain element is measured simultaneously on different spectrometers to increase total sensitivity (see below) this has to be taken into account in an extended function of error propagation. Nevertheless, EMP analysis is limited by the generally observed high background signal of the bremsstrahlung and thus low net-signal–background ratios at low concentration levels. Accordingly, it is difficult to decrease errors or LOD significantly once a certain low value has already been reached. Figure 9.13a displays the detection limits of single spectrometer analyses for Al, and Ti (on two spectrometers having different sensitivity)

**Table 9.4** Analysis protocol for trace elements (Al, K, Ti, Fe) in quartz: recommended beam conditions are: Beam-current/Faraday): 80–120 nA; Accelerating voltage: 20 kV; beam diameter: 7–14 μm (if appropriate)

| Spectrometer | 1 | 2 | 3 | 4 | 5 |
|---|---|---|---|---|---|
| (a) Analysis protocol for analysing the elements (AL, K, Ti, Fe) in quartz | | | | | |
| 1. Sequence: | | | | | |
| Element | Si | Si | K | Ti | Fe |
| X-ray line | Kβ | Kβ | Kα | Kα | Kα |
| Crystal | TAP | TAP | PET | PET | LIFH |
| Peaktime | 60 | 60 | 360 | 360 | 360 |
| Backgr.-time | 30 | 30 | 180 | 180 | 180 |
| Standard | Quartz | Quartz | Sanidine | $TiO_2$, synth | Hematite |
| 2. to 6. Sequence: | | | | | |
| Element | Al | Al | | | |
| X-ray line | Kα | Kα | | | |
| Crystal | TAP | TAP | | | |
| Peaktime | 60 | 60 | | | |
| Backgr.-time | 30 | 30 | | | |
| Standard | $Al_2O_3$, synth | $Al_2O_3$, synth | | | |
| LOD (μg $g^{-1}$) | | Al-tot: 5 | K: 11 | Ti: 18 | Fe: 16 |
| (b) Analysis protocol for Ti(Al) in quartz, used for the TitaniQ thermometer (Wark and Watson 2006; Thomas et al. 2010) | | | | | |
| 1. Sequence: | | | | | |
| Element | Ti | Si | Ti | Ti | Ti |
| X-ray line | Kα | Kβ | Kα | Kα | Kα |
| Crystal | PET | TAP | PET | PET | PETH |
| Peaktime | 360 | 60 | 360 | 360 | 360 |
| Backgr.-time | 180 | 30 | 180 | 180 | 180 |
| Standard | $TiO_2$, synth | Quartz | $TiO_2$, synth | $TiO_2$, synth | $TiO_2$, synth |
| 2. to 6. Sequence: | | | | | |
| Element | | Al | | | |
| X-ray line | | Kα | | | |
| Crystal | | TAP | | | |
| Peaktime | | 60 | | | |
| Backgr.-time | | 30 | | | |
| Standard | | $Al_2O_3$, synth | | | |
| LOD (μg $g^{-1}$) | | Al: 9 | | | Ti-tot: 4 |

dependant on counting time for the EMP in Göttingen. In Fig. 9.13b relative errors are calculated by error propagation of peak and background measurements for single measurements of Al and Ti at different concentration levels.

## 9.5 Analysis Protocol for Trace Elements in Quartz by EMP

Based on our observations and experiments, we propose the following recommendations for analysing traces in quartz:

Two sets of analysis protocols, a more general, containing the elements Si, Al, K, Ti and Fe and a more specialized for the application of the Ti-in-quartz-thermometer are given in Table 9.4.

It has to be emphasised, that this protocol is optimized for the analyser-crystal–spectrometer configuration of the actual EMP in Göttingen.

- The use of a blank quartz sample is essential. Ideally, natural quartz with trace element concentrations below the detection limits for EMP is needed to evaluate "true zero" concentrations.
- Never use $SiO_2$-glass for standardization or as "blank"-control. The behaviour of quartz during electron irradiation is very different to that of $SiO_2$-glass.
- Evaluation of potential negative net-counts should be performed by multiple analyses of the blank-quartz in each set of analyses, preferably at least once after 10 spots on the unknown.
- Due to the beam sensitivity of quartz, a focussed beam should not be used. Good results are obtained with a beam diameter of 10–15 μm in diameter.
- Beam currents are limited for the same reasons. 80 nA for smaller beam diameter or 120 nA (measured on Faraday) for a larger beam diameter are sufficient.
- The curvature of the background signal in the vicinity of the Al-K$\alpha$ line needs to be carefully monitored. If no built-in software is available, a correction for the effect has to be applied offline to the data set. The amount of correction depends on the distance of the background measurement positions on a certain spectrometer, hence no absolute values can be given here.
- For the analysis of Al we recommended dividing the measurement time into subsets. Peak-signal and each background-signal measurement interval has to be repeated in cycles to adapt the background drift due to potential sample damage effects. If the built-in analysis software is not capable of handling a multiple sequence of a single element, the processing of the raw data needs to be done offline. Because small variations of a trace element concentration in the overall equal matrix quartz leads to perfect linear correlation of the net-raw signal to its concentration, the calibration can be easily performed (Fig. 9.12).
- Highest precision and lowest detection limits will be obtained when a single element is analysed on different spectrometers simultaneously. Especially for the titanium used for the TitaniQ-Thermometer an acquisition of the Ti-K$\alpha$ signal on all available PET analyser crystals is recommended. Again, data treatment has to be performed offline, if the analysis software is not capable of handling multiple measurements of a certain element. Results of analyses on single spectrometers need to be carefully compared with results from a multiple-spectrometer analysis approach. Commercial software for this is already available ("aggregate mode", Donovan et al. 2011).

- It is always useful to plot net-counts (normalized) versus the calculated concentrations for different correction procedures, either $\Phi\rho z$ (Armstrong 1995) or the common ZAF-correction procedures (Fig. 9.12). Because of the strong linear behaviour due to the uniform matrix in quartz, linear offline calibrations can be applied. If calibration factors are determined for an individual spectrometer, multiple counting of a certain element on different spectrometers can be calculated individually: If systematic differences occur, this may point to SF-effects. Systematic negative net raw-counts will point to signal drift effects and could be eliminated by subdividing a measurement sequence into subsets (Fig. 9.12).
- Using large beam currents and high energies in a trace program can cause problems when analysing the major elements like Si in quartz, due to detector overflow or offsets in dead-time correction. For quartz we do not recommend to adopt a calculated $SiO_2$-value (fixed or as difference to 100%) because it obscures the quality of the analysis. Thus, the major elements should be analysed in a separate sequence using a smaller beam current, which is not possible in all microprobe types, or alternatively, a weaker X-ray line needs to be chosen. For Si we recommend using Si-K$\beta$ or the second order diffraction of Si-K$\alpha$.

**Acknowledgments** We are grateful to Herbert Nägele from Windhoek who provided the QzGA1 and QzGA2 quartz crystals from Gamsberg, Namibia. The constructive review by A. Renno helped to improve the manuscript.

## References

Armstrong JT (1995) CITZAF: a package of correction programs for the quantitative electron microbeam X-ray analysis of thick polished materials, thin films, and particles. Microbeam Anal 4:177–200

Bastin GF, Loo FJJ, Vosters PJC, Vrolijk JWGA (1983) A correction procedure for characteristic fluorescence encountered in microprobe analysis near phase boundaries. Scanning 5:172–183

Bastin GF, Loo FJJ, Vosters PJC, Vrolijk JWGA (1984) An iterative procedure for the correction of secondary fluorescence effects in electron-probe microanalysis near phase boundaries. Spectrochimica Acta 39B:1517–1522

Bernet M, Bassett K (2005) Provenance analysis by single-quartz-grain SEM-CL/optical microscopy. J Sed Res 75(3):492–500

Castaing R (1960) Electron-probe microanalysis. Adv Electr Electron Phys 13:317–386

Dalton JA, Lane SJ (1996) Electron microprobe analysis of Ca in olivine close to grain boundaries: the problem of secondary X-ray fluorescence. Am Mineral 81:194–201

Donovan JJ, Lowers HA, Rusk BG (2011) Improved electron probe microanalysis of trace elements in quartz. Am Mineral 96:274–282

Drouin D, Couture AR, Joly D, Tastet X, Aimez V, Gauvin R (2007) CASINO V2.42—a fast and easy-to-use modelling tool for scanning electron microscopy and microanalysis users. Scanning 29:92–101

Escuder JA, Salvat F, Llovet X, Donovan JJ (2010) Numerical correction for secondary fluorescence across phase boundaries in EPMA. IOP Conf. Series: Mater Sci Eng 7:012008. doi:10.1088/1757-899X/7/1/012008

Fialin M, Rémy H, Richard C, Wagner R (1999) Trace element analysis with the electron microprobe: new data and perspectives. Am Mineral 84:70–77
Fournelle JH, Kim S, Perepezko JH (2005) Monte Carlo simulation of Nb Kα secondary fluorescence in EPMA: comparison of PENELOPE simulations with experimental results. Surf Interface Anal 37:1012–1016
Fournelle JH (2007) The problem of secondary fluorescence in EPMA in the application of the Ti-in-Zircon geothermometer and the utility of PENEPMA Monte Carlo program. Microsc Microanal 13(supplement 2):1390–1391
Götze J (2009) Chemistry, textures and physical properties of quartz—geological interpretation and technical application. Mineral Mag 73:645–671
Hayden LA, Watson EB, Wark DA (2008) A thermobarometer for sphene (titanite). Contrib Mineral Petrol 155:529–540
Huang R, Audétat A (2011) A critical look at the titanium-in-quartz (TitaniQ) thermobarometer. Mineral Mag 75:1065
Jercinovic MJ, Williams ML, Lane ED (2008) In situ trace element analysis of monazite and other fine-grained accessory minerals by EPMA. Chem Geol 254(3–4):197–215
Johansson T (1933) Über ein neuartiges, genau fokussierendes Röntgenspektrometer. Z Physik 82:507–529
Kleber W (1985) Einführung in die Kristallographie. VEB Verlag Technik, Berlin
Llovet X, Galan G (2003) Correction of secondary X-ray fluorescence near grain boundaries in electron microprobe analysis: application to thermobarometry of spinel Lherzolithes. Am Mineral 88:121–130
Luvizotto GL, Zack T, Meyer HP, Ludwig T, Triebold S, Kronz A, Münker C, Stockli D, Prowatke S, Klemme S, Jacob DE, von Eynatten H (2008) Rutile crystals as potential trace element and isotope mineral standards for microanalysis. Chem Geol 261:346–369
Müller A, Kronz A, Breiter K (2002) Trace elements and growth patterns in quartz a fingerprint of the evolution of the subvolcanic Podlesí Granite System (Krusné hory Mts., Czech Republic). Bull Czech Geol Surv 77(2):135–145
Müller A, René M, Behr H-J, Kronz A (2003a) Trace elements and cathodoluminescence of igneous quartz in topas granites from the Hub Stock (Slavkovsky Les Mts., Czech Republic). Mineral Petrol 79:167–191
Müller A, Wiedenbeck M, van den Kerkhoff AM, Kronz A, Simon K (2003b) Trace elements in quartz—a combined electron microprobe, secondary ion mass spectrometry, laser-ablation ICP-MS, and cathodoluminescence study. Eur J Mineral 15:747–763
Müller A, Ihlen PM, Kronz A (2005) Potential recources of quartz and feldspar raw material in Sørland IV: relationship between quartz, feldspar and mica chemistry and pegmatite type. Geol Surv Norway. NGU report 2005.075, p 94
Müller A, Seltmann R, Halls C, Siebel W, Dulski P, Jeffries T, Spratt J, Kronz A (2006) The magmatic evolution of the land's end pluton, Cornwall, and associated pre-enrichment of metals. Ore Geol Rev 28:329–367
Müller A, Seltmann R, Kober B, Eklund O, Jeffries T, Kronz A (2008a) Compositional zoning of Rapakivi feldspars and coexisting quartz phenocrysts. Can Mineral 46:1417–1442
Müller A, Ihlen PM, Kronz A (2008b) Quartz chemistry in polygeneration Sveconorwegian pegmatites, Froland, Norway. Norway Eur J Mineral 20:447–463
Müller A, Herringon R, Armstrong R, Seltmann R, Kirwin DJ, Stenina NG, Kronz A (2010a) Trace elements and cathodoluminescence of quartz in stockwork veins of Mongolian porphyry-style deposits. Mineralium Deposita 45:707–727
Müller A, Kerkhof AM, Behr H-J, Kronz A, Koch-Müller M (2010b) The evolution of late-Hercynian granites and rhyolites documented by quartz—a review. Earth and environmental transactions of the Royal Rociety of Edinburgh. Earth Environ Sci 100:185–200
Nasdala L, Kronz A, Hanchar JM, Tichomirowa M, Davis DW, Hofmeister W (2006) Effects of radiation damage on back-scattered electron imaging of natural zircon. Am Mineral 91:1739–1746

Nasdala L, Grambole D, Kronz A, Trullenque G (2007) Effects of irradiation damage on the backscattering of electrons: silicon-implanted silicon. Am Mineral 92:1768–1771
Owen MR (1991) Application of cathodoluminescence to sandstone provenance. In: Barber CE, Kopp OC (eds) SEPM-1991 luminescence microscopy: quantitative and qualitative aspects, SEPM (Society for Sedimentary Geology)
Stevens-Kalceff MA (2000) Electron irradiation induced radiolytic oxygen generation and microsegregation in silicon dioxide polymorphs. Phys Rev Lett 84:3137–3140
Stevens-Kalceff MA (2009) Cathodoluminescence microcharacterization of point defects in a-quartz. Mineral Mag 73:585–605
Stevens-Kalceff MA, Phillips MR (1995) Cathodoluminescence microcharacterization of the defect structure of quartz. Phys Rev B 52:3122–3134
Stevens-Kalceff MA, Phillips MR, Moon AR, Kalceff W (2000) Cathodoluminescence microcharacterization of silicon dioxide polymorphs. In: Pagel M, Barbin V, Blank P, Ohnenstetter D (eds) Cathodoluminescence in geosciences. Springer, Berlin
Takahashi R, Müller A, Hiroharu M, Okrugin VM, Shuji O, van den Kerkhof AM, Kronz A, Andreeva ED (2008) Cathodoluminescence and trace elements in quartz: clues to metal precipitation mechanisms at the Asachinskoe gold deposit in Kamchatka. In: Okada H, Mawatari SF, Suzuki N, Gautam P (eds) Origin and evolution of natural diversity, Proceedings of international symposium "the origin and evolution of natural diversity", Sapporo, 1–5 October 2007, pp 175–184
Thomas JB, Watson EB, Spear FS, Shemella PT, Nayak SK, Lanzirotti A (2010) TitaniQ under pressure: the effect of pressure and temperature on the solubility of Ti in quartz. Contrib Min Pet 160:743–759
Tomkins HS, Powell R, Ellis DJ (2007) The pressure dependence of the zirconium-in-rutile thermometer. J metamorphic Geol 25:703–713
Van den Kerkhof AM, Kronz A, Simon K, Riganti A, Scherer T (2004a) Origin and evolution of Archean quarzites from the Nondweni greenstone belt (South Africa) inferences from a multidisciplinary study. S Afr J Geol 107:559–576
Van den Kerkhof AM, Kronz A, Simon K, Scherer T (2004b) Fluid-controlled quartz recovery in granulite as revealed by cathodoluminescence and trace element analysis (Bamble sector, Norway). Contrib Min Pet 146:637–652
Wark DA, Watson EB (2006) TitaniQ: a titanium-in-quartz geothermometer. Contrib Mineral Petrol 152:743–754
Watson EB, Wark DA, Thomas JB (2006) Crystallization thermometers for zircon and rutile. Contrib Mineral Petrol 151:413–433
Zack T, Moraes R, Kronz A (2004) Temperature dependence of Zr in rutile: empirical calibration of a rutile thermometer. Contrib Min Pet 148:471–488

# Chapter 10
# In Situ Analysis of Trace Elements in Quartz Using Laser Ablation Inductively Coupled Plasma Mass Spectrometry

**Belinda Flem and Axel Müller**

**Abstract** In situ micro analysis of ultra trace element composition of quartz using laser ablation inductively coupled plasma mass spectrometry (LA-ICP-MS) allows rapid screening of lattice bound impurities of potential high-purity quartz resources and samples for petrological research without the need to remove solid and liquid inclusions by expensive dressing techniques prior to chemical analysis. Information of the analysed trace element content can not only be used to determine the economic quality of quartz but also the conditions of quartz crystallisation and the origin of the quartz-forming fluids and melts. The main purpose of this paper is to describe an efficient and precise analytical method for estimating the concentrations of lattice-bound trace elements in quartz. The best choice of instrument is considered to be a double focusing sector field inductively coupled plasma mass spectrometry ICP-MS that provides high sensitivity and a mass resolution high enough to separate K from its interferences. The ICP-MS should be coupled to a 193 nm excimer laser, a femto second laser or a similar ablation system. The following elements are included in the standard analytical protocol applied at the Geological Survey of Norway (NGU): Al, B, Be, Ca, Cr, Fe, Ga, Ge, K, Li, Mg, Mn, Na, P, Rb, Sb, Sr, Ti and Zn. Any element with isotopes that can be ionised in an Ar plasma can easily be included if suitable reference materials are available. External calibration was done using the international reference materials NIST SRM 610, 612, 614, 616 and 1830 from the National Institute of Standards and Technology (NIST), BCS 313/1 from the Bureau of Analysed Samples (BAS) and the certified reference material "pure substance No. 1" silicon dioxide $SiO_2$ from the Federal Institute for Material Research and Testing, Berlin, Germany (BAM). To improve the lower limit of quantification and analytical uncertainty at

B. Flem (✉) · A. Müller
Geological Survey of Norway, Leiv Eirikssons vei 39,
7491 Trondheim, Norway
e-mail: Belinda.Flem@NGU.NO

J. Götze and R. Möckel (eds.), *Quartz: Deposits, Mineralogy and Analytics*,
Springer Geology, DOI: 10.1007/978-3-642-22161-3_10,

low concentrations, it is important to have calibration curves with well defined intercepts. This can be achieved by the use of certified standards, with trace element concentrations lower than the quantification limit. Even better is to use a standard where the analyte is not present, however in laserablation matrixmatcing is usually important and a standard blank is usually not avilable. In this work, BAM no. 1 $SiO_2$ is the reference material used whith the lowest consentration of the analytes. Because of the absence of a $SiO_2$ blank, the BAM no. 1 $SiO_2$ is used for the estimation of detection limits. Detection limits for most of the elements are between 1 and 0.02 μg $g^{-1}$. Analysis time and laser spot size are adjusted to the size of the quartz crystal and the thickness (~300 μm) of the polished thick section. The size of the ablation raster is commonly 100 × 350 μm with a depth of 10–20 μm applying the analysis time of less than 1 min.

## 10.1 Introduction

Although quartz is a common rock-forming mineral in silica-saturated lithologies, the speciation and concentration of lattice-bound trace elements in quartz is potentially informative. However, trace element studies of quartz are limited compared to other rock-forming minerals (e.g., Perny et al. 1992; Larsen et al. 2000a, b; Götze et al. 2001; Müller and Welch 2009 and references therein). Because high-purity quartz—i.e. quartz with less than 50 μg $g^{-1}$ lattice-bound trace elements (e.g., Harben 2002)—is a valuable commodity used in the manufacture of various high technology products, there is an industrial need for its chemical characterisation on the μg $g^{-1}$ down to ng $g^{-1}$ level.

Conventional chemical analysis of quartz requires extensive laboratory preparations. e.g., flotation, handpicking and sample digestion. Even then, contamination during preparations and incomplete removal of microscopic inclusions can lead to spurious results. Laser ablation combined with high resolution inductively coupled plasma mass spectrometry (LA-HR-ICP-MS) provides an approach to the analysis of quartz at the μg $g^{-1}$ to ng $g^{-1}$ level with minimal sample preparation and subsequent contamination. State-of-the art laser ablation equipment allows high resolution optical control of the analysed quartz domain to avoid the analyses of contaminating micro fluid- and mineral inclusions. Alternative micro-analytical methods are, for example, secondary ion mass spectrometry (SIMS) and electron probe micro analysis (EMPA). Advantages and disadvantages of these methods compared to quartz analysis by LA-ICP-MS are described in Müller et al. (2003).

Quartz has an exceptionally strong atomic configuration of Si–O bonds which allow only a minimum of other elements into its structure. However, minute amounts of substitutional and interstitial impurities can be incorporated into the atomic lattice and these elements are classified as lattice-bound impurities. Substitutional impurities replace $Si^{4+}$ in the Si–O tetrahedra which make up the quartz lattice. Interstitial impurities, mostly small monovalent ions that fit into structural channels running parallel to the c-axis, function as charge compensators to balance the substitutional heteroelectronic impurities. For example, $Al^{3+}$ is a common substitutional impurity with $Li^+$ or $Na^+$ in the structural channels balancing the missing positive charge. Elements which are most commonly identified as lattice-bound impurities include Al, B, Ca, Cr, Cu, Fe, Ge, H, K, Li, Mg, Mn, Na, P, Pb, Rb, Ti and U (e.g., Fanderlik 1991), though not all studies agree with this list of elements. Jung (1992) suggests that only Al, B, Ge, Fe, H, K, Li, Na, P and Ti may be regarded as true lattice-bound impurities, and assume that Ca, Cr, Cu, Mg, Mn, Pb, Rb and U result from microscopic solid and liquid inclusions which were not entirely removed before analysis. Furthermore, the structural incorporation of $Ga^{3+}$ in quartz was shown by Weil (1993). Rusk et al. (2011) also detected considerable amounts of Sb in hydrothermal quartz and assumed that $Sb^{5+}$, in combination with $Al^{3+}$ substitutes two $Si^{4+}$.

Because LA-ICP-MS can be used either to analyse single spots or to perform raster analysis, quartz crystals can be screened without the requirement for expensive preparation techniques before analysis. Commonly petrological thick sections (~300 μm) with polished sample surfaces are used. The minimum requirements for sample preparation reduce the possibility of contamination and the analysis is therefore rapid and cost-efficient compared to conventional chemical analyses.

The trace element distribution in quartz provides a rarely used source of primary information that may be applicable to a wide variety of geologic problems including igneous geochemistry, provenance studies and rapid feasibility studies of industrial quartz resources. Natural quartz occurs in silica-rich sediments, igneous and metamorphic rocks and in hydrothermal environments as veins and gangue mineral and concentrations of trace elements are mainly controlled by the crystallization temperature (e.g., Wark and Watson 2006), the fractionation degree of the melt (igneous and pegmatite quartz; e.g., Breiter and Müller 2009; Jacamon and Larsen 2009; Beurlen et al. 2011) and the chemistry and acidity of the fluid in which the quartz crystallized (diagenetic and hydrothermal quartz; e.g., Jourdan et al. 2009, Müller et al. 2010).

## 10.2 Instrumentation

### *10.2.1 Inductively Coupled Mass Spectrometry*

There are several manufacturers of ICP-MS instruments, which all display particular advantages and disadvantages. ICP-MS instruments can be divided in two main groups; high mass resolution instruments (HR-ICP-MS) and quadrupole instruments (Q-ICP-MS). HR-ICP-MS can provide eligible mass resolutions above 300 m/Δm, up to around 9,000 Δm/m, while Q-ICP-MS has a mass resolution below or around 300 m/Δm. Resolution in mass spectrometry is normally given

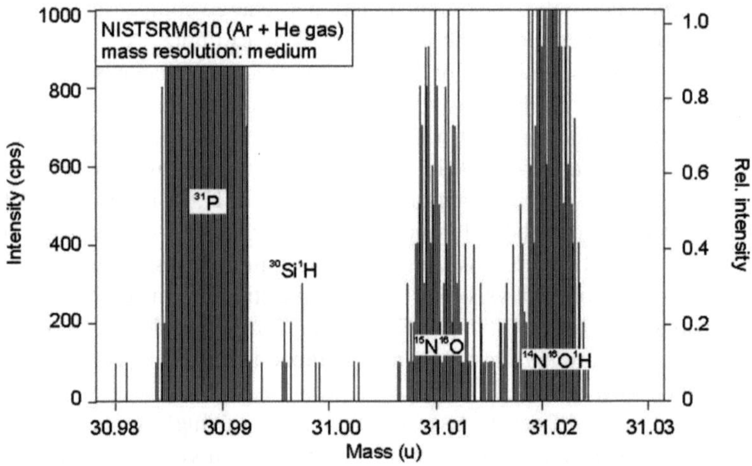


**Fig. 10.1** Analysis of P in NIST SRM 610 (National Institute of Standards and Technology) at medium resolution 3500 m/Δm with adjacent interferences, $^{12}C^{18}O^{1}H$, $^{14}N^{16}O^{1}H$ (shown in figure), $^{28}Si^{1}H_3$, $^{16}O^{15}N$ (shown in figure), $^{29}Si^{1}H_2$ and $^{1}H^{30}Si$ (shown in figure) which need a mass resolution of 933, 967, 1162, 1457, 1840 and 3951 respectively to be completely resolve from the target isotope

according to the 10% valley definition, where IUPAC defines Δm as the mass difference between two resolved peaks, with a 10% valley.

Performing multi-isotope laser ablation on limited sample material, speed is essential and Q-ICP-MS are regarded as the fastest instruments since they can scan the whole mass range (2–260 Dalton) electrically. The Q-ICP-MS mass analyzer consists of four cylindrical rods onto which are applied both RF and DC electrical fields. The ions are separated by velocity determined by their energy and mass. If a different U/V ratio is applied different m/z is allowed to pass through the quadruple rods to the detector. HR-ICP-MS instruments are slower since the magnetic field strength has to be varied when isotopes within a large mass range is measured.

However, Q-ICP-MS has a mass resolution which does not allow interference free measurements of e.g. P, Cr, Fe. A mathematical interference correction has to be applied to the measured intensity of an isotope based upon the measurement of another isotope of the interfering element/species. Using the known natural isotopic abundances of the interfering element/species the contribution to the isotope of interest is calculated, correcting for the isobaric overlap. In most cases this will be successful. However, in some cases, such as P in quartz, where interferences at the mass resolution 300 m/Δm occur with $^{14}N^{16}O^{1}H$, $^{12}C^{18}O^{1}H$, $^{28}Si^{1}H$ and $^{1}H^{30}Si$ (the required mass resolutions are 967, 933, 1162 and 3951 m/Δm, respectively, to resolve completely the target isotope from the interferences; Fig. 10.1) many corrections (with some being sample-dependent) have to be performed making it very difficult to achieve good reproducible analysis. In cases like P the application HR-ICP-MS has a major advantage to perform analyses at high mass resolution to avoid the most serious interferences, which cannot be corrected mathematically without losing precision. The disadvantage in choosing higher mass resolution is loss in sensitivity; however HR-ICP-MS has, in general, ten times higher sensitivity than Q-ICP-MS instruments.

### 10.2.2 Laser Ablation Systems

There are two main types of UV lasers that are widely used in earth sciences: (i) solid state lasers such as the frequency-quadrupled Nd:YAG and (ii) gas lasers such as excimer laser where ArF is commonly used as active laser media. Several authors studied the advantages and disadvantages pertaining to different lasers systems. Russo et al. (2000) studied the influence of wavelength on fractionation (non-stoichiometric generation of vapour species) in laser ablation. They compared three UV wavelengths (157, 213 and 266 nm) and they concluded that the wavelength has a dramatic effect both on the quantity of the mass ablated and on the ablation rate. Fractionation can be observed for all wavelengths, depending in each case on the laser-beam irradiance and the number of laser pulses at each sample-surface location. The transparency of the sample influences the amount of sample ablated at each wavelength, and the extent of fractionation. Jefferies et al. (1998) have compared time resolved signals for analyses of several materials using the 213 and 266 nm wavelengths for ablation. They demonstrate that the use of the 213 nm wavelength reduces the incidence of catastrophic ablation of strongly cleaved minerals in thin sections due to higher adsorption and that the 213 nm produces a longer, flatter and higher intensity signal than the 266 nm laser ablation. They also conclude that the inter-element fractionation is reduced during analysis using the 213 nm laser ablation system.

The shorter the wavelength of the laser, the more controlled or reproducible is the ablation rate and that were the reasons why a 193 nm laser system was introduced at NGU in 2009. The 193 nm excimer laser has been described by Günther et al. (1997). So far, the analysis of quartz has mostly been successful with excimer lasers with a 193 nm wavelength (Thermo Scientific 2011). Fractionation with the 193 nm excimer laser was found to be insignificant if a carefully selected combination of intermediate energy density (8.5–14 J $cm^{-2}$) and relatively low repetition rate (<20 Hz) were used. However, the ablation of quartz is particularly challenging since the natural quartz crystals are typically transparent. Due to the transparency and the specific absorption properties of quartz the ablation is not always ideal. The ablation characteristics and coupling of the laser with the sample is different for each quartz crystal and is not predictable. Even crystals in the same sample show different ablation behaviour (Fig. 10.2). Based on our observations the ablation behaviour of quartz is partially dependent on the crystallographic orientation crystal in addition to the transparency.

In addition to the wavelength of the laser, the laser's pulse is also important for the laser ablation performance. Gonzalez et al. (2004) compared a nanosecond

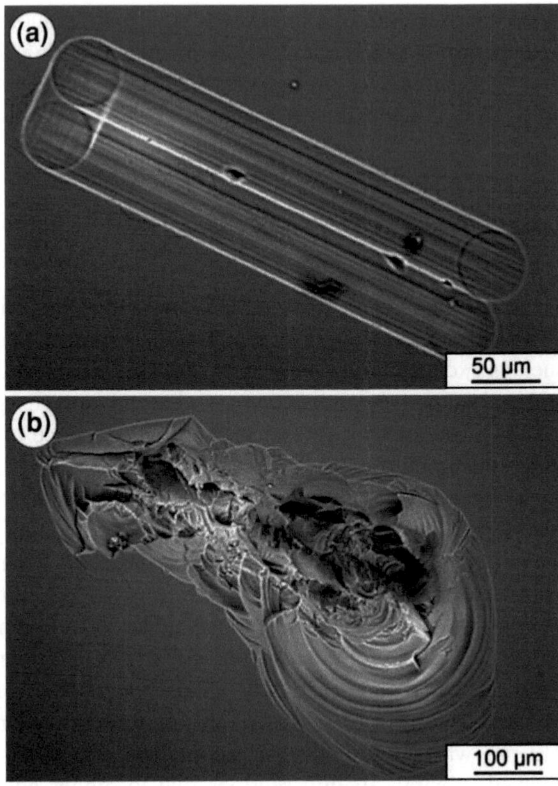


**Fig. 10.2** Backscattered electron images of a perfect laser ablation raster (**a**) and an imperfect laser ablation raster with large outbursts (**b**) in quartz from the Nedre Øyvollen pegmatite, northern Norway. The laser parameters were the same for both analyses. The different ablation behaviour is presumably caused by different crystallographic orientation of the analysed quartz crystals

laser with a femtosecond laser and concluded that by using a femtosecond laser the calibration could be extended due to reduced matrix dependence and because it grants the ability to use non-matrix matched standards. In Fernádez et al. (2007) an overview is given to show the state of art in fs-LA-ICP-MS.

Improved detection limits and ablation performance may also be achieved by performing the laser ablation under a He atmosphere, using a He–Ar mixture as the aerosol carrier (Günther and Heinrich 1999). Different gases can be added to the coolant gas or to the sample gas to improve sensitivity and the stability of the plasma e.g. a small quantity of $N_2$ into the sample gas has the effect of enhancing the sensitivity of the heavy elements (after Hirata and Nesbitt 1995).

At NGU we have experienced that the addition of small amounts methane ($CH_4$) to the sample gas during liquid analysis increased the sensitivity and the plasma stability. The same observation has been made by Rodushkin et al. (2005). Recently, we have also experimented in adding methane to the sample gas during laser ablation analysis and experienced a significant improvement in the analysis of some elements, e.g. P (Fig. 10.3). Comparing the two spectra in Fig. 10.3, one with and one without methane in the sample gas, one can see that the sensitivity is doubled by adding methane. However, adding nitrogen, helium, methane or other

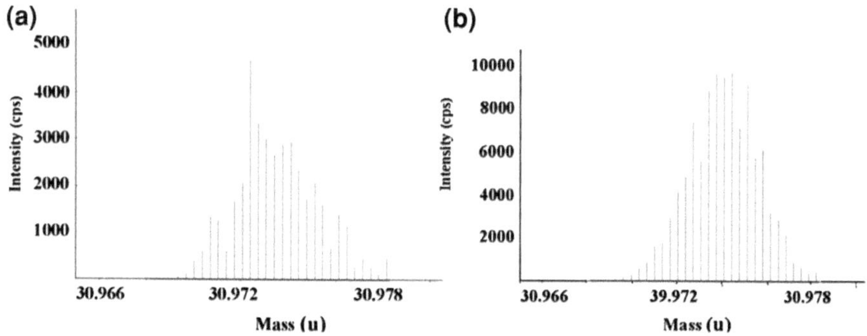


**Fig. 10.3** In the diagram on the *left* the ablation was performed with He as transport gas in the sample chamber and Ar as makeup sample gas. In the diagram on *right* a small amount of methane $CH_4$ was added to the Ar sample gas resulting in higher signal intensity. Please note the difference in scaling between the two diagrams

gases to the sample gas may also add new interferences that have to be considered e.g. adding nitrogen to the sample gas may cause interference from $^{14}N^{++}$ on $^{7}Li^{+}$.

## 10.3 Experimental Set-Up at NGU

A double focusing sector field HR-ICP-MS (Finnigan MAT, model-ELEMENT 1) instrument is used for the LA-ICP-MS analyses carried out at NGU. The instrument has been described in detail elsewhere (e.g., Gießmann and Greb 1994, Feldmann et al. 1994). The instrumental configuration includes the CD-2 Guard Electrode (which is a torch insert module that replaces the standard ELEMENT 1 torch holder) and a grounded platinum electrode (GuardElectrode™) inserted between the quartz ICP torch and the RF load coil (to prevent capacitive coupling from the load coil into the ICP). The ICP is then sustained by «pure» inductive coupling, whereby the secondary discharge between the ICP and the mass spectrometer sampling cone is eliminated.

The laser used for ablation is an UP193 FX (Fast Excimer) laser ablation system from ESI New Wave Research, UK. Helium is used as carrier gas to enhance transport efficiency of ablated material (Günther and Heinrich 1999). The helium carrier gas is mixed with argon as a makeup gas before entering the ICP-MS to maintain stable and optimum excitation conditions. Methane may be used to improve sensitivity and stability of the plasma. Typical key operating parameters of the ICP-MS and the laser probe are summarized in Table 10.1. Different sample chambers are available, but for the sake of convenience the large format cell, New Wave, is used so that several samples and standards can be placed in the ablation chamber at the same time. In order to entrain the aerosol from spreading in the sample chamber a "roving"cup is situated above the ablation

**Table 10.1** Operating parameters of the ICP-MS and key method parameters

| | |
|---|---|
| *Laser parameters* | |
| Wavelength | 193 nm |
| Pulse width | <5 ns |
| Irradiance (power density) | 1.22 GW/cm$^2$ |
| Fluence (energy density) | 6.11 J/cm$^2$ |
| Spot size | 30–100 μm |
| Laser repetition rate/pulse rate | 10–20 Hz |
| Sample helium flow rate | 0.6–1.1 l/min |
| *ICP-MS parameters* | |
| Plasma power | 974–1100 W |
| Auxiliary gas flow | 1.07–1.2 l/min |
| Sample gas flow | 0.9–1.1 l/min |
| Methane (0.5% $CH_4$ in Ar) | 0.03–0.06 l/min |
| Cone | High performance Ni |
| CD-2 guard electrode | Yes |
| *Data collection* | |
| Scan type | E-scan |
| No. of scans | 15–20 |
| Mass range | 20% of isotope peak |
| Low mass resolution (LR) | 80–100% of isotope peak |
| Medium mass resolution (MR) | 60 |
| Samples per peak | 30 |
| Low mass resolution (LR) | 0.024 s |
| Medium mass resolution (MR) | 0.048 s |
| Segment duration (analysis time per isotope per scan) | |
| Low mass resolution (LR) | |
| Medium mass resolution (MR) | |

point and the aerosole is transported from this cup through a tube to the ICPMS plasma. Commonly, raster ablation with a laser diameter of 50 μm is performed. The resulting raster is about 100 × 350 μm in size, applying the measuring parameters listed in Table 10.1. In the case of optimal ablation the depth of the raster is 10–20 μm (Fig. 10.2a). In the case of sample outbreak the crater can be up to 100 μm in depth (Fig. 10.2b). Detailed petrographic examination of the 300-μm thick, polished sections prior to analysis and high-resolution optical control during laser ablation allowed the selection of ablation areas free of mineral and fluid micro inclusions (>0.1 μm). In addition the samples are examined by scanning electron microscope cathodoluminescence (SEM-CL) in order to check the chemical homogeneity of the ablation area. Natural quartz may show small-scale (<500 μm wide) growth zoning, alteration structures or healed fractures with variable trace element content and, therefore, are visualised by SEM-CL. If these structures lie in the sampling area and are smaller than the ablation raster, the trace element concentrations of the structures will be averaged.

The ICP-MS instrumental operating parameters were tuned on $^{139}$La by ablation of the reference material NIST SRM 612 (National Institute of Standards and Technology) to give maximum sensitivity and stability. The oxide formation levels

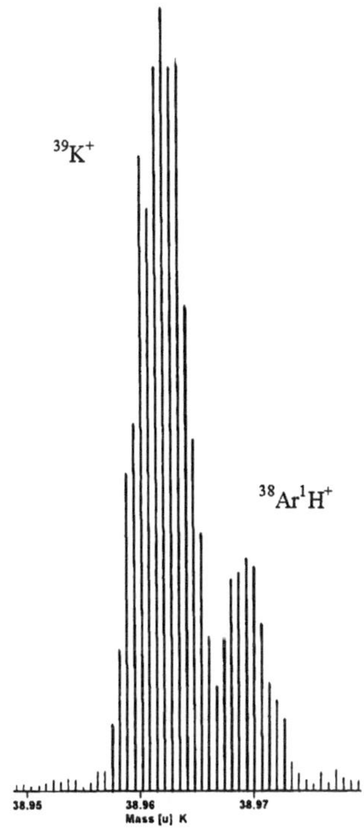


**Fig. 10.4** Spectra of K in medium mass resolution during ablation of the NIST SRM 612 glass standard

were monitored by the ratio of thorium oxide and thorium, as this is one of the most easily formed oxides. The measurement conditions were therefore optimised for the oxide ion ratio formation less than 0.7% ($ThO^+/Th^+ < 0.7\%$).

The following elements are included in the standard analytical protocol at NGU: Al, B, Be, Ca, Cr, Fe, Ga, Ge, K, Li, Mg, Mn, Na, P, Rb, Sb, Sr, Ti and Zn. Other elements like e.g. Th and U can easily be included the only limitation of which elements that can be included is the availability of suitable reference materials and the ionization energy of the element should be below 12 eV. The existence of spectroscopic interferences required the use of variable mass resolutions. Li, Be, B, Mn, Ge, Rb, Sr and Sb were analysed at low mass resolution ($m/\Delta m = 300$), but Na, Mg, Al, P, Ti, Ca, Cr, Ti, Fe, Zn and Ga required medium mass resolution ($m/\Delta m \approx 3500$) for interference free measurements. For the analysis of potassium a resolution of 5689 ($m/\Delta m$) is necessary to avoid the interference from $^{38}Ar^1H^+$ which means that K usually should be measured in high mass resolution mode. However, by careful tuning of the instrument and selection of the integration area of the K peak, it is also possible to achieve appropriate measurements also in medium mass resolution mode (Fig. 10.4). This was done to save time and make it possible to perform the analysis on smaller ablation areas. Typical analysis time per sample, including the elements given above, is less than 50 s.

The purpose of the internal standard is to average out variations in ablation and plasma instabilities. It is difficult (and in most cases impossible) to find suitable standards that have the same matrix as the samples. It is therefore always necessary to use an internal standard during laser ablation work. Usually the concentrations of the major elements are known or can be determined through other analytical techniques. During quartz analysis, it is usual to choose one of the three silicon isotopes as an internal standard. The isotope $^{29}Si$ was chosen as an internal standard at low mass resolution and $^{30}Si$ was used at medium and high mass resolution. $^{30}Si$ and $^{29}Si$ have the lowest abundance of the three silicon isotopes, which is an advantage in this context to avoid tripping of the detector. However, $^{30}Si$ has a spectroscopic interference with the molecular ion $^{14}N^{16}O^+$ at low mass resolution, which can be produced in significant amounts in the plasma. In medium and high mass resolution $^{30}Si$ is not affected by any spectroscopic interference. The interferences on isotope $^{29}Si$ are the less abundant molecular ions $^{13}C^{16}O^+$ and $^{15}N^{14}N^+$ and it is therefore more suitable as an internal standard at low mass resolution. If methane is used to enhance sensitivity and stability of the plasma $^{30}Si$ will be the best suitable internal standard in low mass resolution since it is not affected by any carbon containing interferences.

## 10.4 Standards and Standard Preparation

Since the ablation process is material and matrix dependent, the common opinion is that it is preferable to match the matrix of the standards to that of the samples. The availability of reference materials is however limited. External calibration is commonly achieved by using the synthetic glasses NIST SRM 610, 612, 614 and 616 from the National Institute of Standards and Technology. In addition International Association of Geoanalysts (IAG) arranges once or twice a year “round robin tests” on solid materials. The so-called G-probe program combines traditional natural and synthetic geological glasses with pressed powder samples, polymetallic sulphides, oxides, corals, bones, and organic materials have so far been represented. However, the synthetic glasses made available through the G-probe program have not been particularly suitable for external calibration for quartz analysis. In Jochum et al. 2000 presented eight geological MPI-DING reference glasses for in situ microanalysis; a useful contribution for the development of reference materials for probe analysis. Still there is a need of better caracterised reference materials for quartz analysis, especially for the elements Al, Na and Ca. There is, however an ongoing process of developing new reference materials in several microprobe groups worldwide, which hopefully will make it easier to achieve matrix matching standards in the future.

At NGU, external calibration is achieved by using four silicate glass reference materials produced by National Institute of Standards and Technology (NIST); NIST SRM 610, 612, 614 and 616. In addition the standard reference material NIST SRM 1830 (soda-lime float glass (0.1% $Al_2O_3$) also from NIST), the high purity silica BCS 313/1 reference sample from Bureau of Analysed Samples, UK

(BAS), and the certified reference material "pure substance no. 1" Silicon dioxide $SiO_2$ from BAM are also used.

The NIST SRM 1830 was cut into four sections of approximately 10 × 10 mm. The surface was abraded with diamond paste to enhance absorption of the laser light and to remove any contaminants at the outer surface. BCS 313/1 and BAM no.1 $SiO_2$ were melted in a tungsten crucible under He/$H_2$ atmosphere at 2000°C. Both standards were then mounted in epoxy and polished until a planar surface was obtained. Subsequently, the samples were rinsed with de-ionised water.

## 10.5 Measurement Procedure, Calibration and Data Evaluation

The measurement time choosen for each isotope is dependent on the expected element concentration and resolution needed to remove significant interferences. Each isotope is scanned 15–20 times sequencially, but the measurement time of each isotope can be manyfold by the number of channels allotted to the chosen mass range (number of samples), the time used to measure on each channel and the mass range (mass window). Details on data acquisition are given in Table 10.1.

In order to avoid problems associated with possible outliers (spikes) caused by unstable ablation, sub-microscopic inclusions and spurious single channel spikes of unknown origin (e.g. possible originating in the detector or the electronic circuits) a MAD (median absolute deviation) method, i.e. robust estimation method, was used for identification (Wilcox 1997). The MAD is robust against up to 50% of extreme values in the data set because it is based soly on medians. Due to the robustness of the MAD considerable deviations from the normal distribution are tolerable; Reimann et al. (2008). The big advantage of the test is its high break down point, equal 0.5, which makes it attractive for e.g. outlier identification (Staudte and Sheather 1990). The MAD is based on the median of all the absolute deviation from the series median i.e.:

$$MAD = MED\{|X_1 - M|, \ldots, |X_n - M|\} \tag{10.1}$$

where M is the usual sample median. If observations are randomly sampled from a normal distribution, MAD does not estimate $\sigma$, the standard deviation, but $z_{0.75}\sigma$, where $z_{0.75}$ is the 0.75 quantile of the standard normal distribution. MAD can be rescaled so that it estimates $\sigma$ by (Wilcox 1997):

$$MADN = \frac{MAD}{z_{0.75}} \approx \frac{MAD}{0.6745} \tag{10.2}$$

The outlier is then identified by:

$$z_i = \frac{x_i - M}{MADN} \tag{10.3}$$

If $z_i > z_{limit}$, where $z_{limit}$ usually is chosen equal to 2.5 or 3, then $x_i$ is an outlier.

An Ar-blank is run before each standard and sample measurement. The background signal is subtracted from the instrumental response of the standard before normalisation against the internal standard. This is done to avoid memory effects between samples. Outliers were identified and removed from the background signal by the MAD method described above. After normalisation against the appropriate internal standard, the data was controlled for outliers as described above. Otherwise, outliers (spikes) could have dramatic effects on the results, especially if they occur in the intensity data for the internal standard. This can for instance be illustrated for the measurement of Sr in NIST SRM 1830. In one case, the result without outlier removal gave a normalised intensity equal to 8100 (corresponding to a concentration of 76 µg g$^{-1}$ Sr) with a relative standard deviation of 25% (based on the 15 scans), while the results with outliers removed yielded a normalised intensity equal to 5670 (corresponding to a concentration of 52 µg g$^{-1}$ Sr) with a relative standard deviation of 2.4%.

The actual number of standards used for calibration of each element varied somewhat from element to element, depending on available certified or reference values. The calibration was based on a linear relationship between normalised intensities ($Y_i$) and corresponding concentrations ($X_i$), where $Y_i$ was given by:

$$Y_i = \frac{I_i}{I_{Si}} X_{Si} \tag{10.4}$$

In Eq. (10.4) $I_i$ and $I_{Si}$ are corresponding intensities for the analysed element and reference element (Si), and $X_{Si}$ is the concentration of the reference element. A weighted linear regression model was used for the calculation of the curve parameters. The weighted factors, $w_i$, were based on the square root of intensity for both analysed and reference element (Si), see Eqs. (10.5) and (10.6).

$$s_{Y_i} \propto \tilde{s}_{Y_i} = Y_i \sqrt{\frac{1}{I_i} + \frac{1}{I_{Si}}} \tag{10.5}$$

$$w_i = \frac{1/\tilde{s}^2_{Y_i}}{\sum \left[1/\tilde{s}^2_{Y_i}\right]/n} \tag{10.6}$$

In Eqs. (10.5) and (10.6), $s_Y$ is the uncertainty in $Y_i$, $\tilde{s}_Y$ represents the "expected" relationship for uncertainties in $Y_i$, $w_i$ is the weighted factor, and $n$ is the number of standards used in the regression analysis. The regression analyses are similar to those used by Ødegård et al. (1998); further details can be found therein.

**Table 10.2** Typical detection limits based on three times the standard deviation of ten measurements of BAM no. 1 $SiO_2$, together with certificated concentrations or recommended values were available

| Isotopes | Detection limit, LOD ($\mu g\ g^{-1}$) | BAM no. 1 $SiO_2$ ($\mu g\ g^{-1}$) |
|---|---|---|
| $^{7}Li$ | 0.4 | 0.25 |
| $^{9}Be$ | 0.15 | (<0.1) |
| $^{11}B$ | 0.35 | (<1) |
| $^{55}Mn$ | 0.06 | <0.2 |
| $^{74}Ge$ | 0.1 | <1 |
| $^{85}Rb$ | 0.05 | (<0.007) |
| $^{88}Sr$ | 0.02 | (0.013) |
| $^{121}Sb$ | 0.05 | (0.021) |
| $^{23}Na$ | 14 | <2 |
| $^{24}Mg$ | 3 | <0.5 |
| $^{27}Al$ | 6 | 8.7 |
| $^{31}P$ | 1.4 | (<1) |
| $^{39}K$ | 5.3 | 0.48 |
| $^{44}Ca$ | 35 | 0.42 |
| $^{47}Ti$ | 1.4 | 1.3 |
| $^{56}Fe$ | 0.33 | 0.62 |
| $^{68}Zn$ | 1.2 | <1.3 |
| $^{69}Ga$ | 0.07 | (<0.002) |

## 10.6 Limits of Detection

Due to a lack of a $SiO_2$ blank, 10–15 successive measurements on the BAM no.1 $SiO_2$ were done at each analytical sequence to estimate detection limits. The determination of the limits of detection (LOD) are based on three times the standard deviation of the 10 or 15 measurements divided by the sensitivity, *S*:

$$LOD = \frac{3\hat{\sigma}}{S} \tag{10.7}$$

Typical detection limits are given in Table 10.2. To improve the analytical uncertainty at low concentrations, it is important to have calibration curves with well defined intercepts. This can be achieved by using certified standards with trace element concentrations lower than the BAM no. 1 $SiO_2$ or a sample blank, if available.

The use of He as carrier gas instead of Ar may give a decrease in LOD by a factor of up to two (Jacob 2006). Typical detection limits for Ba,Pb, Th, U and Cr are given by Flem et al. (2002).

## 10.7 Precision

Routine control of precision and accuracy using independent standards is an important part of the analytical protocol. In this case the international reference materials, BR-K1 and BR-FR2 from Breitländer, Germany, have been used

(Flem et al. 2002). In addition NIST SRM 612 is monitored repeatedly throughout the sequence to document instrumental drift.

## 10.8 Discussion

For international standards and reference materials, the concentration of the internal standard is provided on the accompanying certificate. In samples, however, the concentration of the internal standard is unknown. If the samples are high-purity quartz with <50 μg $g^{-1}$ total trace element content, it may be assumed that the silicon oxide concentration is equal to approximately 99.995 wt.% and relate all the elements analysed to this concentration. Otherwise, the concentration of the internal standard can be found prior to analysis by the use of electron probe micro analyses (EPMA; e.g., Müller et al. 2003).

The Ti isotope with the highest abundance, $^{48}$Ti, could not be used for the analysis because of isobaric interference from $^{48}$Ca, present in the matrix of the reference materials NIST SRM 610, 612, 614 and 616 (~12 wt.% CaO). NIST SRM 612 is certified with 59.1 μg $g^{-1}$ Ti equal to 37 μg $g^{-1}$ $^{48}$Ti (73.8% abundance) while it contains 163.4 μg $g^{-1}$ $^{48}$Ca (0.19% abundance). Therefore, we have chosen to use $^{47}$Ti, despite the low abundance of 7.28%. The measurements were done in medium mass resolution (MR) because interferences at low mass resolution occur with the molecular ions $^{7}Li^{40}Ar^{+}$ and $^{15}N^{16}O^{16}O^{+}$. Due to the high CaO content in the reference materials NIST SRM 610–616, the isotope $^{24}$Mg produces a signal which interferes with that of $^{48}Ca^{2+}$, which would affect the calibration curve. The samples, however, do not have the same matrix as the NIST SRM 612–616 reference materials, and the lattice-bound impurity level of Ca in natural quartz is usually <5 μg $g^{-1}$. It was therefore necessary to avoid the interference of $^{48}Ca^{2+}$ on $^{24}$Mg by analysing Mg in MR. The main interferences on $^{52}$Cr are the common argides $^{12}C^{40}Ar^{+}$ and $^{36}Ar^{16}O^{+}$, and $^{40}Ar^{16}O^{+}$ is the main interference on $^{56}$Fe, all of which are resolved at MR. The main advantage with medium and high mass resolution is that interferences like these can be avoided. This advantage has to be considered against the disadvantage in less sensitivity, possibilities of mass drift and the need for a larger ablation volume (Flem et al. 2002).

In Table 10.3, examples of in situ trace element analysis performed at NGU is presented for the elements Li, Be, B, Mn, Ge, Rb, Sr, Na, Al, P, K, Ca, Ti, Fe, Ga for pegmatite quartz applying the analytical parameters described above. The analysis were performed in connection with a mapping project of regional quartz occurrences in Troms, northern Norway. Two different crystals (A and B) were analysed in each polished thick section in order to check the consistency of the results. Consentration spices of Na (analytical results above detection limit) indicate that NaCl-bearing fluid inclusions were hit during the ablation process and that it was too large to be removed by the outlier identification procedure. Also the lower limit of detection at that specific analysis-day is given in the Table 10.3, if compared with the LOD's given in Table 10.2 most of the elements has similar

**Table 10.3** In situ laser ablation ICPMS analysis (LA-ICP-MS) of pegmatite quartz samples from Troms, Norway

| | | $^{7}$Li (LR) μg g$^{-1}$ | $^{9}$Be (LR) μg g$^{-1}$ | $^{11}$B (LR) μg g$^{-1}$ | $^{55}$Mn (LR) μg g$^{-1}$ | $^{74}$Ge (LR) μg g$^{-1}$ | $^{85}$Rb (LR) μg g$^{-1}$ | $^{88}$Sr (LR) μg g$^{-1}$ | $^{23}$Na (MR) μg g$^{-1}$ | $^{27}$Al (MR) μg g$^{-1}$ | $^{31}$P (MR) μg g$^{-1}$ | $^{39}$K (MR) μg g$^{-1}$ | $^{44}$Ca (MR) μg g$^{-1}$ | $^{47}$Ti (MR) μg g$^{-1}$ | $^{56}$Fe (MR) μg g$^{-1}$ | $^{69}$Ga (MR) μg g$^{-1}$ |
|---|---|---|---|---|---|---|---|---|---|---|---|---|---|---|---|---|
| | LOD | 0.22 | 0.08 | 1.28 | 0.07 | 0.07 | 0.03 | 0.02 | 6.4 | 6.0 | 1.0 | 2.3 | 29 | 0.66 | 0.52 | 0.01 |
| *Sample ID* | | | | | | | | | | | | | | | | |
| 60,447-A | Holmevatnet | 11.0 | 0.17 | 1.44 | <0.07 | 1.62 | 0.03 | 0.05 | <6.4 | 34.6 | 1.31 | <2.3 | <29 | 6.90 | <0.52 | <0.01 |
| 60,447-B | | 12.6 | 0.30 | 5.30 | <0.07 | 2.75 | <0.03 | 0.03 | <6.4 | 49.7 | 2.05 | <2.3 | <29 | 10.4 | <0.52 | <0.01 |
| 60,449-A | Nieidajavri | 14.4 | 0.31 | 3.08 | <0.07 | 2.96 | <0.03 | <0.02 | <6.4 | 44.5 | 1.33 | <2.3 | <29 | 1.97 | <0.52 | <0.01 |
| 60,449-E | | 13.9 | 0.12 | 2.77 | <0.07 | 2.37 | <0.03 | <0.02 | 8.5 | 54.2 | 1.38 | <2.3 | <29 | 3.17 | <0.52 | <0.01 |
| 60,450-A | Nieidajavri | 12.7 | 0.16 | 3.78 | 0.33 | 2.14 | 0.04 | 0.06 | 15.9 | 61.4 | 3.13 | <2.3 | <29 | 9.09 | <0.52 | <0.01 |
| 60,450-B | | 11.6 | <0.08 | 3.26 | <0.07 | 2.08 | <0.03 | <0.02 | <6.4 | 49.1 | <1.0 | <2.3 | <29 | 5.52 | <0.52 | <0.01 |
| 60,468-A | Kåfjordbergan | 1.95 | <0.08 | 2.25 | <0.07 | 1.75 | <0.03 | 0.06 | 21.2 | 45.4 | 4.28 | <2.3 | <29 | 8.43 | <0.52 | <0.01 |
| 60,468-B | | 3.49 | 0.16 | 2.02 | <0.07 | 1.66 | <0.03 | <0.02 | <6.4 | 22.7 | 3.78 | <2.3 | <29 | 7.59 | <0.52 | <0.01 |
| 60,470-A | Istinden | 5.62 | 0.11 | 2.82 | 0.38 | 2.13 | <0.03 | 0.04 | <6.4 | 38.7 | 1.62 | <2.3 | <29 | 9.84 | <0.52 | <0.01 |
| 60,470-B | | 6.53 | <0.08 | 2.50 | <0.07 | 2.02 | <0.03 | 0.04 | <6.4 | 33.9 | 6.71 | <2.3 | <29 | 11.1 | <0.52 | <0.01 |
| 60,471-A | Signalhøyda | 6.11 | <0.08 | 1.40 | 0.08 | 2.40 | 0.04 | <0.02 | <6.4 | 16.1 | 2.39 | <2.3 | <29 | 5.38 | <0.52 | <0.01 |
| 60,471-B | | 6.35 | <0.08 | <1.28 | 0.08 | 2.47 | <0.03 | <0.02 | <6.4 | 17.8 | 1.44 | <2.3 | <29 | 4.79 | <0.52 | <0.01 |
| 60,472-A | Tårnvatnet | 0.76 | 0.10 | <1.28 | <0.07 | 0.22 | <0.03 | 0.02 | <6.4 | 30.2 | 3.90 | <2.3 | <29 | 6.79 | <0.52 | <0.01 |
| 60,472-B | | 0.69 | <0.08 | <1.28 | 0.07 | 0.42 | 0.03 | 0.08 | <6.4 | 17.2 | 3.70 | <2.3 | <29 | 5.98 | <0.52 | <0.01 |
| 60,473-A | Tårnvatnet | 2.79 | 0.24 | <1.28 | 0.14 | 0.69 | 0.05 | <0.02 | <6.4 | 14.9 | 2.38 | <2.3 | <29 | 5.03 | <0.52 | <0.01 |
| 60,473-B | | 3.25 | 0.32 | <1.28 | 0.17 | 0.75 | <0.03 | 0.03 | <6.4 | 14.0 | <1.0 | <2.3 | <29 | 4.42 | <0.52 | <0.01 |

Several elements are analysed, both in low mass resolution (LR) and medium mass resolution (MR). Lower limit of detection (LOD) at that specific analysis day is also given

LOD's but some like B is considerably worse in Table 10.3. The reason for this can be found in instrumental choices of e.g. laser spot size and laser energy at the laser and plasma power and gas flows at the ICPMS.

To get good-quality in situ analysis of quartz several criteria have to be considered; primarily all of the hardware equipment, the choice of ICP and laser ablation unit, but as important is also the well defined and matrix-matching standards and reference materials. At NGU multi-standard calibration is applied permanently. Since the concentration level of most elements in natural quartz samples is in a range close to the detection limit and only a very few silicate glass and quartz standards with concentrations in this range are available, a weighted regression model is used to give more importance to the lower standards when calculating the regression line. For these reasons single-point calibration (gas-blank + a standard with high element concentrations) is in our opinion not suitable for precise quartz analysis and, therefore, has not been applied at NGU (see also Müller et al. 2008). Since the concentration level of most elements in natural quartz samples is in a range close to the detection limit, a weighted regression model is used to give more importance to the lower standards when calculating the regression line.

## 10.9 Conclusion

We have shown that the ICP-MS with an attached laser system is very suitable method for the in situ analysis of lattice-bound elements at ultra trace level in single quartz crystals. In our experience the best choice of instrument is considered to be a double focusing sector field ICP-MS that provides high sensitivity and mass resolution high enough to separate most of the possible interferences. The ICP should be coupled to an 193-nm excimer laser, a femto second laser or a similar ablation system. The micro optical system of the laser enables the optical control of the sampling area to be ablated in order to avoid the contamination of the analysis by fluid and mineral micro inclusions. By using the LA-ICP-MS, most elements in the periodic table may be analysed in solid quartz samples at $\mu g\ g^{-1}$ to $ng\ g^{-1}$ levels. In most cases, the availability of suitable standard materials comprises the main limitation in the analyses of certain elements.

**Acknowledgments** We are grateful to Ben Snook and Ian Henderson who improved the English language of the manuscript.

## References

Beurlen H, Müller A, Silva D, Da Silva MRR (2011) Petrogenetic significance of trace-element data analysed with LA-ICP-MS in quartz from the borborema pegmatite province, Northeastern brazil. Mineral Mag 75(5):2703–2719

Breiter K, Müller A (2009) Evolution of rare-metal granitic magmas documented by quartz chemistry. Eur J Mineral 21:335–346

Fanderlik I (1991) Silica glass and its application. Elsevier, Amsterdam

Feldmann I, Tittes W, Jakubowski N, Stuewer D, Giessmann U (1994) Performance characteristics of inductively coupled plasma mass spectrometry with high mass resolution. J Anal At Spectrom 9:1007–1014
Fernandes B, Claverie F, Pecheyran C, Donard OFX (2007) Direct analysis of solid samples by fs-LA-ICP-MS. Trends Anal Chem 26(10):951–966
Flem B, Larsen RB, Grimstvedt A, Mansfeld J (2002) In situ analysis of trace elements in quartz by using laser ablation inductively coupled plasma mass spectrometry. Chem Geol 182:237–247
Gießmann U, Greb U (1994) High resolution ICP-MS a new concept for elemental mass spectrometry. Fresenius' J Anal Chem 350:186–193
Gonzalez J, Liu C, Mao X, Russo RE (2004) UV-femtosecond laser ablation-ICP-MS for analysis of alloy samples. J Anal At Spectrom 19:1165–1168
Götze J, Plötze M, Habermann D (2001) Origin, spectral characteristics and practical applications of the cathodoluminescence (CL) of quartz—a review. Mineral Petrol 71:225–250
Günther D, Heinrich CA (1999) Enhanced sensitivity in laser ablation-ICP mass spectrometry using helium–argon mixtures as aerosol carrier. J Anal At Spectrom 14:1363–1368
Günther D, Frischknecht R, Heinrich CA, Kahlert HJ (1997) Capabilities of an argon fluoride 193 nm excimer laser for laser ablation inductively coupled plasma mass spectrometry microanalysis of geological materials. J Anal At Spectrom 12:939–944
Harben PW (2002) The industrial mineral handybook—a guide to markets, specifications and prices, 4th edn. Industrial Mineral Information.Worcester Park, United Kingdom, p 412
Hirata T, Nesbitt RW (1995) U-Pb isotope geochronology of zircon: evaluation of the laser probe-inductively coupled plasma mass spectrometry technique. Geochim Cosmochim Acta 59:2491–2500
Jacamon F, Larsen RB (2009) Trace element evolution of quartz in the charnockitic Kleivan granite, SW Norway: the Ge/Ti ratio of quartz as an index of igneous differentiation. Lithos 107:281–291
Jacob DE (2006) High sensitivity analysis of trace element-poor geological reference glasses by laser ablation-inductively coupled plasma mass spectrometry (LA-ICP-MS). Geostand Geoanal Res 30:221–235
Jeffries T, Jackson SE, Longerich HP (1998) Application of a frequency quintupled Nd: YAG source ($\lambda = 213$ nm) for laser ablation inductively coupled plasma mass spectrometric analysis of minerals. J Anal At Spectrom 13:935–940
Jochum KP, Dingwell DB, Rocholl A, Stoll B, Hofmann AW et al (2000) The preparation and preliminary characterisation of eight geological MPI-DING reference glasses for in situ microanalysis. Geostand Newsl 24:87–133
Jourdan A-L, Vennemann TW, Mullis J, Ramseyer K (2009) Oxygen isotope sector zoning in natural hydrothermal quartz. Mineral Mag 73:615–632
Jung L (1992) High-purity natural quartz. Part 1: high-purity natural quartz for industrial use. Library of Congress-in-Publication Data, New Jersey, p 538
Larsen RB, Polvé M, Juve G (2000a) Granite pegmatite quartz from Evje-Iveland: trace element chemistry and implications for high-purity quartz formation. Bull Geol Surv Norw 436:57–65
Larsen RB, Flem B, Dundas S, Lahaye Y, Mansfeld J (2000b) LA-HR-ICP-MS analysis of quartz and principles governing the distribution and speciation of structural impurities in igneous quartz. Report Geological Survey of Norway, 2000.081
Müller A, Welch MD (eds) (2009) Frontiers in quartz research. Mineral Mag themat issue 73(4)
Müller A, Wiedenbeck M, van den Kerkhof AM, Kronz A, Simon K (2003) Trace elements in quartz—a combined electron microprobe, secondary ion mass spectrometry, laser-ablation ICP-MS, and cathodoluminescence study. Eur J Mineral 15:747–763
Müller A, Wiedenbeck M, Flem B, Schiellerup H (2008) Refinement of phosphorus determination in quartz by LA-ICP-MS through defining new reference material values. Geostand Geoanalytical Res 32:361–376

Müller A, Herrington R, Armstrong R, Seltmann R, Kirwin G, Stenina N, Kronz A (2010) Trace elements and cathodoluminescence of quartz in stockwork veins of Mongolian porphyry-style deposits. Miner Deposita 45:707–727

Ødegård M, Dundas SH, Flem B, Grimstvedt A (1998) Application of a double-focusing magnetic sector inductively coupled plasma mass spectrometer with laser ablation for the bulk analysis of rare earth elements in rocks fused with Li2B4O7. Fresenius' J Anal Chem 362:477–482

Perny B, Eberhardt P, Ramseyer K, Mullis J, Pankrath R (1992) Microdistribution of Al, Li and Na in alpha-quartz: possible causes and correlation with short-lived cathodoluminescence. Am Mineral 77:534–544

Reimann C, Filzmoser P, Garrett RG, Dutter R (2008) Statistical data analysis explained applies enviromental statistics. Wiley,Chichester

Rodushkin I, Nordlund P, Engström E, Baxter D (2005) Improved multi-elemental analyses by inductively coupled plasma-sector field mass spectrometry through methane addition to the plasma. J Anal At Spectrom 20:1250–1255

Rusk B, Koenig A, Lowers H (2011) Visualizing trace element distribution in quartz using cathodoluminescence, electron microprobe, and laser ablation inductively coupled mass spectrometry. Am Mineral 96:703–708

Russo RE, Mao XL, Borisov OV, Liu H (2000) Influence of wavelength on fractionation in laser ablation ICP-MS. J Anal At Spectrom 15:1115–1120

Staudte RG, Sheather SJ (1990) Robust estimation and testing. Wiley, New York, p 350

Thermo Scientific (2011) <http://www.thermo.com/eThermo/CMA/PDFs/Various/File_265.pdf>. Accessed 30 May 2011

Wark DA, Watson EB (2006) TitaniQ: a titanium-in-quartz geothermometer. Contrib Mineral Petrol 152:743–754

Weil JA (1993) A review of the EPR spectroscopy of point defects in α-quartz: the decade 1982–1992. In: Helms CR, Deal BE (eds) Physics and chemistry of $SiO_2$ and the Si–$SiO_2$ interface. Plenum Press, New York, pp 131–144

Wilcox RR (1997) Introduction to robust estimation and hypothesis testing. Academic Press, San Diego, p 296

# Chapter 11
# Cathodoluminescence Microanalysis of the Defect Microstructures of Bulk and Nanoscale Ultrapure Silicon Dioxide Polymorphs for Device Applications

**Marion A. Stevens-Kalceff**

**Abstract** Cathodoluminescence (CL) techniques have been used to investigate the defect structures of a variety of bulk and nanoscale ultrapure synthetic silicon dioxide ($SiO_2$) polymorphs. CL microanalysis in an electron microscope enables the detection of defect centers with high sensitivity and high spatial resolution. The defect microstructures of a variety of pure silicon dioxide polymorphs have been systematically analyzed using CL microanalysis techniques: The CL microanalysis of $SiO_2$ polymorphs can be challenging as their defect structure may be modified by electron irradiation. Bulk synthetic crystal and amorphous $SiO_2$, $SiO_2$ surface layers (ranging between 50–900 nm thick) on silicon and in situ buried $SiO_2$ in silicon (synthesized using oxygen ion implantation and high temperature annealing processes) have been investigated and their characteristic defects have been determined and compared. The CL emission from pure $SiO_2$ polymorphs is generally related to local point defects in the tetrahedrally coordinated $SiO_2$ host lattice. CL emissions associated with non bridging oxygen defects, oxygen deficient defects and the radiative recombination of the self trapped exciton are observed from bulk and thermal thin films of $SiO_2$ polymorphs. CL emission associated with very low concentrations of substitutional Aluminium impurities is also observed from bulk specimens of Type I and II a-$SiO_2$. In contrast, the CL emission from buried $SiO_2$ in silicon is dominated by emission associated with Si–$SiO_2$ interfaces and Si nanoparticles which form during synthesis

M. A. Stevens-Kalceff (✉)
School of Physics and Electron Microscope Unit,
University of New South Wales, Sydney, NSW 2052, Australia
e-mail: Marion.Stevens-Kalceff@unsw.edu.au

J. Götze and R. Möckel (eds.), *Quartz: Deposits, Mineralogy and Analytics*,
Springer Geology, DOI: 10.1007/978-3-642-22161-3_11,

## 11.1 Introduction

Silicon dioxide is an important material in many technologically important applications including for example, optics, optical fiber and silicon semiconductor device technology, etc. The optical, electrical and mechanical properties are dependent on the presence of defects (imperfections and impurities). Characterization of the defect microstructure of silicon dioxide ($SiO_2$) allows the performance of these applications to be optimised. In addition the defect structure of natural silicon dioxide polymorphs (e.g. crystalline and metamict quartz) can give useful insight into the geological processes associated with their genesis. Despite efforts over many years, aspects of the defect structure of silicon dioxide polymorphs remain controversial.

Cathodoluminescence (CL) is the non incandescent luminescent emission from a material during electron irradiation. The emission of photons in luminescence processes is due to electronic transitions involving localised band gap states and/or the conduction band and/or valence band. In an electron microscope, the focused beam of electrons enables spatially resolved information about the defect structure of a specimen to be obtained. CL techniques provide nondestructive, sensitive, high resolution methods of assessing the relative concentration and distribution of defects in $SiO_2$ polymorphs. The systematic analysis of a range of pure silicon dioxide polymorphs and specimens for which the impurity concentrations are known provides a good foundation for the characterization of the defect microstructure of more complex specimens such as natural specimens containing impurities (e.g. Al, Ti, Fe, Ge). Cathodoluminescence (CL) techniques have been successfully used for many years in mineralogy and petrology applications, Pagel et al. (2000), Gotze and Kempe (2008) but no single technique can provide a complete analysis. For example, some diamagnetic defects associated with luminescent emissions (e.g. neutral oxygen vacancy Skuja (2000)) are not directly detectable using Electron Paramagnetic Resonance (EPR) techniques. Conversely some paramagnetic defects, which have been characterised by EPR, are not associated with luminescent processes (e.g. peroxy radical Nishikawa et al. (1990)), although their presence may sometimes be inferred as a precursor to an observed emission. A range of complementary analyses are therefore necessary for a more comprehensive understanding of the defect structure of silicon dioxide polymorphs.

Silicon dioxide polymorphs are known to be sensitive to ionizing and energetic irradiation. In particular, defect centers may be generated and/or modified by electron irradiation. Atomic displacements from the tetrahedral silicon dioxide lattice sites can result from *elastic scattering via momentum transfer* processes (Hobbs et al. 1994). For example electron beam energies of 200 or 65 keV are necessary for the removal of a silicon or oxygen atom respectively (Pfeffer 1985). Electron-beam induced atomic displacements and modification of the low pressure tetrahedral silicon dioxide microstructure can also occur as a result of *inelastic scattering via radiolytic* processes. *Radiolysis* is an electronic process that results in the formation of stable defects due to the non-radiative decay of electron irradiation-induced electron–hole pairs (Tsai and Griscom 1991), (Hobbs et al. 1994). While elastic displacement occurs above a threshold incident electron energy, radiolysis may depend on a threshold current density and can be reduced by sample cooling (Egerton et al. 2004). Radiolysis has been observed to occur in monovalent ionic compounds (e.g. alkali halides) and covalently bonded oxides (tetrahedral silicon dioxide) Williams et al. (1986), Williams and Song (1990), Song and Williams (1992), Hobbs et al. (1994). Electrostatic charging may also occur due to the poor electrical conductivity of low pressure silicon dioxide polymorphs. The trapped charge will induce a localised induced electric field which may cause micro-segregation of mobile charged defect species within the irradiated specimen (Cazaux 1986), (Stevens-Kalceff 2000, 2001). The changes observed in the CL spectra as a function of electron beam irradiation exposure may therefore also be dependent on localised electromigration of mobile charged defects species (e.g. interstitial ions $H^+$ $OH^-$ $O^{2-}$ etc.).

The electron beam irradiation-induced modification of the CL emission from $SiO_2$ due to radiolysis and charging effects must be assessed during the collection of CL data from silicon dioxide polymorphs. A range of synthetic silicon dioxide polymorphs typically used in device applications have been systematically investigated using cathodoluminescence microanalysis.

## 11.2 Materials and Methods

Cathodoluminescence data were measured in a *JEOL 7001F* FESEM (Schottky emitter) equipped with a *Gatan XiCLone* cathodoluminescence imaging and spectral analysis system. CL spectra were excited at 293 K with a normal incidence electron beam, and collected using a retractable diamond machined paraboloidal mirror collector. Spectra over the wavelength range, ~225–825 nm (i.e. with equivalent energy range ~1.5–5.5 eV) were collected using a cooled *Princeton Instruments* Pixis 100 UV optimised CCD with gratings blazed at 500 nm/2.5 eV.

A diverse range of synthetic silicon dioxide polymorphs typically used in device applications have been systematically investigated using cathodoluminescence microanalysis. The low pressure crystal and amorphous forms of silicon dioxide have similar corner sharing $SiO_4$ tetrahedra (Hobbs et al. 1998). The basic local defect-free tetrahedral $SiO_2$ bonding structure can be simply denoted as $O_3{\equiv}Si{-}O{-}Si{\equiv}O_3$ where (−) represents silicon-oxygen bonds. Each silicon atom is surrounded by four oxygen atoms in an approximately tetrahedral arrangement and adjacent silicon atoms are bridge bonded through a single oxygen atom. The bond angles and lengths are well defined in $\alpha$-$SiO_2$ ($\alpha$ quartz), but in amorphous polymorphs of $SiO_2$ (a-$SiO_2$) a continuous distribution of bond angles and lengths are possible. For example, the O–Si–O tetrahedral angle is 109.5° in $\alpha$-$SiO_2$ but can range between 106–113° in a-$SiO_2$. Similarly the Si–O–Si bridge angle is 144°

**Table 11.1** Maximum concentration of major impurities in the "as-received" Silicon dioxide polymorph specimens investigated by Cathodoluminescence spectroscopy

| $SiO_2$ polymorph | –OH | Al | Cl | Other (e.g. Ca Na K Li) |
|---|---|---|---|---|
| Synthetic high purity $\alpha$-$SiO_2$ (Sawyer z cut $\alpha$-quartz) | <300 ppm | <2 ppm | – | <2 ppm |
| Type I a-$SiO_2$ (Infrasil) | <5 ppm | <15 ppm | – | <1 ppm |
| Type II a-$SiO_2$ (Herasil) | <180 ppm | <50 ppm | – | <4 ppm |
| Type III a-$SiO_2$ (Spectrosil) | ~1000 ppm | ~0.1 ppm | ~50 ppm | <1 ppm |
| Type IV a-$SiO_2$ (Spectrosil WF) | ~10 | ~0.1 ppm | ~180 | <1 |
| 50, 300 and 900 nm thick Dry Thermal $SiO_2$ on Si(100) | Nominally pure (Impurity concentrations depend on substrate) | | | |
| 400 nm thick buried $SiO_2$ sandwiched between the 220 nm Si top layer and the Si(100) substrate. (SIMOX 400) | Si nanoparticles and clusters (form during oxygen ion implantation and high temperature annealing) Impurity concentrations also depend on substrate | | | |

(Sawyer Research Products Inc. 1994), (Heraeus Quarzglas 1995), (Saint-Gobain Quartz Ltd 1997), (MTI Corporation 2010), (IbIS Technology Corporation 2000) Impurity and native defects may provide precursor states for irradiation-induced defects. Substitutional and interstitial species may exist in the specimens as a result of the method of synthesis. Depending on the specimen temperature, interstitial species may include both atomic and molecular/ion species of O, H, Cl, (e.g. $O_2^-$ , $ClO_2$, $OH^-$ , $H^+$ , HCl etc.) depending on the specimen type

in $\alpha$-$SiO_2$ but ranges between 120–180° in the a-$SiO_2$ (Wright 2000). The density of $\alpha$-$SiO_2$ is 2.7 g.cm$^{-3}$ while the density of the more open structured amorphous a-$SiO_2$ is typically 2.2 g.cm$^{-3}$. A range of crystal and amorphous silicon dioxide specimens were investigated and are described below. Their impurity concentrations are compared and summarised in Table 11.1.

Single crystal bulk synthetic z-cut premium ultra-pure $\alpha$-quartz produced by hydrothermal processes was investigated (Sawyer $\alpha$-quartz (Sawyer Research Products Inc.)). A range of homogeneous bulk amorphous silicon dioxide (a-$SiO_2$) specimens were also investigated including anhydrous (Type I) and hydrated fused quartz (Type II), and hydrous (Type III) and anhydrated (Type IV) silica glass. Pure fused quartz $SiO_2$ glasses are produced by melting natural quartz. Type I a-$SiO_2$ is obtained by electric melting of natural quartz crystals under vacuum, while Type II a-$SiO_2$ is produced by $H_2$–$O_2$ flame fusion of natural quartz, which results in significant residual –OH concentration. Type III a-$SiO_2$ and Type IV a-$SiO_2$ synthetic fused silica are produced by chemical oxidation of $SiCl_4$ to form $SiO_2$ via vapor-phase hydrolysis or electrically fused respectively. The cation impurities in Types III and IV are substantially lower than in Types I and II because of the high purity of $SiCl_4$ used in the processing (Hench and Vasconcelos 1990). However, –OH concentrations of up to 1,000 parts per million (ppm) may be present in Type III silica due to flame hydrolysis, while unreacted residual chlorine ions in concentrations of up to a few hundred ppm are retained in both Types III and IV a-$SiO_2$. Particular methods of manufacture are proprietary processes and the descriptions above are general, however commercially available specimens with representative impurity characteristics typical of each type of amorphous silicon dioxide were investigated including Type I a-$SiO_2$ (Infrasil (Heraeus Quarzglas 1995)), Type II a-$SiO_2$ (Herasil(Heraeus Quarzglas 1995)), Type III a-$SiO_2$ (Spectrosil (Saint-Gobain Quartz Ltd 1997)) and Type IV a-$SiO_2$ (Spectrosil WF (Saint-Gobain Quartz Ltd 1997)).

Silicon wafers exposed to air at room temperature will grow up to ~2 nm of "native oxide". A high temperature, oxidation process can be used to produce thermal oxide layers on silicon. Thermal oxide incorporates pure oxygen gas into the silicon substrate at high temperature to form a "dry" thermal oxide. Dry oxidation generally produces a more uniform and denser thermal oxide than that produced by the "wet" oxidation process which uses a pyrogenic steam as the oxidation gas. The quality of thermal oxides is also dependent on defects in the silicon substrate. A small percentage of chlorine may also be added to the oxidizing ambient to minimise the incorporation of metal ions. Dry thermal oxides of thickness 50, 300 and 900 nm, produced at 1000°C on Si(100) substrates have been investigated (MTI Corporation 2010).

Finally an in situ buried oxide layer in silicon has been investigated. A buried oxide layer of uniform thickness can be fabricated in silicon using the *Separation by IMplantation of OXygen* (SIMOX) process: Oxygen ions are implanted into the silicon substrate which is then annealed at high temperature to form a uniform buried silicon dioxide layer which ideally has a low defect density. The high dose, single implant specimen has a 396 ± 10 nm thick buried oxide layer of silicon dioxide sandwiched between a 219 ± 5 nm thick silicon top layer and the silicon (100) substrate wafer. (SIMOX 400 (IbIS Technology Corporation 2000)) The silicon top layer is often removed to facilitate experimental investigations of the properties of the buried oxide. For example, the silicon top layer can absorb photons preventing the efficient excitation and detection of photoluminescence from the buried oxide layer. CL microanalysis allows the defect microstructure of buried $SiO_2$ layers to be investigated in situ without removal of the silicon top layer. Significant residual strain associated with buried oxide layers in silicon has been reported (Camassel et al. 2001), (Afanas'ev et al. 1996b). Removal of the silicon top layer will modify these residual strain fields and therefore it is useful to investigate the defect microstructure of the in situ buried oxide layer in silicon. Following suitable correction for optical absorption, CL microanalysis allows the defect microstructure of the buried oxide layer to be investigated in situ without removal of the silicon-on-insulator top layer (Stevens-Kalceff 2011).

The polished bulk α-quartz and Types I-IV a-$SiO_2$ specimens were coated with a thin grounded conducting layer of amorphous carbon prior to the CL investigations to minimise surface charging and subsequent deflection and/or deceleration of the incident electron beam. The thin film and buried a-$SiO_2$ layers do not require a conductive coating, as radiation induced conductivity (Hobbs et al. 1994) through the surface and interface is sufficient to minimise charging under the relatively low dose conditions of this experiment. Modulations of the CL emission due to microcavity effects associated with the thin films or layered structures were not observed during these experiments (Billeb et al. 1997).

## 11.3 Results

Tetrahedral silicon dioxide polymorphs may be modified by electron irradiation (Stevens-Kalceff and Phillips 1995a), (Stevens-Kalceff 2000), (Stevens-Kalceff 1998). It is therefore necessary to monitor the effect of electron irradiation on the CL spectra from $SiO_2$ polymorphs during data collection. Irradiation-induced effects can be minimised by reducing the electron power density delivered to the electron beam-specimen interaction volume. This can be achieved by limiting the electron beam current, increasing the volumes of the specimen from which the CL is generated and/or exposing the specimen intermittently to the electron beam. To minimise electron irradiation effects, the CL data presented in this report were generated by a 10 keV, 4.2 nA electron beam scanned over 2500 $\mu m^2$ areas of the surfaces of the range of synthetic silicon dioxide polymorphs described in Table 11.1, at a rate of 0.2 s/frame ($\sim 1.23 \times 10^6$ pixels/frame).

CL spectra have been collected as a function of wavelength $\lambda$ (nm), and following subtraction of dark current have been corrected for instrument response. The response of the components within the CL system (e.g. gratings, detectors, etc.) is not uniform across the detected wavelengths. For example, the sensitivity of the CCD camera used to detect the CL spectra presented in this report is maximised at wavelengths $\sim$650–750 nm, but reduces to <30% of the maximum sensitivity for wavelengths >950 and <450 nm. Thus instrument response corrections are important prior to componentizing spectra when CL emission wavelengths are at the low sensitivity limits of CL detection systems and in particular for materials with broad emissions such as $SiO_2$ polymorphs. As the instrument broadening function is negligible with respect to the natural width of the $SiO_2$ emission components, a simple spectral decomposition rather than deconvolution is sufficient (Stevens-Kalceff 2009). Cathodoluminescence spectra must also be converted from wavelength $\lambda$ (nm) to energy E (eV) prior to peak fitting/componentization, as $SiO_2$ CL emission components are approximately Gaussian in energy space:

$$E = hc/\lambda \text{ and } I_E = \left(\lambda^2/hc\right) I_\lambda) \tag{11.1}$$

Figure 11.1a shows typical CL spectra from bulk synthetic $\alpha$-$SiO_2$ (quartz) excited with a normal incidence electron beam of energy 10 keV and current 4.2 nA scanned over 2500 $\mu m^2$ surface regions as a function of increasing electron beam irradiation exposure time (i.e. dose) ranging from 50 to 500 s. The 10 keV electrons penetrate up to $\sim$1 $\mu$m into silicon dioxide, and therefore the CL is generated from a volume of up to $\sim$2500 $\mu m^3$. Similarly, CL spectra from the bulk, thin film and buried amorphous silicon dioxide specimens have been collected under identical conditions (10 keV, 4.2 nA electron beam scanned over 2500 $\mu m^2$ regions as a function of increasing electron beam irradiation exposure ranging from 50 to 500 s). All spectra have been corrected for instrument response and converted from wavelength to energy space to enable fitting of Gaussian profiles to the spectral components. Figures 11.1b, c, d, e show CL spectra from bulk wafer specimens of Type I, II, III and IV a-$SiO_2$ respectively.

The contributions to the CL emission from the silicon (001) substrates and/or top layer must be determined in the case of thin a-$SiO_2$ films of dry oxide on bulk silicon (001) wafer substrates and in situ buried oxide layer between the ~220 nm thick silicon top layer and the silicon (001) wafer substrate. A silicon (001) wafer was therefore investigated under identical conditions and no CL emission was detected over the visible energy range 1.5–5.5 eV (i.e. equivalent wavelength range of ~225–825 nm). Bulk silicon has a band gap of ~1.1 eV at 293 K and therefore emits CL at infrared rather than visible wavelengths. It is noted that silicon has an exciton Bohr radius of 4.9 nm (Pauc et al. 2005) and therefore quantum confinement effects will contribute to luminescent emission in the visible wavelength range only when the silicon dimensions are of order 15 nm and less (Pauc et al. 2005). Thus the silicon substrates and/or top layers do not contribute to the CL spectra presented in this report. Figures 11.1f, g, h show CL spectra associated with the thin a-$SiO_2$ films of dry thermal oxide of thickness 50, 300 and 900 nm respectively, grown on bulk silicon (001) wafer substrates. Similarly, Fig. 11.1i shows typical CL spectra collected under the same conditions associated with the in situ buried oxide layer in SIMOX (Stevens-Kalceff 2011).

The spectra in Fig. 11.1i from the buried oxide layer have also been corrected for instrument response (see Eq. 11.1). Silicon absorbs light in the visible wavelength range (i.e. the transmission and reflection of light through silicon are wavelength dependent Green (2008)). The CL spectra from the buried oxide layer should therefore also be corrected for the effects of optical absorption and reflection by the silicon top layer prior to componentization. The CL emitted from the in situ buried oxide layer has passed through a minimum silicon top layer thickness of ~220 nm depending on the angle of the emitted CL. The transmission T($\lambda$) of the *normal component* of the CL generated in the buried oxide layer through the silicon top layer of thickness $t$, can be estimated from the absorption coefficient $\alpha(\lambda)$

$$T(\lambda) = \frac{(1-R)^2 \exp[-\alpha(\lambda)t]}{1 - R^2 \exp[-2\alpha(\lambda)t]} \tag{11.2}$$

where R($\lambda$) is the reflectivity calculated from the wavelength dependent refractive index n($\lambda$):

$$R = \left(\frac{n-1}{n+1}\right)^2 \tag{11.3}$$

These formulae assume a number of simplifications. It is assumed for example that the interfaces and surface are smooth with no reflective or absorptive losses. Equations 11.2 and 11.3 (e.g. see Bhat et al. (2008)) and the optical properties of silicon Green (2008) have been used to calculate the transmission of photons through a 220 nm thick silicon layer (see Fig. 11.2). More than 99% of photons of

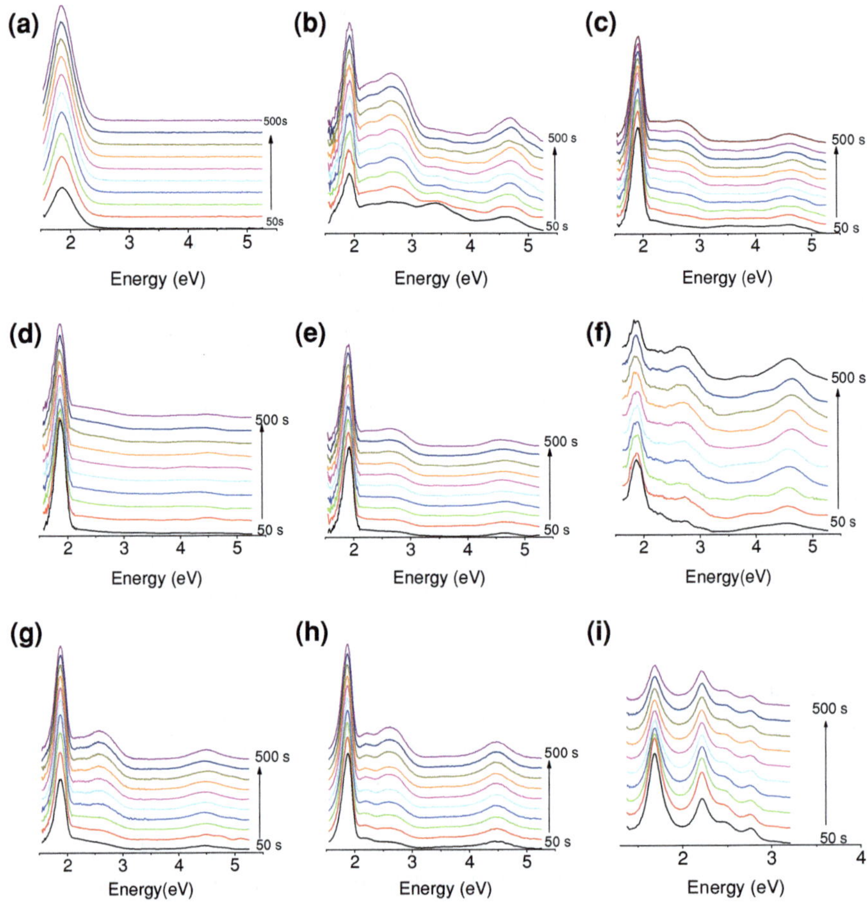


**Fig. 11.1** Typical CL spectra from **a** single crystal z-cut $\alpha$-$SiO_2$; **b** bulk Type I a-$SiO_2$; **c** bulk Type II a-$SiO_2$; **d** bulk Type III a-$SiO_2$; **e** bulk Type IV a-$SiO_2$; **f** 50 nm thick dry thermal a-$SiO_2$ on Si (001); **g** 300 nm thick dry thermal a-$SiO_2$ on Si (001); **h** 900 nm thick dry thermal a-$SiO_2$ on Si (001); **i** in situ 400 nm thick buried oxide layer in Si (001): SIMOX. These $SiO_2$ polymorphs have been irradiated with a 10 keV, 4.2 nA electron beam as a function of increasing irradiation dose for t = 50, 100, 150, 200, 150, 300, 350, 400, 450 and 500 s. The spectra have been corrected for instrument response, and in the case of (i) the in situ 400 nm thick buried oxide layer in Si (001), have also been corrected for the effects of optical absorption by the silicon top layer for energies <3.2eV

energy <1.35 eV (i.e. >920 nm), will be transmitted through the 220 nm thick silicon top layer (Stevens-Kalceff 2011). However photons of energy >3.2 eV (i.e. <390 nm) will be absorbed by the 220 nm silicon top layer and therefore will not be detected (Stevens-Kalceff 2011). Normal components of the emergent CL emission are attenuated by 220 nm of the silicon top layer, as shown in Fig. 11.2.

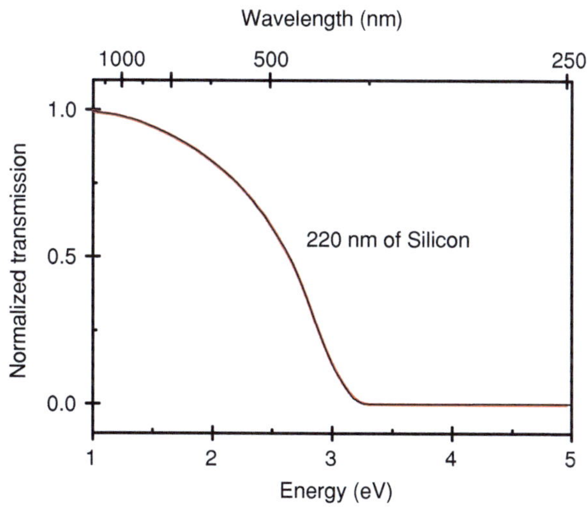


**Fig. 11.2** Calculated normalised transmission of 1.0–5.0 eV photons through a 220 nm thick layer of silicon

Non-normal components of the emergent CL emission must also be considered, but their contribution will be limited because of the effects of total internal reflection within the silicon top layer. Dependent on the angle of incidence and energy of the CL photon, multiple internal reflections will occur at the interfaces. In the case of the silicon-buried $SiO_2$ interface, the refractive index of Si is greater than the refractive index of $SiO_2$ and therefore, over the photon energy range of interest (1.4–3.2 eV), the majority of the non-normal CL emission will be totally internal reflected and/or attenuated within the Si top layer. It is noted that this difference between the refractive indices of Si and $SiO_2$ and the resultant multiple internal reflections at the high quality low roughness silicon on insulator interfaces is the basis for the development of low loss Si infrared optical waveguides fabricated from SIMOX structures (Reed 2008). In the case of the silicon-vacuum interface, at the low energy limit of the detected CL spectra (i.e. 1.5 eV/825 nm) the refractive index of silicon is $n_{Si} = 3.6$ (Green 2008) and therefore the corresponding critical angle is 16.1°, which corresponds to a maximum path length through the silicon top layer of ~229 nm. Similarly, at the high energy limit of the detected CL spectra at 3.2 eV, the refractive index of silicon is $n_{Si} = 6.6$ (Green 2008) and the corresponding critical angle is 8.7°, which corresponds to a maximum path length through the silicon top layer of ~223 nm. Thus, taking into account the multiple internal reflections, the maximum path lengths through the silicon top layer of the lower and higher energy *non-normal* components of the *emergent* CL photons are within 5% of the normal component path length of ~220 nm and therefore within the uncertainty limits of this experiment. The CL spectra from the buried oxide layer in Fig. 11.1i have therefore been corrected for the effects of optical absorption by the silicon top layer by normalizing to the normal component only (Stevens-Kalceff 2011).

## 11.4 Discussion

Consistent with previous studies, CL emission from tetrahedral silicon dioxide polymorphs is observed to change as a function of electron beam exposure due to irradiation-induced defect generation (radiolysis) and transformation (see Fig. 11.1). Electron irradiation can also result in electrostatic charging effects and temperature dependent effects, and CL microanalysis allows the in situ monitoring of the formation and transformation of irradiation sensitive defects.

The radiative transition between two different linearly coupled electronic states of a point defect in a vibrating lattice produces a characteristic emission profile at low temperatures. The shapes of luminescence bands are dependent on the magnitude of electron–phonon coupling. If there is weak defect-host lattice interaction, the spectral profile consists of a purely electronic transition (zero phonon line; ZPL) and a series of equally spaced, lower energy, overlapping phonon replicas related to fundamental lattice vibration frequencies. The lower energy phonon replicas broaden successively and overlap to form an asymmetric sideband which is described as a Pekarian envelope (Henderson and Imbusch 1989). In the case of strong coupling the resultant emission bands are broad and approximately Gaussian in profile (Henderson and Imbusch 1989). Many $SiO_2$ defects are characterised by strong defect-lattice coupling resulting in relatively large Stokes shifts (e.g. self trapped exciton (Itoh et al. 1994) with $\sim$6 eV Stokes shift in $\alpha$-$SiO_2$ Ismail-Beigi and Louie (2005)) or larger Huang Rhys factors (e.g. E′ centre with S = 13.2, Palma et al. (1996) and homogeneous broadening which can range up to 1 eV (Skuja 2000). Homogeneous broadening results from strong coupling between electron–phonon transitions and is the same for all related defects throughout the specimen. CL emissions associated with defects with strong electron–phonon coupling, are therefore fundamentally broad, Gaussian in profile and generally without fine structure even at cryogenic temperatures, due to homogeneous broadening. Inhomogeneous broadening results from local disorder in the $SiO_2$ host lattice at the defect site. Inhomogeneous broadening is minimal for ultrapure synthetic $\alpha$-$SiO_2$, but can become significant in natural and metamict quartz. For defects in amorphous silicon dioxide $SiO_2$ the main contribution to the peak widths of luminescence emissions is usually associated with homogeneous broadening of $\sim$0.2–1 eV (Skuja 2000) while the magnitude of inhomogeneous broadening is typically up to $\sim$0.1 eV (Skuja 2000). CL is produced by relatively high energy non-selective excitation, which may be associated with radiolytic radiation damage of $SiO_2$ and enhanced inhomogeneous broadening of CL emissions. The corrected experimental CL spectra from the $SiO_2$ polymorphs in Fig.11.1 have broad overlapping emission components which have been fitted with multi-parameter Gaussian functions using a non linear least squares curve fitting algorithm. The minimum numbers of statistically significant components have been fitted to each spectrum comparing $\chi^2$ factors (Stevens-Kalceff 2009).

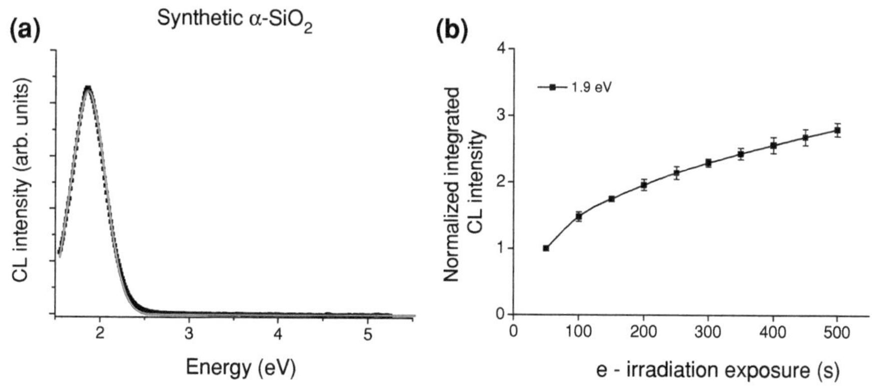


**Fig. 11.3** **a** Typical example of a componentised CL spectrum from single crystal z-cut $\alpha$-$SiO_2$ showing the fitted Gaussian component for an electron irradiation exposure of t = 50 s, from Figs. 11.1a. **b** Normalised integrated CL emission intensities of the fitted Gaussian spectral component from Fig. 11.1a as a function of electron irradiation dose for single crystal z-cut $\alpha$-$SiO_2$. The integrated CL intensity is normalised to t = 50 s electron irradiation exposure

## 11.5 Bulk Single Crystal $\alpha$-$SiO_2$ (Quartz)

Figure 11.3a shows a typical example of a componentised CL spectrum from Fig. 11.1a (i.e. for t = 50 s) for bulk pure single crystal z-cut $\alpha$-$SiO_2$ (quartz). Under low dose electron irradiation conditions, a single broad CL emission at ~1.9 eV(~650 nm) is observed. The CL emission is due to defect centers in the silicon dioxide tetrahedral structure ($O_3{\equiv}Si{-}O{-}Si{\equiv}O_3$). This CL spectrum can be fitted with a single Gaussian component at an energy of 1.9 eV and full width at half maximum (FWHM) of 0.4 eV. While fitting of any more than one spectral component to the low dose CL data in Fig. 11.1a is not statistically justified, it is known from previous investigations that this broad CL emission is comprised of a number of components (Stevens-Kalceff and Phillips 1995a), (Stevens-Kalceff 2000, 2009). These previous investigations for which the electron beam power density/dose was at least 3 orders of magnitude higher, have shown that the broad ~1.9 eV emission is a composite of *at least two overlapping emission components*. For example, the ~1.9 eV CL emission from ultra pure synthetic quartz can be resolved into components at 1.9 eV (FWHM ~0.15 eV) and 1.95 eV (FWHM ~0.4 eV) (Stevens-Kalceff and Phillips 1995a). The 1.9 eV (FWHM ~0.15 eV) CL emission is associated with the non-bridging oxygen center (NBOHC) defect which is a hole trapped in a pure $2p\pi$ orbital of a single (non-bridging) oxygen bonded to a three-fold coordinated silicon atom (i.e. $O_3{\equiv}Si{-}O\bullet$ where the "•" represents an unpaired electron). The higher dose electron beam irradiation results in the irradiation-induced attenuation of the broader ~1.95 eV (~635 nm) emission due to the dissociation of hydrogen from the non-bridging hydroxyl precursor defect ($O_3{\equiv}Si{-}O{-}H$). Higher dose electron irradiation also induces other CL emissions associated with electron irradiation-induced defects in $\alpha$-$SiO_2$ (quartz) including the Oxygen Deficient Centers (e.g. at ~4.3 eV/290 nm), and the radiative recombination of Self Trapped Excitons involving $E_1'$ centers (e.g. at ~2.7 eV/460 nm). The quartz CL emission energy, peak width and identifications are summarised in Table 11.2.

The spectra in Fig. 11.1a confirm that $\alpha$-$SiO_2$ is sensitive to electron beam irradiation, even for low dose electron beam irradiation. The observation of changes in the CL spectrum due to the modification of electron irradiation sensitive defects can provide additional useful information. The integrated intensities of each Gaussian emission component may be plotted as a function of irradiation exposure. These emission response plots summarise the irradiation-induced changes in the intensities of the CL emission bands in a spectrum. In Fig. 11.3b the integrated CL emission intensity of the fitted Gaussian spectral component normalised to $t = 50$ s (see Fig. 11.3a) from each of the spectra in Fig. 11.1a is plotted as a function of electron beam irradiation time for bulk synthetic $\alpha$-$SiO_2$ (quartz). The unresolved multi-component ~1.9 eV emission increases in intensity consistent with electron beam induced modification of the defect structure of $\alpha$-quartz.

The data in Figs. 11.1a and 11.3 and previous studies show that electron irradiation induces defects in $\alpha$-quartz (Hobbs and Pascucci 1980) and natural quartz (Rusk et al. 2006). At higher doses the irradiation-induced disorder can locally transform the crystal to an amorphous environment (i.e. bond lengths and bond angles are distorted in the local vicinity of irradiation-induced defects). For example stationary electron beam irradiation can produce localised amorphised outgrowths on the surface of quartz (Stevens-Kalceff and Phillips 1995b), (Stevens-Kalceff et al. 1996), (Stevens-Kalceff 2000). Localised amorphous environments may occur within natural crystal quartz due to the presence of defects. The defect structure and associated CL spectra from amorphous silicon dioxide polymorphs may give insight into the microstructure of amorphised and metamict natural quartz.

## 11.6 Bulk Amorphous a-$SiO_2$

The visible CL spectra in Fig. 11.1b, c, d, e from bulk amorphous Type I, II, III and IV $SiO_2$ have been fitted with a minimum of 4 or 5 Gaussian components. The integrated intensities of each of the fitted Gaussian emission components have then been plotted as a function of irradiation exposure for each of the bulk amorphous $SiO_2$ specimens. The componentised spectrum at $t = 50$ s and the emission response plots of the fitted components for bulk amorphous Type I, II, III and IV $SiO_2$ are plotted in Figs. 11.4(i)(a, b), Figs. 11.4(ii)(a, b), Figs. 11.4(iii)(a, b) and Figs. 11.4(iv)(a, b), respectively. Type I, II, III and IV $SiO_2$ exhibit four common CL emissions at ~1.9 eV (~ 650 nm), ~2.2 eV (~ 565 nm), ~2.7 eV (~460 nm) and ~4.5 eV (~280 nm) which are associated with point defects in the a-$SiO_2$ tetrahedral structure. CL emission energies and peak widths are similar

**Table 11.2** Experimental emission energies and full width at half maximum of *low dose* CL emission components observed from synthetic pure crystal and amorphous silicon dioxide polymorphs in comparison with in situ buried oxide layer in silicon

| Wave-length (nm) | CL energy (eV) FWHM, eV | | | | Association/Identification | |
|---|---|---|---|---|---|---|
| | Bulk $\alpha$-$SiO_2$ | Bulk a-$SiO_2$ | Dry thermal $SiO_2$ | Buried a-$SiO_2$ layer | | |
| | | | | | $O_3\equiv Si-O-Si\equiv O_3$ | Silicon dioxide (defect free) |
| 760 | | | | 1.65 ± 0.01(0.18 ± 0.01) | Si/H<br>Si/$SiO_2$ | Passivated silicon nano-clusters and/or unpassivated $SiO_2$–Si nano-particle interface states |
| 650 | | | | 1.9 ± 0.03 (0.17 ± 0.02) | $O_3\equiv Si-O$ | NBOHC ? unresolved component in $\alpha$-$SiO_2$[a] |
| | 1.87 ± 0.01[a] (0.39 ± 0.01) | 1.89 ± 0.02[b] (0.18 ± 0.01) | 1.86 ± 0.01 (0.17–0.22) | | | NBOHC unresolved component in $\alpha$-$SiO_2$[a] |
| | | | | | $O_3\equiv Si-OH$ | NBOHC with hydroxyl precursor |
| 590 | | | | 2.15 ± 0.01 (0.20 ± 0.01) | Si/$SiO_2$ | $SiO_2$–Si nanoparticle interface states? |
| 565 | | 2.2 ± 0.1 (0.43 ± 0.02) | 2.2 ± 0.15 (0.4 ± 0.05) | | e.g. $O_3\equiv Si-O\bullet\ \bullet Si\equiv O_3$ $\rightarrow O_3\equiv Si-O-Si\equiv O_3+h\nu$ | Radiative recombination of STE |
| 535 | | | | 2.35 ± 0.01 (0.18 ± 0.01) | Si | Si nano-clusters |
| 460 | | | | 2.7 ± 0.01 (0.23 ± 0.01) | Si/O | Oxidised silicon nano-cluster |
| | | 2.69 ± 0.03 (0.37 ± 0.03) | 2.7 ± 0.92 (0.35 ± 0.01) | | $O_2{=}Si:$ | ODC(II): 2-fold coordinated Si |

(continued)

**Table 11.2** (continued)

| Wave-length (nm) | CL energy (eV)FWHM, eV | | | | Association/Identification | |
|---|---|---|---|---|---|---|
| | Bulk$\alpha$-$SiO_2$ | Bulka-$SiO_2$ | Dry thermal $SiO_2$ | Buried a-$SiO_2$ layer | | |
| | | | | | $O_3\equiv Si-O-Si\equiv O_3$ | Silicon dioxide (defect free) |
| 415 | | | | $\sim 3$[c] | | Silicon excess/oxygen vacancy defect and/or coesite crystalline inclusions near the $SiO_2$–Si substrate interface |
| 365 | | $3.45 \pm 0.10$ $(0.8 \pm 0.05)$ | | | $O_3\equiv Al{:}M-O-Si\equiv O_3$ | Charge compensated substitutional $Al^{3+}$ $(Al^{3+}\text{-}M^{+})$ where $M^{+}$ is $Li^{+}$, $Na^{+}$, $K^{+}$ or $H^{+}$ |
| 275 | | $4.5 \pm 0.15$ $(0.5–0.7)$ | $4.45 \pm 0.10$ $(0.6–0.75)$ | | $O_2{=}Si{:}$ $O_3\equiv Si-Si\equiv O_3$ | ODC(II); 2-fold coordinated Si (possible additional contribution from the relaxed O vacancy; ODC(I)) |

2500 $\mu m^2$ regions of each $SiO_2$ polymorph have been irradiated with a 10 keV, 4.2 nA electron beam for between t = 50–500 s. In the case of the buried oxide layer, photons of energy greater than $\sim$3.2 eV are not transmitted through the $\sim$220 nm silicon top layer. The maximum uncertainties (experimental and fitting) are indicated

[a] 2+ unresolved components at low dose (Stevens-Kalceff 2009)

[b] Evidence for extra component in Type II and Type III (hydrated) a-$SiO_2$ at low doses

[c] Emission profile is truncated for energies >3.2 eV due to reduction in transmission through Si

for Type I-IV a-$SiO_2$, but the relative and absolute emission intensities are different.

The four common CL emissions observed from bulk amorphous $SiO_2$ (see Fig. 11.4) are due to native point defect centers associated with the silicon dioxide tetrahedral structure ($O_3{\equiv}Si{-}O{-}Si{\equiv}O_3$) and include non-bridging oxygen hole centers at 1.9 ± 0.02 eV (NBOHC: $O_3{\equiv}Si{-}O\cdot$) where ($\cdot$) represents an unpaired electron. The homogenous broadening of the NBOHC in amorphous $SiO_2$ has been investigated using site selective photoluminescence excitation and transient spectral hole-burning techniques revealing spectral fine structure corresponding to a Huang-Rhys factor of ~1.5 (Skuja et al. 1995). This spectral fine structure is not resolved in the CL experiments which reveal an approximately Gaussian peak shape for the ~1.9 eV emission (FWHM ~0.18 eV). CL is produced by relatively high energy non-selective excitation provided by an energetic (keV) electron beam, and may be associated with radiation damage and enhanced inhomogeneous broadening, resulting in the approximately Gaussian profile of the ~1.9 eV CL emission. The broad CL emission (FWHM ~0.4 eV) observed at 2.2 ± 0.1 eV is associated with the radiative recombination of the self trapped exciton (e.g. STE: $O_3{\equiv}Si{-}O{\cdot\cdot}Si{\equiv}O_3 \rightarrow O_3{\equiv}Si{-}O{-}Si{\equiv}O_3$). The STE is a correlated electron–hole pair localised in a self-induced lattice distortion (Williams and Song 1990). Radiative STE recombination from $SiO_2$ is characterised by lifetimes typically of order ms (Tsai and Griscom 1991), a large Stokes shift, (Ismail-Beigi and Louie 2005) and an intrinsically broad, approximately Gaussian band profile due to relatively strong defect-lattice coupling (Williams and Song 1990). Note, for example that the E′ defect ($\cdot Si{\equiv}O_3$), which is a component of the STE, has a Huang-Rhys factor which has been determined to be ~13.2 (Palma et al. 1996). CL emissions associated with Oxygen Deficient Centers (ODC) are observed at 2.69 ± 0.03 eV and 4.5 ± 0.15 eV. The variation of peak energies (4.4–4.65 eV) and peak widths (0.5–0.7) between the different bulk a-$SiO_2$, are possibly associated with the concentration of defect precursors introduced during synthesis, and resultant sensitivities to energetic ionising radiation and inhomogeneous broadening. A variety of oxygen deficient type defects have been proposed for amorphous $SiO_2$ (Skuja 1998, 2000). Oxygen Deficient Centers (ODC) in amorphous $SiO_2$ are known as ODC(I) and ODC(II). The ODC (I) has been associated with the relaxed neutral oxygen vacancy (i.e. $O_3{\equiv}Si{-}Si{\equiv}O_3$) with a covalent Si–Si bond and reduced atom spacing of ~2.5 Å, and a luminescent emission at ~4.4 eV (Nishikawa et al. 1994), (Pacchioni and Ierano 1997). A possible configuration for the ODC(II) has been proposed, consistent with damage induced by energetic ionizing radiation, and is known as the two-fold coordinated silicon defect ($O_2{=}Si{:}$) (Skuja et al. 1984), (Griscom 1991), (Skuja 1994). Theoretical simulations have predicted luminescent emissions associated with the silica ODC(II) at both 2.6 eV ($T_1 \rightarrow S_0$) and 4.6 eV ($S_1 \rightarrow S_0$) (Pacchioni and Ierano 1998). In addition, to the four common CL emissions, an emission at ~3.4 ± 0.1 eV (~365 nm) is observed in Type I and II a-$SiO_2$ CL spectra (see Figs. 11.1b, c, 11.4(i)a and (ii)a). This radiation sensitive CL emission is due to low concentrations of Al impurities in natural quartz raw material used in the

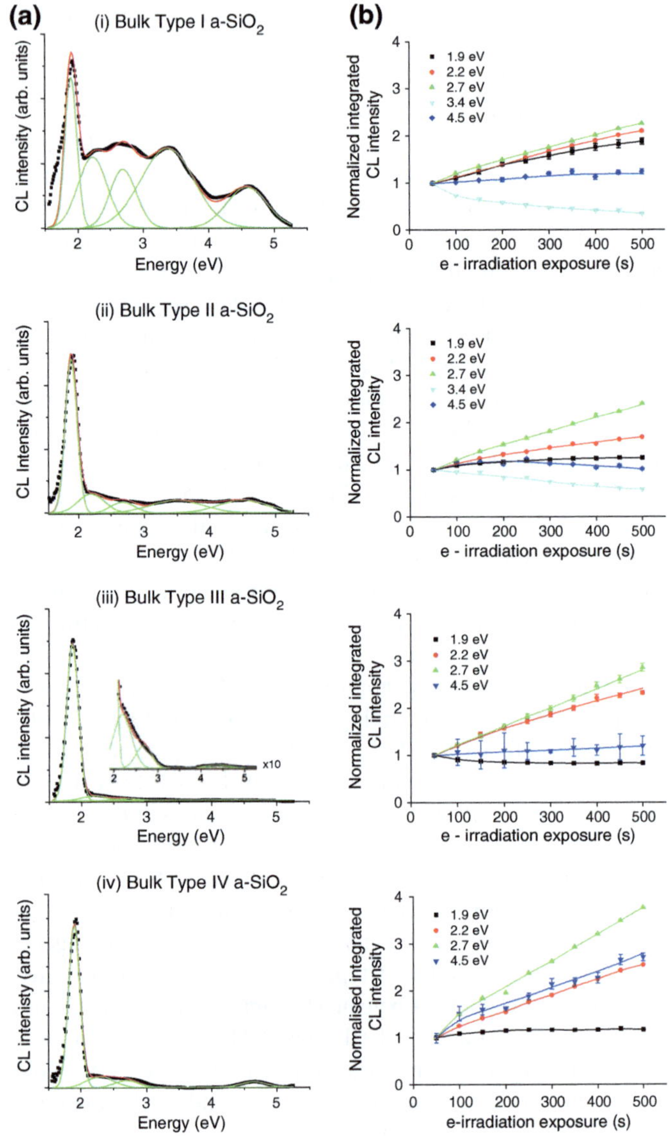


**Fig. 11.4** **a** Typical example of a componentised CL spectrum from amorphous bulk (i) Type I a-$SiO_2$; (ii) Type II a-$SiO_2$; (iii) Type III a-$SiO_2$; and (iv) Type IV a-$SiO_2$ polymorphs showing the fitted Gaussian components for an electron irradiation exposure of t = 50 s, from Figs 11.1b, c, d, e. **b** Normalised integrated CL emission intensities of the fitted Gaussian spectral component from Figs 11.1b, c, d, e as a function of electron irradiation dose for (i) Type I a-$SiO_2$; (ii) Type II a-$SiO_2$; (iii) Type III a-$SiO_2$; and (iv) Type IV a-$SiO_2$ polymorphs. The integrated CL intensity is normalised to t = 50 s electron irradiation exposure

manufacture of Type I and II a-$SiO_2$. The ~3.4 eV may be associated with the charge compensated substitutional $Al^{3+}$ alkali ion defect which has been reported in amorphous and natural crystalline $SiO_2$ polymorphs; $O_3{\equiv}(Al^{3+}{:}M^+){-}O{-}Si{\equiv}O_3$ where $M^+$ is typically $Li^+$, $Na^+$, $K^+$ or $H^+$ which has been characterised using evidence from EPR (Halliburton et al. 1981), thermally stimulated luminescence, (Alonso et al. 1983), and cathodoluminescence experiments (Ramseyer and Mullis 1990), (Stevens-Kalceff 1998). Electron irradiation tends to dissociate the charge compensating hydrogen or alkali ion from substitutional aluminium, attenuating the ~3.4 eV CL emission (Ramseyer and Mullis 1990), (Stevens-Kalceff and Phillips 1995a), (Gorton et al. 1997), (Gotze et al. 2001). The CL emission energies, peak widths and identifications for bulk amorphous $SiO_2$ polymorphs are summarised in Table 11.2.

The CL energies and peak widths are generally similar but the relative CL intensities are different for each bulk a-$SiO_2$ polymorph. For example, in comparison with the 1.9 eV emission the intensities of the 2.2, 2.7, 3.4 and ~4.5 eV CL emissions are greater for Types I and II a-$SiO_2$ than for Type IV and in particular for Type III a-$SiO_2$. This is consistent with the generally higher concentrations of impurities in Type I and Type II a $SiO_2$, noting that chlorine impurities are not directly associated with any of the observed CL emissions from a-$SiO_2$. The peak widths of the 2.2 and 2.7 eV CL emissions from Type I a-$SiO_2$ are generally slightly broader (i.e. by ~0.05 eV) than the peak widths of CL emissions from the other bulk a-$SiO_2$ polymorphs. The broader emissions from Type I a-$SiO_2$ are likely to be due to enhanced inhomogeneous broadening resulting from local disorder in the $SiO_2$ host lattice. The electron beam radiation responses of each emission component in each spectrum are different consistent with their association with a range of independent defect centers (see Fig. 11.4(i)b, Fig. 11.4(ii)b, Fig. 11.4(iii)b and Fig. 11.4(iv)b). Electron irradiation of bulk a-$SiO_2$ generally results in enhancement of the CL emissions because of the increase in defect concentration. For example the ~2.7 eV emission associated with oxygen deficient defects is observed to increase significantly in all cases as a function of dose. The ~4.5 eV CL emission which is also associated with oxygen deficient defects, is not as sensitive to electron irradiation in Types II and III a-$SiO_2$ (that have significant concentrations of OH impurities) in comparison with the ~4.5 eV CL emission in Types I and IV a-$SiO_2$ (with negligible OH concentration). See Table 11.1. Note that for bulk Type III a-$SiO_2$, the very low intensity of the ~4.5 eV emission results in a large uncertainty in the estimation of the associated normalised integrated CL intensities: The modification of the ~4.5 eV emission from Type III a-$SiO_2$ by the low dose electron irradiation is not significant within experimental uncertainty (see Fig 11.4(iii)b). Mobile OH interstitials may anneal some types of oxygen deficient defects. The electron irradiation response of Type I-IV a-$SiO_2$ polymorphs is influenced by trace levels of impurities including hydrogen (e.g. H, OH), aluminium, alkali ions and chlorine. The ~3.4 eV CL emission associated with aluminium impurities in Type I and II a-$SiO_2$ is attenuated due to the irradiation-induced disassociation of the charge compensating hydrogen or alkali ion from substitutional aluminium ($Al^{3+}$).

Types III and IV a-$SiO_2$ contain significant concentrations of chlorine however no CL emissions that can be directly associated with chlorine impurities are observed in the spectra. See Table 11.1. Because of the irradiation sensitivity of silicon dioxide, irradiation exposure experiments may allow overlapping peaks to be resolved and therefore fitted with less ambiguity. Defect generation/modification can be monitored in situ allowing a more nuanced analysis of the luminescence from $SiO_2$. The presence of precursor defects can be indirectly deduced (e.g. the $E_1'$ precursor for the radiative recombination of the STE, and the non-bridging hydroxyl precursor for the NBOHC).

## 11.7 Dry Thermal Oxide (Amorphous $SiO_2$) on Si (001)

Similarly Gaussian components have been fitted using the non-linear least square fitting procedure described above, to the observed visible CL spectra in Figs. 11.1f, g, h from 50, 300 and 900 nm thick dry thermal oxide (amorphous $SiO_2$) on Si (001), which have been collected under the same low dose conditions (see Figs. 11.5i, ii, iii respectively). The CL spectra from the dry thermal oxide thin films have been fitted with a minimum of four Gaussian components. The integrated intensities of each of the fitted Gaussian emission components have then been plotted as a function of irradiation exposure for each specimen. For example, a typical componentised spectrum (for $t = 50$ s) and the corresponding emission response plots of the fitted components from 50, 300 and 900 nm thick dry thermal oxide are plotted in Figs. 11.5(i):a and b, Figs. 11.5(ii):a and b, and Figs. 11.5(iii):a and b, respectively. The CL spectra and irradiation response plots from the dry thermal a-$SiO_2$ thin films are similar to the CL spectra from bulk low OH ("water free") Type IV a-$SiO_2$ with a minimum of four common CL emissions at ~1.9 eV (~650 nm), ~2.2 eV (~565 nm), ~2.7 eV (~460 nm) and ~4.5 eV (~280 nm). The CL emissions from the dry oxide thin films can be identified with analogous native point defects in bulk a-$SiO_2$, due to their similar peak positions, peak widths and electron irradiation response. Thus the $1.86 \pm 0.01$ eV emission is associated with the non-bridging oxygen hole centre (NBOHC). The $2.1 \pm 0.15$ eV CL emission is associated with the radiative recombination of the self trapped exciton (STE), and CL emissions observed at $2.69 \pm 0.02$ eV and $4.45 \pm 0.1$ eV are associated with Oxygen Deficient Centers (ODC). The $SiO_2$ CL emission energies, peak widths and identifications from dry thermal amorphous $SiO_2$ thin film are summarised in Table 11.2.

CL emission energies from the 50, 300 and 900 nm thick dry thermal a-$SiO_2$ thin films are similar, but the peak widths, relative and absolute emission intensities vary. For example, the peak widths of the 1.9 eV emission increases from 0.17 to 0.20–0.22 eV for the 50, 300 and 900 nm thick dry thermal a-$SiO_2$ thin films respectively. The increase in the peak widths of the CL emissions from dry thermal a-$SiO_2$ thin films as the film thickness decreases, indicates an increase in inhomogeneous broadening due to the different local environments experienced by the NBOHC defect (associated with the ~1.9 eV emission) located at different

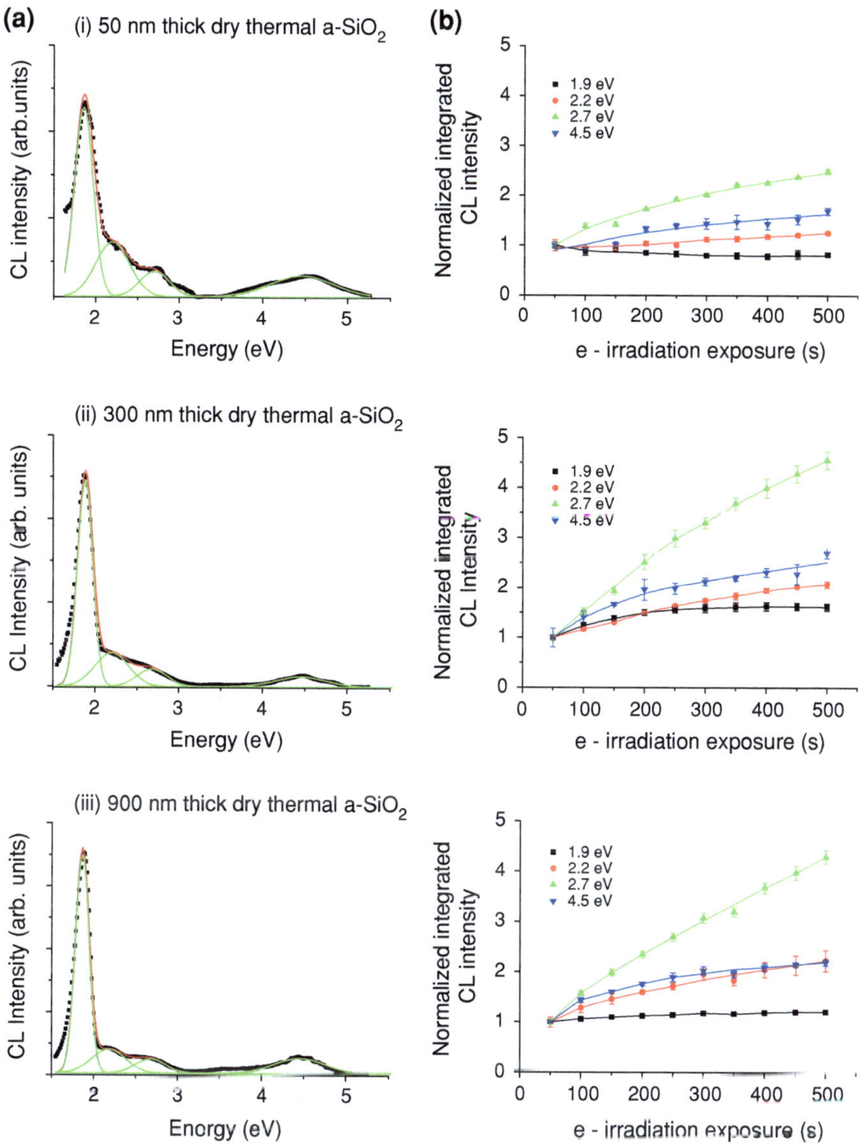


**Fig. 11.5** **a** Typical example of a componentised CL spectrum from (i) 50 nm (ii) 300 nm and (iii) 900 nm thick dry thermal a-$SiO_2$ on Si (001) showing the fitted Gaussian components for an electron irradiation exposure of t = 50 s, from Figs. 11.1f, g, h. **b** Normalised integrated CL emission intensities of the fitted Gaussian spectral component from Figs. 11.1f, g, h as a function of electron irradiation dose for (i) 50 nm (ii) 300 nm and (iii) 900 nm thick dry thermal a-$SiO_2$ on Si (001). The integrated CL intensity is normalised to t = 50 s electron irradiation exposure

sites. Disorder can be localised in the vicinity of defects (point defects, interfaces, surfaces, etc.) and at sites of strain and irradiation damage. Inhomogeneous broadening is particularly evident in CL from the 50 nm thick dry thermal a-$SiO_2$, where the effects of interface and surface states are more significant in comparison with defects in the volume of the thin film. In contrast, the CL peak energies, peak widths and electron irradiation responses of the 300 nm and in particular the 900 nm thick dry thermal a-$SiO_2$, approach that of the bulk Type IV a-$SiO_2$. It is also noted that in comparison with the 1.9 eV CL emission, the intensities of the 2.2, 2.7, and ~4.5 eV CL emissions are greater for the 50 nm thick dry thermal a-$SiO_2$ than for the 300 or 900 nm thick dry thermal a-$SiO_2$. This is consistent with generally higher concentrations of defects (and higher surface/volume and interface/volume ratio) in the thinner dry oxide. It is also noted that the variation in the energy and peak width of the ~4.5 eV emission observed in the CL spectra from the bulk a-$SiO_2$ Type I-IV a-$SiO_2$ specimens is not observed in the dry Thermal $SiO_2$ specimens which are synthesised using the same process and have the same impurity concentrations. The variety of responses of CL emission intensities to electron irradiation shown in Figs. 11.4, 11.5, 11.6 are consistent with defect generation and, in particular, oxygen vacancy generation in electron irradiated amorphous $SiO_2$.

## 11.8 Buried Amorphous $SiO_2$ in Si (001)

Gaussian components have also been fitted to the observed visible CL spectra in Fig. 11.1i collected under the same low dose conditions from SIMOX (i.e. 400 nm thick layer of amorphous $SiO_2$ buried 220 nm below the surface of a Si (001) wafer). The SIMOX CL spectra have been corrected for instrument response (see Eq. 11.1) and for absorption by the silicon top layer as described above (see Eqs. 11.2 and 11.3). A minimum of six CL emission components are observed from the in situ buried oxide at ~1.65 eV (~760 nm), ~1.9 eV (~650 nm), ~2.15 eV (~590 nm), ~2.35 eV (~535 nm), ~2.7 eV (~460 nm) and ~3 eV (~415 nm). CL emissions at energies greater than ~3.2 eV are not transmitted through the 220 nm silicon top layer (see Fig. 11.2), therefore any CL emission at ~4.5 eV associated with ODCs in the buried oxide layer will not be detected in the current experiment. A typical example of a componentised spectrum (for t = 50 s) and the corresponding integrated intensities of each of the fitted Gaussian emission components has been plotted as a function of irradiation exposure from the 400 nm thick buried oxide layer and is shown in Figs. 11.6a and b, respectively. In general, CL emissions from the buried oxide cannot simply be identified with analogous native point defect centers associated with the silicon dioxide tetrahedral structure because of their different emission energies, peak widths and irradiation responses (Stevens-Kalceff 2011).

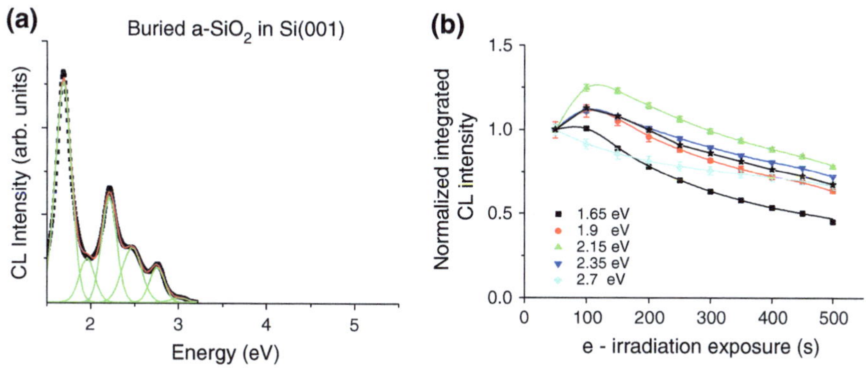


**Fig. 11.6** **a** Typical example of a componentised CL spectrum from in situ 400 nm thick buried oxide layer in SIMOX on Si (001) showing the fitted Gaussian components for an electron irradiation exposure of t = 50 s, from Fig. 11.1(i). **b** Normalised integrated CL emission intensities of the fitted Gaussian spectral component from Fig. 11.1(i) as a function of electron irradiation dose for in situ 400 nm thick buried oxide layer in SIMOX on Si (001). The integrated CL intensity is normalised to t = 50 s electron irradiation exposure

Oxygen vacancies are known to be incorporated into the buried oxide during the fabrication process. Following oxygen ion implantation, the subsequent high-temperature anneal results in an oxygen reduced oxide due to out-diffusion of oxygen into the silicon and/or in-diffusion of a reducing species from the silicon (Nishikawa et al. 1999). Silicon nanoparticles are formed within the buried oxide layer during the post implantation, high-temperature anneal (Afanas'ev and Stesmans 1999), (Nishikawa et al. 1999). Investigations of the charge trapping properties indicate smaller Si clusters (<1 nm diameter) are found to be approximately uniformly distributed through the buried oxide and are associated with confined exciton states in Si nanocrystals (Revesz and Hughes 1997), (Afanas'ev et al. 1996a). In comparison, larger ( ~ 3–4 nm diameter) Si clusters have been found near the interface between the buried oxide—silicon substrate (Afanas'ev et al. 1996a) and silicon top layer—the buried oxide (Nishikawa et al. 1999).

The fabrication of a buried oxide structure via the SIMOX process produces significant residual strain which is maximised at the upper and lower Si-$SiO_2$ interfaces (Camassel et al. 2001). Residual compressive strain is observed in the buried oxide layer between the two Si–$SiO_2$ interfaces. Residual tensile strain is observed across the silicon top layer from the upper Si-$SiO_2$ interface and extending from the lower Si–$SiO_2$ interface into the silicon substrate(Camassel et al. 2001). The formation or transformation of $SiO_2$ defects which involve the structural rearrangement or relaxation of the surrounding oxide network may be hindered in the confined strained oxygen deficient buried oxide layer. The strained confined buried oxide is also likely to contain significant concentrations of surface or interface defects associated with the Si–$SiO_2$ interfaces and the Si nanoclusters (which have large surface to volume ratios).

The CL data in Figs. 11.1i, 11.6a and b have been collected under low dose conditions. High dose CL spectroscopy of the buried oxide also reveals a minimum of six Gaussian CL emissions components with similar wavelengths and peak widths, but different relative intensities. The origins of these emissions have previously been described in detail (Stevens-Kalceff 2011). Silicon nanoparticles formed within the buried oxide layer during the post implantation, high temperature anneal are associated with the ~1.65 eV (~750 nm) see Figs. 11.1i and 11.6a. Luminescence from nanostructured silicon is believed to be strongly dependent on surface states and surface passivation (Saeta and Gallagher 1997), (Wolkin et al. 1999), (Fauchet et al. 1998). The ~1.65 eV CL emission is related to either Si nanoparticle–$SiO_2$ interface defects or passivated quantum confined silicon nanoparticles (Afanas'ev and Stesmans 1999), (Saeta and Gallagher 1997), (Wolkin et al. 1999), (Vasiliev et al. 2002) or competition between these two recombination mechanisms (Godefroo et al. 2008). During electron irradiation, radiolytic dissociation of the passivating species from the silicon nanoparticles will result in the attenuation of the 1.65 eV CL emission.

An unresolved low intensity emission component is observed at 1.9 eV (FWHM ~0.17 eV) from the buried oxide layer: The 1.9 eV (650 nm) CL emission may be partially resolved in high dose experiments only when the adjacent CL components have been sufficiently attenuated by the electron beam irradiation (Stevens-Kalceff 2011). The fitted 1.9 eV CL emission energy and peak width are consistent with the 1.9 eV CL emission (full width half maximum; FWHM ~0.18 eV) observed from bulk a-$SiO_2$ observed under the same conditions and associated with the native NBOHC defect. CL emissions at 2.15 eV (FWHM 0.20 eV) and 2.35 eV (FWHM 0.18 eV) are observed in CL spectra from the buried oxide layer (see Figs. 11.1i and 11.6a). The peak widths of both the 2.15 eV (590 nm) and 2.35 eV (535 nm) emissions from the buried oxide layer are half that observed under identical excitation conditions for the 2.2 eV (~565 nm) CL emission from bulk $SiO_2$ which is attributed to the radiative combination of the STE (Stevens-Kalceff 2011). Thus neither the 2.15 or 2.35 eV CL emissions from the buried oxide layer are likely to be associated with the characteristically broad STE. In addition the residual strain within the buried oxide layer will reduce the probability of radiative relaxation of the $SiO_2$ tetrahedral structure and therefore radiative recombination of the STE is unlikely (Stevens-Kalceff 2011). The absence of a CL component associated with the STE is also consistent with complementary data showing enhanced concentrations of E' type defect centers (e.g. $\cdot Si\equiv O_3$) in SIMOX (Revesz and Hughes 1997), (Warren et al. 1993) as the radiative STE recombination would reduce the concentration of E' defects (e.g. $O_3\equiv Si{-}O{\cdot\cdot}Si\equiv O_3 \rightarrow O_3\equiv Si{-}O{-}Si\equiv O_3{+}h\nu_{STE}$) (Stevens-Kalceff 2011). The 2.15 eV emission has been observed only in CL spectra from buried $SiO_2$ to date (Stevens-Kalceff 2011). It is most sensitive to electron irradiation and may possibly be associated with (hydrogen) passivated silicon nano-clusters. Electron beam induced dissociation of the passivating species from the silicon nanoparticles will result in the attenuation of the 2.15 eV CL emission. Previous photoluminescence (PL) investigations have removed the silicon top layer to enable efficient excitation and detection of luminescence from the buried oxide: Etch-back photoluminescence experiments have revealed a 2.4 eV PL emission associated with silicon nanoclusters localised at the interface between the silicon top layer and the buried oxide. The CL emission observed at 2.35 eV (FWHM 0.18 eV) has a similar emission energy and peak width to the 2.4 eV PL emission.

The peak width of the CL emission from the buried oxide layer at 2.7 eV (FWHM 0.23 eV) is significantly less than the 2.7 eV (FWHM 0.35 eV) emission observed from bulk a-$SiO_2$ under identical conditions. Due to strong defect lattice coupling, the peak width of the ~2.7 eV emission from bulk a-$SiO_2$ is intrinsically broad (~0.30 eV) even at 5 K (Stevens-Kalceff 1998). (Reducing specimen temperature assists in the resolution of spectral components due to reduction in the CL peak widths with a decrease in thermal broadening and enhancement of CL intensities associated with reduced diffusion of competitive non-radiative centers). In addition, the formation of ODC (II) defects which is associated with significant structural rearrangement is likely to be hindered in the strained confined buried oxide layers (Nishikawa et al. 1999). Therefore in contrast to the situation observed in bulk a-$SiO_2$ e-irradiated under the same conditions, the 2.7 eV CL emission from the buried oxide layer is unlikely to be associated with the ODC (II). Insight into the possible origin of the CL emission observed at ~2.7 eV from the buried oxide layer may be obtained from complementary investigations. Photoionization spectroscopy of silicon nanoparticles in $SiO_2$ attribute defects at oxidised clusters of silicon to luminescent emission observed at 2.4–2.8 eV (Afanas'ev and Stesmans 1999). This is consistent with photoluminescence data from etch-back experiments of buried oxide in silicon which attribute a 2.6–2.8 eV PL emission to Si clusters found near the buried oxide—silicon interface (Nishikawa et al. 1999). The 2.7 eV CL emission is therefore likely to be associated with oxidised silicon clusters located in strained buried oxide near the $SiO_2$–Si interface. The 2.7 eV CL emission is also attenuated by electron irradiation but both low and high dose (Stevens-Kalceff 2011) experiments show, that it is not as irradiation-sensitive as the other CL emissions associated with Si clusters in the buried oxide (i.e. 1.65, 2.15 and 2.35 eV emissions; see Fig. 11.6b).

The peak profile of the low intensity CL emission observed at ~3 eV (415 nm) is strongly influenced by reduced transmission of photons through the Si top layer (e.g. see Fig. 11.2). Despite the correction of the SIMOX CL spectra over the wavelength range 1.4–3.2 eV, for transmission through the silicon top layer, it is difficult to accurately determine the energy, peak width or irradiation response of the ~3 eV (415 nm) CL emission due to uncertainties introduced by the steep drop off in photon transmission through the silicon top layer at these energies. Insight into the possible origins of the ~3 eV CL is again provided by comparison with other investigations. Photoionization experiments of Si particles in $SiO_2$ attribute energy levels 3.1 eV below the $SiO_2$ conduction band to a hydrogen complexed oxygen vacancy in $SiO_2$ (Afanas'ev and Stesmans 1999) while a 3.1 eV PL emission is attributed to excess silicon near the $SiO_2$–Si substrate interface (Nishikawa et al. 1999). In addition, inclusions of a dense coesite-like crystalline phase have also been reported near the $SiO_2$–Si substrate interface (Afanas'ev et al. 1996b; 1997) which is a location of high residual compressive strain (Camassel et al. 2001). Natural coesite can form as a result of high pressure. Coesite crystallites exhibit a broad CL emission at ~3 eV (Trukhin et al. 2003). Thus the ~3 eV CL emission may be associated with excess silicon defects and/or crystalline coesite inclusions near the $SiO_2$ –Si substrate interface. The CL emission energies, peak widths and identifications from the buried oxide layer are summarised in Table 11.2.

During low dose irradiation, the CL emissions associated with defects in the buried oxide initially increase indicating an electron irradiation-induced increase in defect concentrations, however as dose increases, all CL emissions associated with the buried oxide layer are attenuated. This is in contrast to the (low and high dose) situation in bulk a-$SiO_2$ where generally the CL emission components increase in intensity as a function of irradiation. This attenuation of all CL emissions associated with the buried oxide layer corresponds to the electron beam induced modification of the defects within the buried oxide and at the Si–$SiO_2$ interfaces rather than an annealing or reduction in the concentration of the defects. At higher doses all CL emissions from the buried oxide attenuate and volume loss occurs from the irradiated specimen, which is consistent with the breakdown of the $SiO_2$ structure in the buried oxide layer (Stevens-Kalceff 2011). In bulk, thin film and buried silicon dioxide, charge trapping at existing and radiation induced defects will result in an irradiation-induced electric field. Charge trapping in irradiated buried oxides is enhanced in comparison with other forms of $SiO_2$ (Paillet et al. 1995), due to the higher defect density in the strained buried oxide layer and at the Si-$SiO_2$ interfaces. The irradiation-induced localised electric field is enhanced therefore contributing to the electron irradiation-induced breakdown of the buried oxide layer.

## 11.9 Conclusions

A range of synthetic pure silicon dioxide polymorphs typically used in device applications have been systematically investigated using cathodoluminescence spectroscopy techniques. Bulk single crystal silicon dioxide (quartz), bulk Type I–IV amorphous silicon dioxide, dry amorphous thermal silicon dioxide thin films on silicon (001) and buried strained amorphous silicon dioxide layer in silicon (001) have been investigated (see Table 11.1) under the same low dose excitation conditions. CL emissions associated with local $SiO_2$ defects are fundamentally broad due to strong electron–phonon coupling. The influence of homogeneous broadening may become significant in amorphous polymorphs of silicon dioxide.

A single broad multicomponent emission (FWHM 0.4 eV) at approximately 1.9 eV (650 nm) is observed from $\alpha$-$SiO_2$ and is attributed to the non-bridging oxygen hole centre (NBOHC) with contribution from at least two precursors.

CL emissions from bulk Type I–IV amorphous $SiO_2$ and dry thermal amorphous $SiO_2$ thin films are identified with a range of native defect centers associated with point defects in the a-$SiO_2$ tetrahedral structure. These characteristic CL emissions are associated with oxygen deficient defects, (e.g. oxygen vacancies), non bridging oxygen defects and self trapped excitons, however the relative intensities, peak widths and/or irradiation kinetics differ between each bulk and thin film a-$SiO_2$ polymorph. The CL emissions from the bulk and thin film amorphous $SiO_2$ polymorphs include the non-bridging oxygen hole centre (NBOHC) at 1.9 eV (650 nm); the radiative recombination of the self trapped exciton (STE) at 2.2 eV (565 nm); and Oxygen Deficient Centers (ODC) at 2.7 eV (460 nm) and ~4.5 eV (275 nm). A CL emission at ~3.4 eV (365 nm) is observed from Type I and II a-$SiO_2$ and is attributed to be charged compensated substitutional $Al^{3+}$:$M^+$ defect (where $M^+$ is typically $Li^+$, $Na^+$, $K^+$ or $H^+$). See Table 11.2.

The CL emission from buried amorphous $SiO_2$ is significantly different from bulk and thin film amorphous $SiO_2$ polymorphs. In general, the CL emissions from the buried oxide layer cannot simply be associated with analogous amorphous silicon dioxide native point defects. The Separation by IMplantation of OXygen (SIMOX) fabrication process produces buried oxides confined within the silicon wafer with residual compressive strain maximised at the two Si-$SiO_2$ planar interfaces. The concentration of defects is enhanced at surfaces and interfaces. The formation of the native $SiO_2$ defects which involve the relaxation of the surrounding oxide network is hindered in the confined strained buried oxide, while the formation of silicon nano-clusters and crystalline coesite platelets is facilitated. CL emission from the confined strained buried oxide is dominated by defects associated with large surface to volume ratio of nanoscale silicon clusters and their interfaces. Following suitable corrections for optical absorption, CL microanalysis allows the defect microstructure of the buried oxide layer to be investigated in situ without removal of the silicon top layer. A minimum of six visible CL emission components from the in situ buried oxide are observed (see Table 11.2). The 1.65 eV (760 nm) emission is associated with passivated silicon nano-clusters and $SiO_2$-Si particle surface states. Similarly the 2.15 eV (590 nm) and 2.35 eV (535 nm) emissions are associated with surface states of silicon nano-clusters near the Si—buried oxide interfaces. The 2.7 eV (460 nm) emission is associated with the oxidised silicon nano-clusters. The peak profile of the nominally ~3 eV (415 nm) emission is distorted by reduced transmission through the silicon top layer, preventing the peak width and energy from being more accurately determined. The ~3 eV emission is associated with a silicon excess defect (e.g. H complexed oxygen vacancy) and/or coesite crystalline platelets near the $SiO_2$ Si substrate interface. The unresolved 1.9 eV CL emission is possibly associated with the NBOHC (See Table 11.2).

The most significant physical processes contributing to the changes in the CL spectra are defect generation via radiolysis and local modification due to highly localised electric fields produced by charge trapping at defects within the $SiO_2$. The rate of generation/modification of electron irradiation-induced effects can be reduced by decreasing the electron beam power density. Electron irradiation-induced dose dependent changes in the CL emission intensities as shown in Figs. 11.4, 11.5, 11.6 are consistent with defect generation and transformation.

In particular, electron irradiation results in oxygen vacancy generation in the bulk and thin film amorphous $SiO_2$ polymorphs (see Figs. 11.3,11.5, 11.6). Electron irradiation can produce localised trapped-charge-induced electric fields within the irradiated micro-volume. This can result in local electromigration of charged mobile defect species with local modification of the defect structure and/or chemical composition. Charge trapping is enhanced within the strained confined buried oxide layer at interfaces and nanoscale silicon cluster defects. These enhanced electric fields can ultimately contribute to volume loss and breakdown of the irradiated confined buried oxide.

**Acknowledgments** Support from the Australian Research Council and the Australian Microscopy and Microanalysis Research Facility in the Electron Microscope Unit at the University of New South Wales is gratefully acknowledged.

## References

Afanas'ev VV, Revesz AG, Hughes HL (1996a) Confinement phenomena in buried oxides of SIMOX structures as affected by processing. J Electrochem Soc 143:695–700

Afanas'ev VV, Stesmans A (1999) Photoionization of silicon particles in $SiO_2$. Phys Rev B Condens Matter 59:2025–2034

Afanas'ev VV, Stesmans A, Revesz AG, Hughes HL (1997) Structural inhomogeneity and silicon enrichment of buried $SiO_2$ layers formed by oxygen ion implantation in silicon. J Appl Phys 82:2184–2199

Afanas'ev VV, Stesmans A, Twigg ME (1996b) Epitaxial growth of $SiO_2$ produced in silicon by oxygen ion implantation. Phys Rev Lett 77:4206

Alonso PJ, Halliburton LE, Kohnke EE, Bossol RB (1983) X-ray induced luminescence in crystalline $SiO_2$. J Appl Phys 54:5369–5375

Bhat R, Dutta PS, Guha S (2008) Crystal growth and below-bandgap optical absorption studies in InAs for non-linear optic applications. J Cryst Growth 310:1910–1916

Billeb A, Grieshaber W, Stocker D, Schubert EF (1997) Microcavity effects in GaN epitaxial films and in Ag/GaN/sapphire structures. Appl Phys Lett 70:2790–2792

Camassel J, Falkovsky LA, Planes N (2001) Strain effect in silicon-on-insulator materials: Investigation with optical phonons. Phys Rev B Condens Matter 63:1880

Cazaux J (1986) Some considerations on the electric field induced in insulators by electron bombardment. J Appl Phys 59:1418–1430

Egerton RF, Li P, Malac M (2004) Radiation damage in the TEM and SEM. Micron 35:399–409

Fauchet PM, Tsybeskov L, Zacharias M, Hirschman K (1998) Nanocrystalline silicon/amorphous silicon dioxide superlattices. Mater Res Soc Symp Proc 485:49–59

Godefroo S, Hayne M, Jivanescu M, Stesmans A, Zacharias M, Lebedev OI, van Tendeloo G, Moshchalkov VV (2008) Classification and control of the origin of photoluminescence from Si nanocrystals. Nat Nanotechnol 3:174–178

Gorton NT, Walker G, Burley SD (1997) Experimental analysis of the composite blue cathodoluminescence emission in quartz. J Lumin 72:669–671

Gotze J, Kempe U (2008) A comparison of optical microscope- and scanning electron microscope-based cathodoluminescence (CL) imaging and spectroscopy applied to geosciences. Mineral Mag 72:909–924

Gotze J, Plotze M, Habermann D (2001) Origin, spectral characteristics and practical applications of the cathodolumimescence (CL) of quartz—a review. Mineral Petrol 71:225–250

Green MA (2008) Self-consistent optical parameters of intrinsic silicon at 300 K including temperature coefficients. Sol Energy Mater Sol Cells 92:1305–1310
Griscom DL (1991) Optical properties and structure of defects in silica glass. J Ceram Soc Jpn 99:923–942
Halliburton LE, Koumvakalis N, Markes ME, Martin JJ (1981) Radiation effects in crystalline $SiO_2$: The role of aluminum. J Appl Phys 52:3565–3574
Hench LL, Vasconcelos W (1990) Gel-silica science. Annu Rev Mat Sci 20:269–298
Henderson B, Imbusch GF (1989) Optical spectroscopy of inorganic solids. Clarendon, Oxford
Heraeus Quarzglas (1995) Hanau, Germany
Hobbs LW, Clinard FW Jr, Zinkle SJ, Ewing RC (1994) Radiation effects in ceramics. J Nucl Mater 216:291–321
Hobbs LW, Jesurum CE, Pulim V, Berger B (1998) Local topology of silica networks. Philos Mag A 78:679–712
Hobbs LW, Pascucci MR (1980) Radiolysis and defect structure in electron-irradiated a-quartz. Journal de Physique 7:C6-237–C6-242
IBIS Technology Corporation (2000) Danvers, USA
Ismail-Beigi S, Louie SG (2005) Self-trapped excitons in silicon dioxide: mechanism and properties. Phys Rev Lett 95:156401-4
Itoh N, Shimizu-Iwayama T, Fujita T (1994) Excitons in crystalline and amorphous $SiO_2$ : formation, relaxation and conversion to Frenkel pairs. J Non-Cryst Sol 179:194
MTI Corporation (2010) Richmond, CA 94804, USA
Nishikawa H, Nakamura R, Tohmon R, Ohki Y, Sakurai Y, Nagasawa K, Hama Y (1990) Generation mechanism of photoinduced paramagnetic centers from preexisting precursors in high-purity silicas. Phys Rev B Condens Matt 41:7828
Nishikawa H, Stahlbush RE, Stathis JH (1999) Oxygen-deficient centers and excess Si in buried oxide using photoluminescence spectroscopy. Phys Rev B Condens Matt 60:15910–15918
Nishikawa H, Watanabe E, Ito D, Ohki Y (1994) Decay kinetics of the 4.4 eV photoluminescence associated with the 2 states of oxygen-deficient-type defect in $SiO_2$. Phys Rev Lett 72:2101
Pacchioni G, Ierano G (1997) Computed optical absorption and photoluminescence spectra of neutral oxygen vacancies in alpha -quartz. Phys Rev Lett 79:753–756
Pacchioni G, Ierano G (1998) Ab Initio theory of optical transitions of point defects in $SiO_2$. Phys Rev B Condens Matt 57:818–832
Pagel M, Barbin V, Blanc P, Ohnenstetter D (2000) Cathodoluminescence in geosciences. Springer Verlag, Berlin Heidelberg. ISBN 3-540-659870-0
Paillet P, Autran JL, Leray JL, Aspar B, Auberton-Herve AJ (1995) Trapping-detrapping properties of irradiated ultra-thin SIMOX buried oxides. IEEE Trans Nucl Sci 42:2108–2113
Palma A, Lopez-Villanueva JA, Carceller JE (1996) Electric field dependence of the electron capture cross section of neutral traps in $SiO_2$. J Electrochem Soc 143:2687–2690
Pauc N, Calvo V, Eymery J, Fournel F, Magnea N (2005) Electronic and optical properties of Si/ Si O2 nanostructures. I. Electron-hole collective processes in single Si/Si O2 quantum wells. Phys Rev B Condens Matt 72:205324
Pfeffer RL (1985) Damage center formation in $SiO_2$ thin films by fast electron irradiation. J Appl Phys 57:5176–5180
Ramseyer K, Mullis J (1990) Factors influencing short-lived blue cathodoluminescence of alpha-quartz. Am Mineral 75:791–800
Reed GT (ed) (2008) Silicon photonics: the state of the art. Wiley, Chichester
Revesz AG, Hughes HL (1997) Properties of the buried oxide layer in SIMOX structures. Microelectron Eng 36:343–350
Rusk BG, Reed MH, Dilles JH, Kent AJR (2006) Intensity of quartz cathodoluminescence and trace-element content in quartz from the porphyry copper deposit at Butte, Montana. Am Mineral 91:1300–1312
Saeta PN, Gallagher AC (1997) Photoluminescence properties of silicon quantum-well layers. Phys Rev B Condens Matt 55:4563
Saint-Gobain Quartz Ltd (1997) Wallsend, UK

Sawyer Research Products Inc (1994) Eastlake, USA
Skuja L (1994) Direct singlet-to-triplet optical absorption and luminescence excitation band of the twofold-coordinated silicon center in oxygen-deficient glass. J Non-Cryst Solids 167: 229–238
Skuja L (1998) The nature of optically active oxygen deficiency related centers in amorphous silicon dioxide. J Non-Cryst Solids 239:16–48
Skuja L (2000) Optical properties of defects in Silica. In: Pacchioni G, Skuja L, Griscom DL (eds) Defects in $SiO_2$ and related dielectrics: science and technology. Kluwer, Dordrecht
Skuja L, Suzuki T, Tanimura K (1995) Site-selective laser-spectroscopy studies of the intrinsic 1.9-eV luminescence center in glassy $SiO_2$. Phys Rev B 52:15208–15216
Skuja LN, Streletsky AN, Pakovich AB (1984) A new intrinsic defect in amorphous $SiO_2$ twofold coordinated silicon. Solid-State Commun 50:1069
Song KS, Williams RT (1992) Self-trapped excitons. Springer, Berlin
Stevens-Kalceff MA (2011) Cathodoluminescence microcharacterization of the radiation-sensitive defect microstructure of in situ buried oxide in silicon. J Phys D Appl Phys 44:255–402
Stevens-Kalceff MA (1998) Cathodoluminescence microcharacterization of the defect structure of irradiated hydrated and anhydrous fused silicon dioxide. Phys Rev B. Condens Matt 57:5674–5683
Stevens-Kalceff MA (2000) Electron irradiation induced radiolytic oxygen generation and micro-segregation in silicon dioxide polymorphs. Phys Rev Lett 84:3137–3140
Stevens-Kalceff MA (2001) Micromodification of silicon dioxide in a variable pressure/ environmental scanning electron microscope. Appl Phys Lett 79:3050–3052
Stevens-Kalceff MA (2009) Cathodoluminescence microcharacterization of point defects in a-Quartz. Mineral Mag 73:521–541
Stevens-Kalceff MA, Philips MR, Moon AR (1996) Electron irradiation induced changes in the surface topography of silicon dioxide. J Appl Phys 80:4308
Stevens-Kalceff MA, Phillips M (1995a) Cathodoluminescence microcharacterization of the defect structure of quartz. Phys Rev B Condens Matt 52:3122–3134
Stevens-Kalceff MA, Phillips M (1995b) Electron irradiation induced outgrowths from quartz. J Appl Phys 77:4125–4127
Trukhin AN, Jansons JL, Dyuzheva TI, Lityagina LM, Bendeliani NA (2003) Luminescence of different modifications of crystalline silicon dioxide: Stishovite and coesite. Solid State Commun 127:415–418
Tsai TE, Griscom DL (1991) Experimental evidence for excitonic mechanism of defect generation in high-purity silica. Phys Rev Lett 67:2517–2520
Vasiliev I, Chelikowsky JR, Martin RM (2002) Surface oxidation effects on the optical properties of silicon nanocrystals - art. no. 121302. Phys Rev B Condens Matt 6512:1302
Warren WL, Shaneyfelt MR, Schwank JR, Fleetwood DM, Winokur PS, Devine RAB, Maszara WP, McKitterick JB (1993) Paramagnetic defect centers in BESOI and SIMOX buried oxides. IEEE Trans on Nucl Sci 40:1755–1764
Williams RT, Song KS (1990) The self-trapped exciton. J Phys Chem Solids 51:679
Williams RT, Song KS, Faust WL, Leung CH (1986) Off-center self-trapped excitons and creation of lattice defects in alkali halide crystals. Phys Rev B Condens Matt 33:7232–7240
Wolkin MV, Jorne J, Fauchet PM, Allan G, Delerue C (1999) Electronic states and luminescence in porous silicon quantum dots: the role of oxygen. Phys Rev Lett 82:197
Wright AC (2000) Defect-free vitreous networks The idealised structure of $SiO_2$ and related glasses. In: Pacchioni G, Skuja L, Griscom DL (eds) Defects in $SiO_2$ and related dielectrics: science and technology. Kluwer, Dordrecht

# Chapter 12
# Trace Element Characteristics, Luminescence Properties and Real Structure of Quartz

**Thomas Götte and Karl Ramseyer**

**Abstract** Recent results on the cathodoluminescence (CL) and the trace element composition of quartz are the starting point to review the properties of quartz from different origin. CL-spectroscopy revealed five emission bands to be important in quartz additionally to one at approx. 340 nm which has been reported in the literature: the first one in the near-UV at 395 nm, the second in the blue range of the spectrum at 450 nm, the third at 505 nm (greenish blue), the forth at 570 nm (greenish yellow) and the last one in the red range of the spectrum at 650 nm. The bands at 395 and 505 nm are characterised by a strong decrease of intensities during irradiation while the band at 650 nm increases with increasing dose. This phenomenon is very common in quartz grown from aqueous solutions while magmatic quartz may show more stable luminescence emission. Trace element analyses display also differences in the composition between these two groups of quartz. Aluminium, Li and H have been found to be most important in authigenic, hydrothermal and metamorphic quartz but magmatic quartz is generally enriched in Ti. Germanium, Fe, B and Na is present at low levels in all quartz samples. A strong linear correlation between Al and Li indicates combined incorporation in $[AlO_4|Li^+]$-defects. A high unstable intensity at 395 nm has been observed especially in Al-rich quartz. In these samples, the luminescence commonly attenuates completely. However, different quartz samples show different correlation with Al. This result puts doubt on the interpretation that Al-related centres are the only reason for the near-UV emission. The emission band at 505 nm which also shows

T. Götte (✉)
Institute of Geosciences, Goethe-University Frankfurt,
Altenhöferallee 1, 60438 Frankfurt, Germany
e-mail: goette@em.uni-frankfurt.de

K. Ramseyer
Institute of Geological Sciences, University Bern, Baltzerstr. 1+3,
3012 Bern, Switzerland

J. Götze and R. Möckel (eds.), *Quartz: Deposits, Mineralogy and Analytics*,
Springer Geology, DOI: 10.1007/978-3-642-22161-3_12,

unstable behaviour in pegmatitic and hydrothermal quartz might also be related to trace elements, but the correlation is not well established. The increasing emission at 650 nm might be influenced by the water content of quartz, but the indicators for this interpretation are even more ambiguous than in the case of the 395 nm-band because SiOH-centres are commonly only present at very low levels. The emission bands at 450 nm, 505 nm (at least partly) and 570 nm are probably of intrinsic origin.

## 12.1 Introduction

Quartz is one of the most important and most abundant minerals in the Earth's crust and forms in a great variety of physicochemical environments in continental rocks (Fig. 12.1). Hence, it has been a matter of great interest to Earth scientists, but also to physicists and chemists. So the properties of quartz have been dealt with in numerous studies in different research fields. Although the luminescence has also been intensively investigated and used for various purposes (e.g., visualisation of growth fabrics, age dating), many details on the defect structure and luminescence centres in quartz are poorly understood.

The emission of light after excitation of solid material by means of high-energy electrons is termed cathodoluminescence (CL). In quartz, the CL ranges from near ultra-violet light (350 nm) to near infrared light (750 nm). The luminescence properties of quartz are related to the physicochemical conditions at the time of crystal growth and the history thereafter (Götze 2001). In general, the defect structure (and also the luminescence properties) of quartz precipitating from an aqueous solution will differ from that of quartz crystallized from a silicate melt or a supercritical fluid. Moreover, quartz which experienced elevated temperatures after crystallization (e.g., metamorphosis between 200 and 800°C for several tens of millions of years or $>10^{14}$ s) probably has different defect structure than unannealed quartz with a more imperfect lattice. Therefore, the luminescence properties may provide useful information about the formation and subsequent history of quartz. It is worthwhile mentioning that irradiation of quartz with high energy particles (e.g., electrons, protons, ions) has an important effect on point defects such as $[AlO_4^-|M^+]^0$ or some intrinsic defects. The excitation of luminescence by particles changes the type and concentration of point defects in quartz. This effect is well shown by EPR taken from quartz as collected in nature compared with those after annealing or irradiation (e.g., Botis et al. 2005; Götze et al. 2005).

Commonly, quartz is nearly pure $SiO_2$ and for that reason little attention has been paid to its geochemical investigation for a long time. However, recent improvements of analytical techniques with high spatial resolution and low limits of detection have stimulated numerous studies on the trace element geochemistry of quartz (Larsen et al. 2004; Landtwing and Pettke 2005; Götze et al. 2004; Larsen et al. 2009; Müller and Koch-Müller 2009; Lehmann et al. 2011;

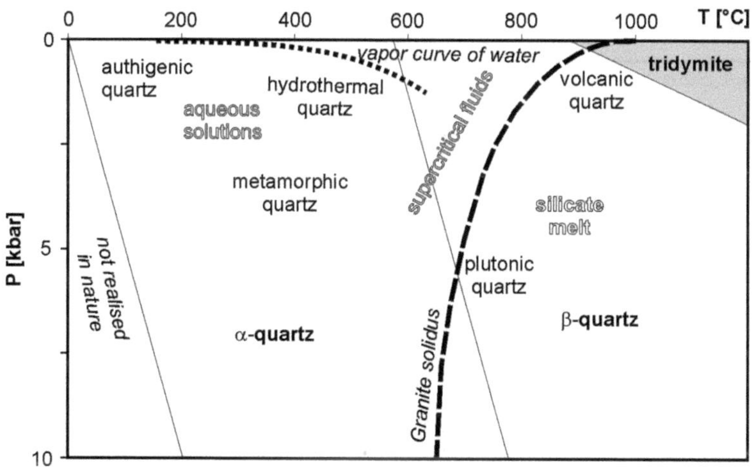


**Fig. 12.1** Phase diagram of quartz (α- and β-quartz) with different characteristic environments of quartz formation. The parental phases which are present in the Earth's crust are shown in outlined text

Götte et al. 2011). The focus has been on magmatic quartz which is commonly enriched in Ti (Larsen et al. 2004; Müller 2000; Götze et al. 2004) and on hydrothermal quartz providing higher Al-, Li-, Na- and K-concentrations (Bambauer et al. 1961; Monecke et al. 2002; Landtwing and Pettke 2005; Götte et al. 2011). But systematic variation in Ge and P-concentrations has also been reported (Larsen et al. 2004, 2008). Different mechanisms influence the trace element composition. Elements that are incompatible to most magmatic minerals such as Ge, P or B are enriched in the melt at the beginning of crystallisation and can be incorporated in quartz formed from a highly differentiated magma (Larsen et al. 2004). Quartz grown from aqueous solutions is more variable in its trace element composition and can be enriched in Al, Li and H, while Ti is commonly absent below temperatures of about 400°C. The trace element incorporation also influences the real structure and vice versa, because impurities provide additional electrons or electron traps.

This study aims at reviewing the spectral characteristics of the luminescence emission of quartz and to compare it to the trace-element composition and to intrinsic point defects. Many results reviewed in this paper are published elsewhere (Müller 2000; Monecke et al. 2002; Larsen et al. 2004; Götze et al. 2004; Landtwing and Pettke 2005; Lehmann et al. 2011; Götte et al. 2011), but some new data are included in this article. A complete list of the samples to which this work refers is given in Table 12.1. A set of seven crystalline samples from New Zealand (five metamorphic rocks and two plutonic rocks, Preusser et al. 2006) have been investigated in detail by trace element analysis (LA-ICP-MS) and cathodoluminescence (SEM-CL). A second set of 14 magmatic rocks was investigated in an optical CL-microscope, and trace elements in these samples

**Table 12.1** List of samples which where analysed for this study

| Sample | Lithology | Location | Reference |
|---|---|---|---|
| *Authigenic quartz* | | | |
| Bo 1–4 | Ruhr-sandstone | Bochum (W-Germany) | |
| WrgI 1–7 | Solling-sandstone | Würgassen (Germany) | |
| Kh | Solling-sandstone | Karlshafen (Germany) | |
| *Hydrothermal quartz* | | | |
| Gig 1b | Rock crystal | Gigerwald (Switzerland) | Götte et al. (2011) |
| Roh 2 | Rock crystal | Rohdenhaus (W Germany) | Götte et al. (2011) |
| *Metamorphic quartz* | | | |
| FJ1 | Biotite gneiss | New Zealand Alps | Preusser et al. (2006) |
| FJ2 | Augengneiss | New Zealand Alps | Preusser et al. (2006) |
| OK2 | Greenshist | New Zealand Alps | Preusser et al. (2006) |
| TAR1 | Quartz segregation | New Zealand Alps | Preusser et al. (2006) |
| TAR2 | Quartz tectonite | New Zealand Alps | Preusser et al. (2006) |
| *Magmatic quartz* | | | |
| Gt22 | Rhyolite | Saar-Nahe (Germany) | Götte and Richter (2006) |
| Gt23 | Rhyolite | Saxony (Germany) | Götte and Richter (2006) |
| Gt24 | Rhyolite | Rosia Montana (Romania) | Götte and Richter (2006) |
| Gt26 | Dacite | Saar-Nahe (Germany) | Götte and Richter (2006) |
| Gt27 | Quartz trachyte | Steiermark (Austria) | Götte and Richter (2006) |
| Gt28 | Two-mica-granite | Tumasvlakte (Namibia) | Götte and Richter (2006) |
| Gt29 | Granite | kl. Spitzkoppe (Namibia) | Götte and Richter (2006) |
| Gt30 | Granite | Jannowice (Poland) | Götte and Richter (2006) |
| Gt31 | Granite | Uludag National Park (Turky) | Götte and Richter (2006) |
| Gt32 | Granite | Günesli (Turky) | Götte and Richter (2006) |
| Gt33 | Granite | Asilomar State Park (USA) | Götte and Richter (2006) |
| Gt34 | Tonalite | Zaroów (Poland) | Götte and Richter (2006) |
| Gt35 | Granodiorite | Saxony (Germany) | Götte and Richter (2006) |
| Gt36 | Quartz diorite | Niemcza (Poland) | Götte and Richter (2006) |
| Gt37 | Diorite | Orhaneli (Turky) | Götte and Richter (2006) |
| Gt38 | Granodiorite | Pirn Creek (USA) | Götte and Richter (2006) |
| Gt40 | Rhyolite | Fréjus (France) | Götte and Richter (2006) |

were obtained by proton micro-probe. In the following sections, the cathodoluminescence properties and the trace-element composition will be briefly discussed, before the correlation of distinct emission bands with the defect structure of quartz will be reviewed.

## 12.2 Methods

### 12.2.1 CL-Microscopy and -Spectroscopy

The SEM-CL-system GATAN CL3 mounted to an EVO 50 Zeiss SEM at the Institute of Geological Sciences at the University of Bern has been used for the CL-spectroscopy of hydrothermal, authigenic and metamorphic quartz. The SEM is equipped with a grating of 150 lines/mm and a PIXIS CCD-detector for the need to detect simultaneously the entire spectral range comprising the visible, near-UV and near-IR light (i.e., 320–860 nm). The wavelength calibration was done with a Hg-lamp. A Faraday cup for accurate measurements of the beam current has been installed. An acceleration voltage of 14 kV and a beam current of about 5 nA have been used for the CL-spectroscopy and twenty spectra each with an exposure time of 10 s were recorded. The electron beam was defocussed to avoid rapid destruction of the crystal lattice. The resulting spot size was about 95 $\mu m^2$ resulting in a dose rate of 0.7 $\mu A\ mm^{-2}$.

The optical CL-microscope HC1-LM at the Institute for Geology, Mineralogy and Geophysics of the Ruhr-University Bochum has been used for the CL-analyses of 14 magmatic samples (for technical details cf. Neuser et al. 1996). The microscope is equipped with a grating spectrograph of EG&G Princeton Research Instruments comprising a grid with 150 lines/mm and a CCD-camera for recording of luminescence spectra. The spatial resolution of the spectroscope is approx. 30 μm and an acceleration voltage of 14 kV, exposure time of 20 s per spectrum and a beam current density of 5 $\mu A\ mm^{-2}$ were used for the analyses. A sequence of 10 spectra was recorded with this instrument.

A computer software was implemented to analyse the CL-spectra and non-linear least square fitting of Gaussian, Lorentzian and Voigtian curves to the recorded data. The procedure ensures optimal deconvolution of overlapping emission bands in quartz spectra. All spectra were converted into the energy-space for fitting but spectra are presented in the wavelength-space in the figures because this is more common in geosciences.

### 12.2.2 Trace Element Analyses

Trace-element composition of the 14 magmatic quartz samples was analysed by the proton micro-probe at the Dynamitron Tandem Laboratory (Ruhr-University Bochum, for technical details cf. Meijer et al. 1994). A proton energy of 1 MeV ensures optimal excitation of "light" elements (up to Fe). The detection limits are below 10 ppm for transition elements (Ti, Fe), but unfortunately much higher (approx. 100 ppm) for Al. Elements lighter than Na ($z = 11$) cannot be analysed due to a Mylar-filter in front of the EDX-detector. In particular, the inability to measure Li-concentrations as a result of this technical limitation restricts the applicability of this method to quartz analysis.

The trace elements of metamorphic, hydrothermal and authigenic quartz samples have been analysed by means of an Elan DRC-e quadrupole mass spectrometer (Perkin Elmer) which is combined with a GeoLas-Pro 2006 193 nm ArF Excimer laser system (Lambda Physik/Coherent) at the Institute of Geological Sciences of the University of Bern. All element concentrations are calibrated with the NIST 612 standard. Pit sizes of 60–120 μm have been found to be optimal for magmatic, metamorphic and hydrothermal quartz which displays homogeneous growth zonations that are large enough. Smaller pit sizes of 16–32 μm, which were chosen for the analysis of authigenic quartz cements in sandstones, result in higher standard deviations, but cannot be avoided due to zonations and small sizes of the overgrowths.

## 12.3 Luminescence and Trace-Element Characterisation of Quartz

### *12.3.1 Cathodoluminescence Spectroscopy*

Cathodoluminescence spectra of quartz are commonly composed of five broad emission bands centred at 395, 450, 505, 570 and 630 nm (in volcanic quartz) to −650 nm (in hydrothermal and metamorphic quartz; Figs. 12.2, 12.3; Götte et al. 2011). Most often, the spectrum is dominated by the emission at 390, 505 and 650 nm. Additionally, emission bands at 570 nm and at 450 nm can occur. The definite relation between the distinct emission bands depend—amongst others—from the genesis of quartz (Götte and Richter 2006; Götze 2009). The results of the fitting of CL-spectra are summarised in Table 12.2. The CL-spectra of the magmatic quartz samples display a predominantly blue emission of the bands at 450 and 505 nm, which is supplemented by an emission band at 630–650 nm (Fig. 12.2). This band increases during irradiation time. The same phenomenon can be observed in the spectra of hydrothermal, metamorphic or authigenic quartz where the increase is even stronger (Fig. 12.3, Götte 2004; Götte and Richter 2006). The intensities of the bands in the blue range of the spectrum substantially decrease with increasing irradiation dose (i.e., short-lived luminescence, Ramseyer et al. 1988; Ramseyer and Mullis 1990; Götze et al. 2005). This effect is also much more pronounced in the CL-spectra of quartz grown from aqueous solutions than in the magmatic samples but Götze et al. (2005) reported an equivalent behaviour of quartz from pegmatites. The intensity changes can be modelled by the equation

$$I = I_{inst}\, e^{-\kappa Rt} + I_{stbl}$$

for decreasing and

$$I = I_{inst}\,(1 - e^{-\lambda Rt}) + I_{stbl}$$

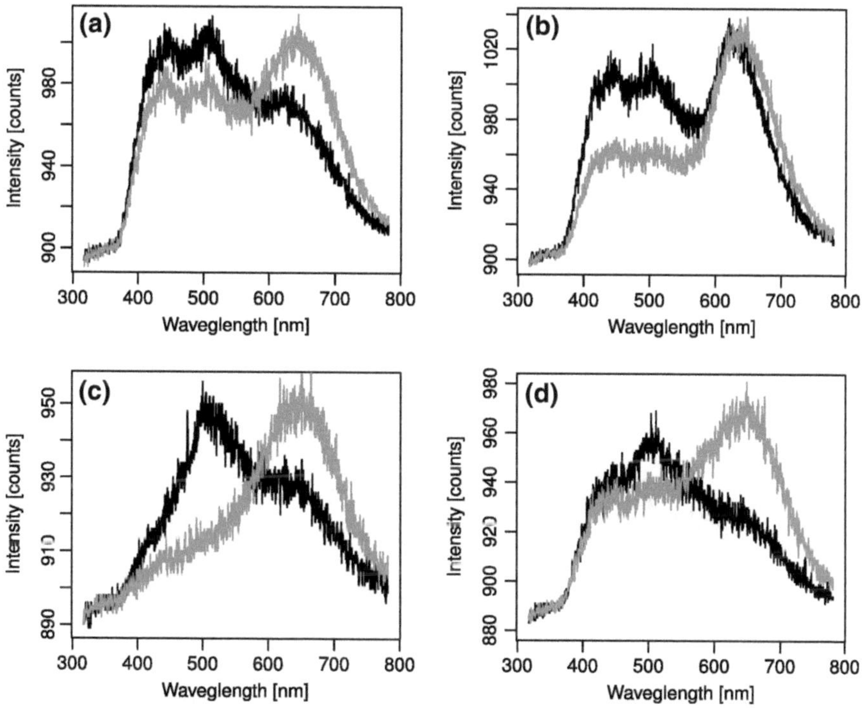


**Fig. 12.2** CL-spectra of various magmatic quartz samples (*black*: initial spectrum, *grey*: final spectrum after 200 s), **a** Quartz trachyte (Gt 27, Steiermark, Austria), **b** Rhyolite (Gt 23: Saxony, Germany), **c** Granite (Gt 29: kl. Spitzkoppe, Namibia), **d** Granite (Gt 33: Asilomar State Park, USA)

for increasing emission, respectively, with the measured intensity I, the modelled unstable part $I_{inst}$, the stable part $I_{stbl}$, the decay constant of defects $\kappa$ and $\lambda$, respectively, the dose rate $R$ (in μA $mm^{-2}$) and the irradiation time $t$ (Fig. 12.4, Götte et al. 2011). The "decay"-constants $\kappa$ and $\lambda$ depend on the technical equipment and need to be determined for each experimental set-up. For the combination of an EVO-Zeiss SEM and a Gatan CL-system, Götte et al. (2011) determined values of 31 ± 12 $mm^2$ $\mu A^{-1}$ $s^{-1}$ and 34 ± 13 $mm^2$ $\mu A^{-1}$ $s^{-1}$ for the 395 nm-band and the 505 nm-band, respectively, and 4.5 ± 3 $mm^2$ $\mu A^{-1}$ $s^{-1}$ for the 650 nm band. The slight difference in the former bands is within the range of error, but the red band increases with a significantly slower rate. This indicates that there is no common mechanism for both phenomena as suggested by King et al. (2010). The emission band at 570 nm has been found to be dominant only in the hydrothermal quartz from Rohdenhaus which shows fabrics of fast crystallisation (Götte 2004; Götte et al. 2011).

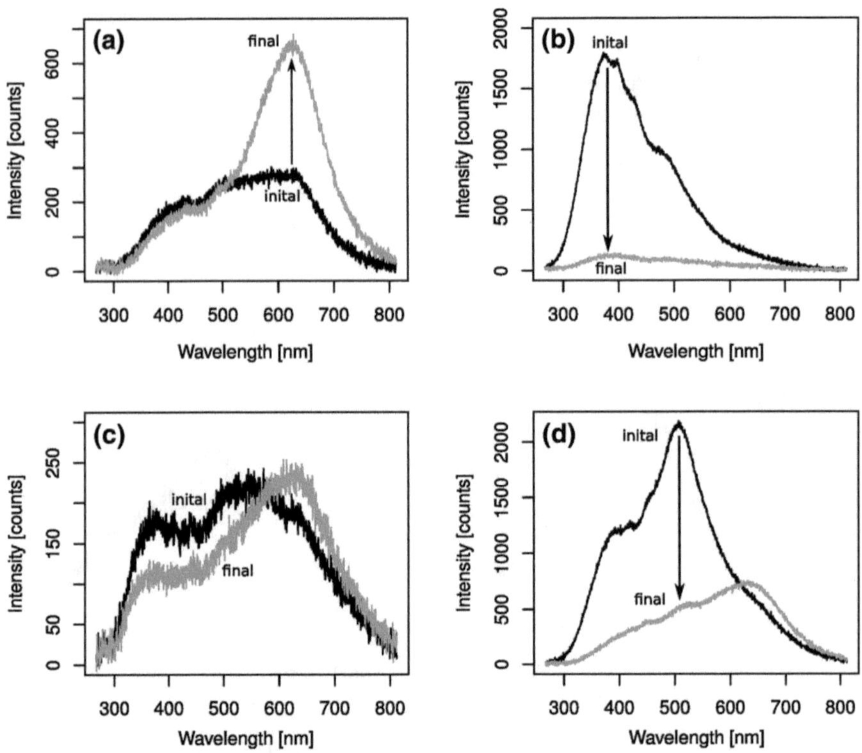


**Fig. 12.3** CL-spectra of quartz formed from aqueous solutions samples (*black*: initial spectrum, *grey*: final spectrum after 200 s), **a** quartz cement in the Finefrau-Sandstone (Upper Carboniferous, W-Germany), **b** Al-rich quartz cement in the Finefrau-Sandstone (Upper Carboniferous, W-Germany), **c** quartz cement in the Solling-Sandstone (Lower Triassic, Germany), **d** metamorphic quartz (Alps, New Zealand)

## 12.3.2 Trace Element Composition

Quartz is known to be relatively pure $SiO_2$, but traces of different mono-, tri-, tetra- or even pentavalent trace elements can be incorporated on interstitial or tetrahedral sites and form additional point defects. The most important trace elements detected in quartz are Al, Li, Na, Ti, Ge and B (Table 12.2). Additionally, H is assumed to be present but could not be analysed by LA-ICP-MS.

Al is strongly linear correlated to the sum of Na and Li in authigenic, metamorphic, hydrothermal and pegmatitic quartz ($R^2 = 0.70$–$0.98$, Fig. 12.5), i.e., in quartz formed from aqueous solutions or from Li-rich, highly differentiated melt. In contrast, the correlation is less obvious in magmatic quartz ($R^2 = 0.11$ for plutonic quartz, Fig. 12.5). Potassium and phosphorous also contribute to the balance of the positive charge deficit resulting from the incorporation of trivalent Al and B for Si (Müller and Koch-Müller 2009), but are commonly only present in small concentrations (if at all). However, it should be noted that the charge balance

**Table 12.2** Results of trace element analyses and CL-spectroscopy for the samples from Table 12.1. Standard deviations are given in brackets. Please note that Al and Li cannot be analysed with the proton microprobe

| | Li | B | Na | Al | P | K | Ti | Fe | Ge | 390 nm | | 505 nm | | 650 nm | |
|---|---|---|---|---|---|---|---|---|---|---|---|---|---|---|---|
| | | | | | | | | | | $I_{inst}$ | $I_{stbl}$ | $I_{inst}$ | $I_{stbl}$ | $I_{inst}$ | $I_{stbl}$ |
| | μmol/mol | μmol/mol | μmol/mol | μmol/mol | μmol/mol | μmol/mol | μmol/mol | μmol/mol | μmol/mol | cts/nA/s | cts/nA/s | cts/nA/s | cts/nA/s | cts/nA/s | cts/nA/s |
| *Authigenic quartz*[a] | | | | | | | | | | | | | | | |
| Finefrau Z2 | 92 (76) | 6.3 (1.1) | 16 (10) | 750 (430) | 23 (4) | n.d. | 57 (50) | 20 (13) | 1.4 (0.6) | 5.7 | 2.6 | 0.5 | 1.1 | 15 | 9 |
| Finefrau Z3 | 430 (170) | 4.7 | 19 (7) | 2,100 (370) | 28 (7) | 410 (220) | 1.4 (0.5) | 65 (27) | 1.9 (0.6) | 65 (38) | 6.5 (2.3) | 13 (8) | 2.1 (0.6) | 1.7 (6.5) | 0 (4.2) |
| Solling Z2 | 17 (10) | n.d. | 29 | 100 (29) | 17 (3) | n.d. | 8.5 | 25 | 0.7 (0.1) | 7.2 (8.1) | 5.4 (3.1) | 5.9 (3.5) | 3.6 (2) | 4.9 (1) | 4.3 (0.9) |
| Solling Z3 | 83 (23) | n.d. | 50 (24) | 300 (35) | 17 (4) | 33 (20) | 8.5 | 8.4 | 0.7 (0.2) | n.d. | n.d. | n.d. | n.d. | 8.5 | 4.7 |
| Karlshafen Z2 | 36 (27) | 20 | 58 (95) | 160 (125) | 25(5) | 128 | 23 (13) | 600 (410) | 1.[illegible] (0.2) | n.d. | n.d. | n.d. | n.d. | n.d. | n.d. |
| *Hydrothermal quartz*[a] | | | | | | | | | | | | | | | |
| Gig1b 1 | 650 (170) | 1.9 (1.1) | 2.5 (1.7) | 1,590 (510) | 95 (55) | n.d. | 0.7 (0.4) | 5 (7) | 4.8 (0.9) | 122 (29) | 12 (2.4) | 30 (6.8) | 3.9 (0.8) | 1.7 | 3.2 |
| Gig1b 2 | 82 (8) | 1.2 (0.3) | 4.5 (0.7) | 140 (19) | 46 (16) | n.d. | 0.5 (0.2) | 2 (1) | 1.2 (0.1) | 9.5 (0.3) | 2.6 (0.2) | 26 (4.8) | 4.3 (0.2) | 6.8 (0.4) | 3.2 (1) |
| Gig1b 3 | 94 (3) | 1.2 (0.2) | 2.5 (1) | 214 (9) | 34 (7) | 38 | 0.7 (0.3) | 2 (1) | 1.2 (0.1) | 37 (5.9) | 7.4 (0.9) | 51 (15) | 9.9 (0.9) | 12 (5.4) | 11 (0.7) |
| Roh2 1 | 360 (180) | 9 (3) | 35 (14) | 520 (245) | 34 | n.d. | n.d. | n.d. | 3.4 (0.9) | 6.9 | 6.6 | 15 | 9.4 | 8.6 | 13 |
| Roh2 2 | 27 (37) | n.d. | 2.7 (3.8) | 39 (49) | 37 (4) | n.d. | n.d. | n.d. | 1.8 (0.1) | 2 (0.8) | 4.8 (1.1) | 3.4 (0.2) | 5.7 (2.4) | 9.6 (4.6) | 8.9 (3.3) |
| Roh2 3 | 3,000 (250) | 40 (5) | 126 (65) | 6,800 (250) | 64 (22) | 51 | 45 | 117 | 15 (10) | 26 (7.5) | 4.1 (0.6) | 3.5 | 2.3 | 0.9 (0.5) | n.d. |
| Roh2 4 | 1,770 (490) | 13 (2.7) | 75 (22) | 2,200 (540) | 34 | 93 (59) | n.d. | n.d. | 4.5 (0.7) | 16 | 17 | 28 | n.d. | 2.7 | n.d. |
| Roh2 5 | 270 (90) | 9.5 (1.5) | 38 (10) | 420 (120) | 34 | 83 (32) | n.d. | n.d. | 4.3 (0.1) | 3.4 (0.6) | 10 (0.2) | n.d. | n.d. | 26 (3.8) | 13 (0.6) |
| *Metamorphic quartz*[a] | | | | | | | | | | | | | | | |
| FJ1 | 21 (4) | 1.8 (0.3) | 3.6 (5.7) | 69 (73) | 13 (0.9) | n.d. | 5.1 (9.8) | 190 | 0.4 (0.1) | 1.9 (0.4) | 1.8 (0.5) | 5.3 (2.3) | 2.6 (0.8) | 2.5 (0.4) | 1.4(0.01) |
| FJ2 | 23 (2) | 0.9 (0.2) | 6.2 (12.7) | 270 (560) | 11 (0.5) | 530 | 2.4 (3.1) | 14 (27) | 0.4 (0.4) | 5.5 (5.5) | 3.8 (0.9) | 15 (13) | 5.1 (1.6) | 10 (16) | 8.7 (13) |
| OK2 | 21 (1) | 5.8 (4.1) | 2.6 (0.9) | 43 (2.6) | 21 (1) | n.d. | 2.3 (0.4) | n.d. | 0.2 (0.05) | 3 (1.6) | 5.5 (1.5) | 2.6 (2.1) | 4.4 (0.8) | 9.4 (3.4) | 37 (32) |
| Tar1 | 30 (14) | 7.2 (5.7) | 3.4 (3.8) | 160 (140) | 23 (0.6) | n.d. | 0.7 (0.9) | 19 (14) | 0.2 (0.1) | 10 (7.3) | 3.4 (0.8) | 18 (12) | 6.6 (1.1) | 7.2 (1.5) | 3.5 (1.3) |
| Tar2 | 21 (16) | 7.9 (1) | 7.9 (7.6) | 57 (38) | 12 (1) | n.d. | 0.23 (0.1) | 4 (1.7) | 0.7 (0.1) | 0.6 (0.1) | 0.4 (0.1) | 0.4 (0.3) | 0.6 (0.2) | 1.6 (0.4) | 0.8 (0.2) |
| *Magmatic quartz*[b] | | | | | | | | | | | | | | | |
| Gt22 | n.d. | n.d. | n.d. | n.d. | n.d. | n.d. | n.d. | n.d. | n.d. | 0.23 | −0.01 | 0.3 | −0.03 | 0.11 | 0.27 |
| Gt23 | n.d. | n.d. | n.d. | n.d. | n.d. | n.d. | 66 | 34 | n.d. | 0.19 | 0.25 | 0.23 | 0.29 | 0.64 | 0.06 |
| Gt24 | n.d. | n.d. | n.d. | n.d. | n.d. | n.d. | 54 | n.d. | n.d. | 0.09 | 0.14 | 0.05 | 0.19 | 0.79 | 0.02 |
| Gt26 | n.d. | n.d. | n.d. | n.d. | n.d. | n.d. | 56 | n.d. | n.d. | 0.12 | −0.01 | 0.2 | 0.14 | 0.2 | 0.28 |
| Gt27 | n.d. | n.d. | n.d. | n.d. | n.d. | n.d. | 69 | 34 | n.d. | 0.21 | 0.13 | 0.34 | 0.18 | 0.29 | 0.22 |
| Gt28 | n.d. | n.d. | n.d. | n.d. | n.d. | n.d. | 23 | 30 | n.d. | 0.05 | 0.07 | 0.14 | 0.06 | 0.1 | 0.21 |
| Gt29 | n.d. | n.d. | n.d. | n.d. | n.d. | n.d. | n.d. | n.d. | n.d. | 0.12 | −0.11 | 0.08 | 0.22 | 0.13 | 0.22 |

(continued)

**Table 12.2** (continued)

| | Li | B | Na | Al | P | K | Ti | Fe | Ge | 390 nm | | 505 nm | | 650 nm | |
|---|---|---|---|---|---|---|---|---|---|---|---|---|---|---|---|
| | | | | | | | | | | $I_{inst}$ | $I_{stbl}$ | $I_{inst}$ | $I_{stbl}$ | $I_{inst}$ | $I_{stbl}$ |
| | μmol/mol | μmol/mol | μmol/mol | μmol/mol | μmol/mol | μmol/mol | μmol/mol | μmol/mol | μmol/mol | cts/nA/s | cts/nA/s | cts/nA/s | cts/nA/s | cts/nA/s | cts/nA/s |
| Gt30 | n.d. | n.d. | n.d. | n.d. | n.d. | n.d. | 29 | n.d. | n.d. | 0.13 | −0.03 | 0.16 | 0.11 | 0.12 | 0.26 |
| Gt31 | n.d. | n.d. | n.d. | n.d. | n.d. | n.d. | 33 | n.d. | n.d. | 0.21 | 0.03 | 0.34 | 0.23 | 0.12 | 0.3 |
| Gt32 | n.d. | n.d. | n.d. | n.d. | n.d. | n.d. | n.d. | n.d. | n.d. | 0.03 | 0.14 | 0.2 | 0.23 | 0.19 | 0.29 |
| Gt33 | n.d. | n.d. | n.d. | n.d. | n.d. | n.d. | 33 | n.d. | n.d. | 0.07 | 0.08 | 0.17 | 0.19 | 0.13 | 0.27 |
| Gt34 | n.d. | n.d. | n.d. | n.d. | n.d. | n.d. | 77 | n.d. | n.d. | 0.15 | −0.06 | 0.23 | −0.08 | 0.16 | 0.14 |
| Gt35 | n.d. | n.d. | n.d. | n.d. | n.d. | n.d. | 92 | n.d. | n.d. | 0.14 | −0.11 | 0.12 | −0.08 | 0.07 | 0.12 |
| Gt37 | n.d. | n.d. | n.d. | n.d. | n.d. | n.d. | 52 | n.d. | n.d. | 0.25 | 0.17 | 0.35 | 0.3 | 0.27 | 0.24 |
| Gt38 | n.d. | n.d. | n.d. | n.d. | n.d. | n.d. | 55 | 18 | n.d. | 0.15 | 0.03 | 0.23 | 0.08 | 0.1 | 0.15 |
| Gt40 | n.d. | n.d. | n.d. | n.d. | n.d. | n.d. | n.d. | n.d. | n.d. | 0.09 | 0.03 | 0.06 | 0.11 | 0.13 | 0.72 |

[a] CL-spectra by SEM-CL, trace elements by LA-ICP-MS
[b] CL-spectroscopy in an optical CL-microscope, trace elements by proton microprobe

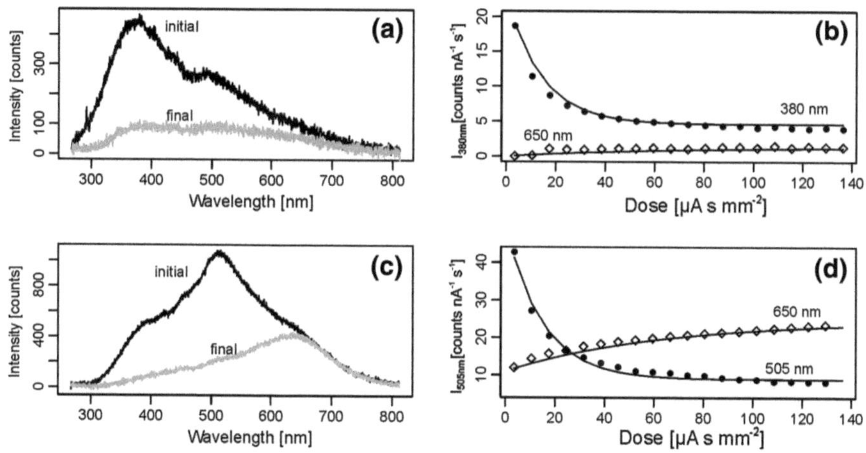


**Fig. 12.4** Decreasing luminescence in the near-UV (**a**, **b**) and *greenish blue* (**c**, **d**) range of the spectrum and increasing luminescence at 650 nm in hydrothermal quartz form Rohdenhaus (**a**, **b**) and Gigerwald (**c**, **d**, modified from Götte et al. 2011)

is not completely satisfied in most samples if only Li and Na are considered. This indicates the presence of significant amounts of AlOH-defects even in magmatic quartz. Li/Al-ratios are high (0.5–0.8) in hydrothermal and pegmatitic quartz, but are significantly lower (0.1–0.5) in magmatic, metamorphic and authigenic quartz.

Boron is commonly low in quartz with maximum values between 5 and 25 μmol mol$^{-1}$. Hydrothermal, metamorphic and pegmatitic quartz are slightly enriched in B, which shows no correlation to Al. Tourmaline, the most important B-mineral in magmatites, is a common constituent of many granitic pegmatites underlining the incompatible character of B. It is therefore enriched in pegmatitic rocks compared to plutonic or volcanic quartz. Enhanced concentrations of B can also be found in agates (Götze, pers. com.). The concentration of Fe in quartz shows non-systematic variation and is assumed to be affected by submicroscopic inclusions. Otherwise, contents as high as 190 μmol mol$^{-1}$ (in sample FJ1) cannot be explained. However, it is probably present on single μmol mol$^{-1}$ level in many magmatic or metamorphic quartz samples. Its occurrence in hydrothermal quartz probably depends on the regional geological setting.

Ti-concentration is generally below 50 μmol mol$^{-1}$ and is highest in the plutonic and volcanic quartz. Trace contents of Ti have been found in hydrothermal quartz formed in Alpine fissures at high temperature (Gig1b, ≤350°C) but it is generally absent in hydrothermal quartz, which formed at significantly lower temperatures (Roh2, ≤130°C, Table 12.2). Titanium detected in authigenic quartz cement is interpreted to indicate submicroscopic mineral inclusions or contamination with detrital quartz during the analysis (cf. Lehmann et al. 2011). It has long been known, that Ti is a common impurity in magmatic quartz but might form exsolutions of rutile during cooling. Recently, it has been found to correlate strongly with temperature (Wark and Watson 2006).

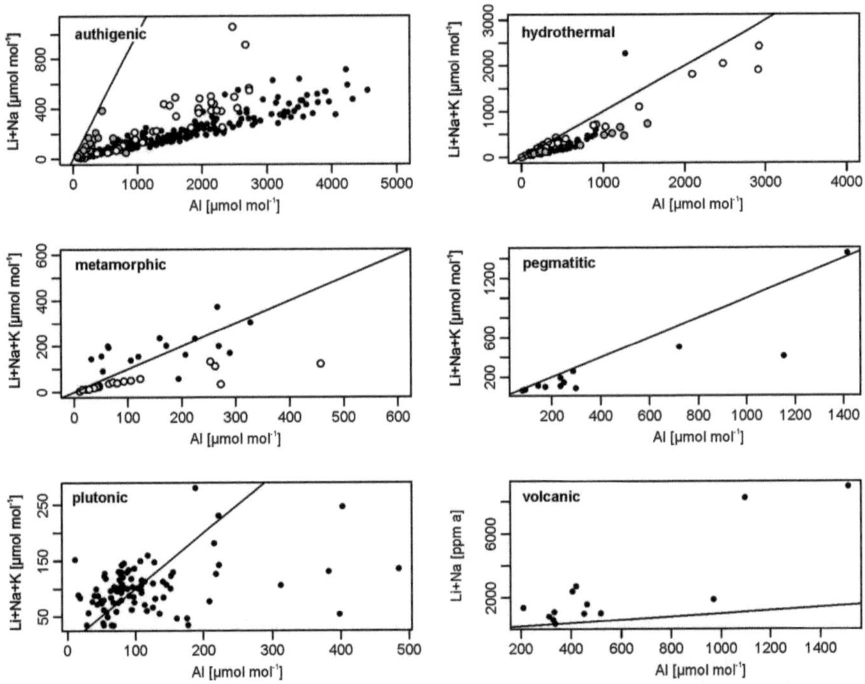


**Fig. 12.5** Correlation of monovalent cations (Li + Na + K) and Al in different quartz (*solid line*: 1:1-correlation). Authigenic quartz: unpublished data from Götte (*open circles*, *grey circles*) and data from Lehmann et al. (2011, *black dots*); hydrothermal quartz: data from Götte et al. (2011, *grey* and *white circles*) and Landtwing and Pettke (2005, *black circles*); metamorphic quartz: this study (*white circles*) and data from Monecke et al. (2002, *black circles*); pegmatitic quartz: from Götze et al. (2005); plutonic quartz: from Larsen et al. (2004); volcanic quartz: from Müller (2000)

The concentration of Ge in metamorphic quartz rarely exceeds 2 µmol mol$^{-1}$, which is close to the average of the continental crust (1.6 µmol mol$^{-1}$, Bernstein 1985). In contrast, it might be higher in hydrothermal quartz ($\leq$25 µmol mol$^{-1}$, Table 12.2). This result underlines the incompatible character of Ge in respect to quartz. In authigenic quartz, between 1.5 and 3 µmol mol$^{-1}$ Ge has been detected. A weak linear correlation exists between Ge and Al in hydrothermal and authigenic quartz while these elements seem to be more independent in pegmatitic, plutonic and metamorphic quartz. Germanium is preferentially absorbed on the surface of Fe-oxides and might be effectively removed from the melt or the solution when Fe-oxides are formed (Bernstein 1985). This might probably be the reason for the low Ge-concentration in many plutonic quartz samples (Larsen et al. 2004).

Although P is the element with the most similar ionic radius to Si, its concentration in quartz rarely exceeds some tens of µmol mol$^{-1}$ and the analysis of P poses some technical problems (Lehmann et al. 2011). Phosphorous concentrations are therefore interpreted as maximum values. Apparently, it is more enriched in pegmatitic and hydrothermal quartz than in their plutonic or volcanic counterparts. This observation is in accord with its incompatible behaviour. However, the P-concentration can be reduced if apatite crystallises contemporaneously from the melt (Larsen et al. 2008).

## 12.4 Correlation of CL-Properties to Point Defects and Trace-Element Composition

Cathodoluminescence properties of quartz have repeatedly been correlated to its trace element composition. However, we should be aware that nearly any impurity in quartz might be present in different "states" some of which are paramagnetic (i.e., with unpaired electrons) and some are not and some of which might be luminescence centres but most will probably not. One reason for the complex real structure may be the relatively strong covalent character of the Si–O-bond which provides the possibility of dissipating electric charge.

The band gap (i.e., conduction to valence band) should be similar in any quartz. Values reported in the literature (e.g., from quantum mechanical band calculations) range from 5.6 (220 nm) to 9.3 eV (135 nm, Xu and Ching 1991) and thus luminescence resulting from the direct band transition is only detectable with special equipment in the vacuum UV. Therefore, all emission in the visible, near UV and near IR is related to point defects that generate additional energy levels within the band gap. The changes in intensity of different emission bands gives strong evidence that the defect structure of quartz is very sensitive to irradiation and indicate that luminescence centres are destroyed (especially between 390 and 520 nm–3.1 and 2.8 eV; Ramseyer et al. 1988; Götze et al. 2005) or generated (at 650 nm–1.91 eV; Stevens-Kalceff and Phillips 1995; Götte and Richter 2006). Other components of the quartz luminescence seem to be stable with respect to electron irradiation. Experiments with high purity synthetic quartz indicates that numerous intrinsic defects are involved in the luminescence emission (e.g., Luff and Townsend 1990; Gorton et al. 1997), but the nature of the distinct luminescence centres is poorly understood. The contribution of trace elements to the luminescence emission of quartz is still unclear; nevertheless, impurities may play an important role for stabilising and charge compensating intrinsic defects (Stevens-Kalceff and Phillips 1995).

### *12.4.1 Ultra-Violet to Blue Luminescence Between 330 and 400 nm*

An emission band at 330–340 nm (3.7 eV) has been observed by Demars et al. (1996) and Botis et al. (2005). Demars et al. (1996) concluded from the relationship between CL intensity and the Li and Al concentration that the $[AlO_4|Li^+]$

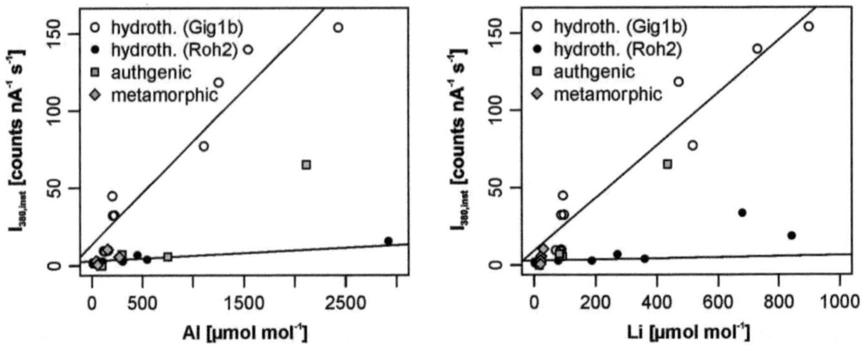


**Fig. 12.6** Linear correlation between the unstable intensity of the 390 nm-band and the Al-concentration (*left*) and the Li-concentration (*right*, *solid lines*: linear regression). Data for hydrothermal quartz from Götte et al. (2011), data of authigenic and metamorphic quartz in Table 12.2

defect centre in quartz is the luminescence centre. But since this emission band is lacking in many quartz samples with high Al and Li concentrations this interpretation is highly questionable. In contrast, Botis et al. (2005) observed a 350 nm (3.5 eV) emission in naturally irradiated quartz from uranium deposits and assumed that a silicon vacancy-hole centre is responsible for this band because the emission band and the centre disappear after annealing the quartz for 24 h at 550–600°C. However, it should be noted that these annealing conditions are similar to those of Al-centres compensated by single charged cations (Ramseyer and Mullis 1990).

Alonso et al. (1983) firstly proposed that the emission at 380 nm (3.1 eV) is attributed to an Al-defect in quartz and, indeed, the unstable part $I_{inst}$ of this band displays a good correlation with the Al-concentration in many cases (Fig. 12.6). Aluminium is incorporated in $[AlO_4|M^+]$-defects in quartz (Rossman 1994) that are destroyed by ionizing irradiation and form $[AlO_4]^0$- and $M^0$-centres, which are no luminescence centres. This model has been the preferred interpretation for short-lived emission in hydrothermal and pegmatitic quartz up to now (Perny et al. 1992; Götze et al. 2005; King et al. 2010). Perny et al. (1992) found short-lived emission (i.e., substantial decrease of the intensity during continuous electron irradiation) in hydrothermal quartz from the Swiss Alps containing more than 50 ppm Al, but without being directly correlated to the Al-concentration. Similarly, Götte et al. (2011) found different correlation for different quartz samples and proposed that only Li-compensated Al-centres may contribute to the short-lived luminescence. However, the uncertainty of the correlation is too high and the discrepancies in published data are too large to accept this correlation as a prove for the physical relationship of the $[AlO_4|Li^+]$-defect and the 380 nm-band. Luff and Townsend (1990) and Gorton et al. (1997) reported a significant emission at 380/390 nm in ultra-pure synthetic quartz. Both authors excluded the relation to an Al-defect but the observed growth related intensity change is a clear evidence of a defect incorporated during growth according to Luff and Townsend (1990).

In essence, the CL emission in the near UV and specifically the short-lived type at 390 nm, shows clear relations with $[AlO_4|M^+]^0$ defects (for more details see Ramseyer and Mullis 1990; Götze et al. 2005; Götte et al. 2011), but other defect centres may also emit in this spectral range. In the case of TL, McKeever et al. (1985) and Yang and McKeever (1990) suggest that the emission at 380 nm for the 110°C TL peak is correlated with the release of electrons from the $[GeO_4]^-$-centres and recombination with holes trapped at $[AlO_4]^0$ and $[H_3O_4]^0$-centres.

### 12.4.2 Blue to Bluish-Green Luminescence: The 450 and 505 nm-Bands

Blue luminescence with maximal emission at 450 nm is commonly observed in magmatic or high-grade metamorphic quartz formed at high temperatures. The emission at 420–490 nm (2.9–2.5 eV) is commonly assigned to irradiation induced self-trapped exciton (Tanimura and Halliburton 1986; Itoh et al. 1988; Gorton et al. 1997; Ismail-Beigi and Louie 2005). The intensity of this defect is effectively quenched at temperatures above −110°C (160 K) and is therefore not important in CL recorded at room temperature (Itoh et al. 1988). Woda et al. (2002) reported a positive relation between the concentration of $[AlO_4]^0$-centres and the radioluminescence (i.e., excitation with 624 keV electrons from a $^{137}Cs$ source) emission at 490 nm. However, $[AlO_4]^0$-centres are not luminescence centres in most CL-studies because the luminescence of Al-rich quartz breaks down completely during irradiation (Fig. 12.3b, Götte 2004; Götze et al. 2005; Götte et al. 2011), and an intrinsic origin of this emission band seems to be more probable. This is in accord with the recognition of a luminescence band at 460 nm in vitreous $SiO_2$ which is proposed to be correlated with distinct oxygen vacancies (ODC(II), Skuja 1998; Stevens-Kalceff 2009).

The luminescence centres that are responsible for the emission at 505 nm are poorly known. Commonly, it is unstable and decays with similar rates as or slightly faster than the 380 nm-band. This short-lived CL emission has been reported from hydrothermal (Ramseyer and Mullis 1990) and pegmatitic quartz (Götze et al. 2005) but can also be found in metamorphic quartz (Fig. 12.3d). The latter authors observed a strong luminescence at 505 nm in pegmatitic quartz which was enriched in trace elements but displayed no significant amounts of lattice defects, and correlated this emission with $[GeO_4^-|M^+]$-defects, but the correlation is poor. However, the unstable character indicates that the corresponding centre is strongly sensitive to ionizing irradiation as it is common for impurity-related defects in quartz.

Most authors propose an intrinsic origin for the (stable) emission at approx. 500 nm (e.g., Stevens-Kalceff and Philipps 1995; Stevens-Kalceff 2009). The 505 nm-band is also assigned to self-trapped excitons by some authors (e.g., Itoh et al. 1988) whose luminescence is quenched at temperatures above −110°C

(160 K) and only plays a minor role at room or elevated temperatures (Itoh et al. 1988). A broad emission band with several side-bands has been observed in luminescence spectra of silica nano-particles. This band has been assigned to SiH-groups at the surface (Glinka et al. 1999). A convincing correlation of these emission bands with a distinct defect in the bulk crystal has not been found up to now and some further research is required in this case. However, it should be noticed that different defects might contribute to the luminescence at approx. 500 nm.

### 12.4.3 Greenish-Yellow Luminescence at 580 nm

A predominant emission band at 580 nm (2.1 eV) is rarely observed and is restricted to uncommon growth conditions in a hydrothermal environment or during silicification of anhydrite, fossil wood and sediments. Götze et al. (1999) reported a yellow luminescence which was associated with an increased emission at 580 nm in agate containing unusual high amounts of $E_1'$ centres. However, Schilles et al. (2001) found no evidence that E′-centres contribute radiatively to this emission. Luff and Townsend (1990) assumed that the 570 nm emission is related to STE-trapping at substituting Ge-impurities. Götte and Richter (2003) and Götte (2004) found high intensities at 570 nm in hydrothermal quartz exhibiting a greenish-yellow luminescence. Due to the internal fabric (e.g., enrichment of primary fluid inclusions, patchy luminescence), they assumed that this quartz has been rapidly formed. In contrast, blue luminescing growth zones with predominant emission at 440 nm in the same quartz crystals are suggested to be formed more slowly.

### 12.4.4 Brown to Reddish Luminescence Between 630 and 650 nm

An emission in the red range of the spectrum (637–650 nm, 1.91–1.94 eV, Stevens-Kalceff and Phillips 1995) is observed in many spectra of quartz of different origin. Commonly, the intensities increase during irradiation and the band is assigned to non-bridging oxygen hole centres (NBOHC, Stevens-Kalceff and Phillips 1995; Götze 2001). It forms from precursor defects such as hydroxyl- (Si–OH) or peroxy- (O–O) linkages by breaking off bonds, diffusing away hydrogen (above 120 K, Skuja and Naber 1997) and forming electron holes at one oxygen ion by ionizing irradiation (Stevens-Kalceff and Phillips 1995; Götze et al. 1999; Glinka et al. 2000). This formation is in accord with the observed increase in the 650 nm peak intensity during irradiation specifically in quartz grown from aqueous solutions in low-grade metamorphic, diagenetic or hydrothermal conditions (Götze et al. 1999; Götte and Richter 2006). This model seems to be

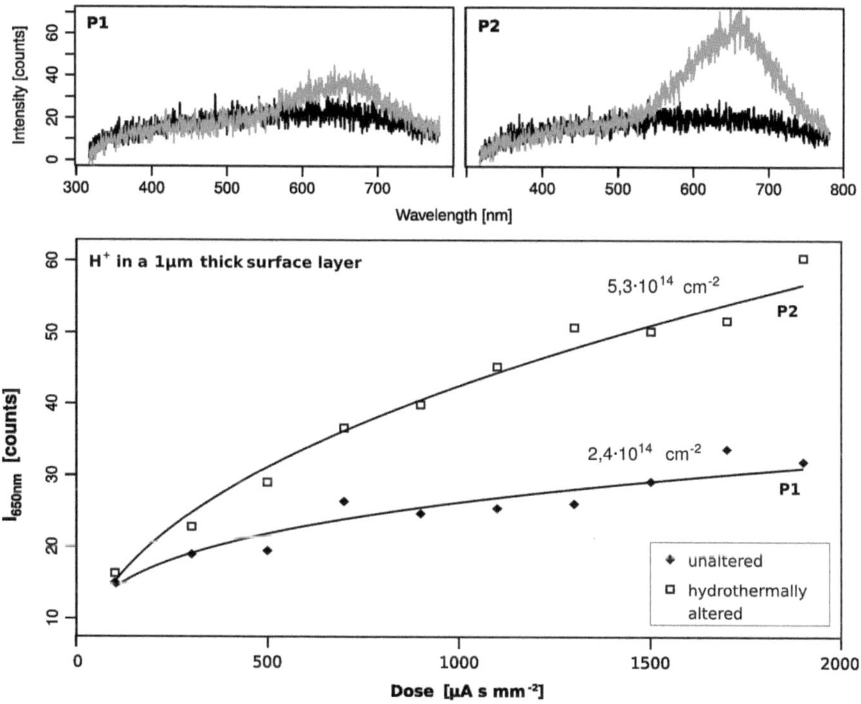


**Fig. 12.7** Increasing intensity at 650 nm with increasing dose in unaltered (P1) and hydrothermally altered synthetic quartz (P2, *black*: initial, *grey*: final). Data from Dersch (2001)

confirmed by diffusion experiments of water in quartz, where hydrothermally altered samples of synthetic quartz display a significantly higher increase than unaltered samples (Fig. 12.7, Dersch 2001; Götte 2004). But, Götte et al. (2011) found no correlation between the concentration of SiOH-defects and the unstable intensity of this particular emission band. The free energy of SiOH-defects is relatively high (Rosa et al. 2005), generally resulting in low concentrations of these defects except in poorly crystallized, fast grown zones where it is probably present in defect clusters (Stenina 2004; Götte et al. 2011). However, it is remarkable that the increase shows distinct differences in different quartz. Al rich quartz, for example, never has a prominent emission in the red range of the spectrum, probably due to quenching of the $[AlO_4]^0$-defects that are formed by irradiation. Strong increase of the 650 nm-band has been observed especially in quartz that formed form aqueous solutions at lower temperatures (e.g., Al-poor hydrothermal, authigenic and low grade metamorphic quartz). In contrast, the luminescence of magmatic quartz is often more or less stable.

In silica nano-particles, the 650 nm band has also been found to increase after heating to 1,100°C in photoluminescence experiments due to thermal degradation of SiOH defects and production of NBOHC at the surface (Glinka et al. 1999,

2000). Kuzuu et al. (1993) assigned the emission at 650 nm and a related absorption band at 260 nm (4.8 eV) in silica glass to defects which are related to oxygen excess (free $O_2$-molecules as well as NBOHC). Recently, Botis et al. (2005) argued that the high thermal stability (up to the α-quartz—tridymite transformation temperature of 867°C) of the centre emitting at 650 nm is unrelated with the presence of NBOHC's as the latter centre is only stable below 500°C whereas the emission band has been detected even in quartz that was heated to 600°C. But however, this is not necessarily contradictory because NBOHC might be generated from peroxy-defects which are still present at this temperature during the measurement.

A strong relationship between the luminescence at 650 nm and irradiation by α-particles has repeatedly been reported. In irradiation experiments, yellowish to reddish luminescing zones or halos are produced which show increased intensities at 650 nm (Komuro et al. 2002; Krickl et al. 2008). In naturally irradiated samples, quartz crystals show "rims" of yellowish to brown luminescence adjacent to the radioactive material (e.g., Götze 2009). However, experiments with proton beams revealed a more complex interaction with the defect structure, because other bands (e.g., at 580 nm) are also affected.

An additional band was reported in some cases in the near infrared at 710 nm (1.74 eV, Kempe et al. 1999). It is likely caused by the incorporation of $Fe^{3+}$.

In essence, CL is a valuable tool to decipher the origin and thermal history of quartz. Furthermore, some of the luminescence centres are partly related to trace elements commonly incorporated in natural quartz (i.e., Al, Li, H, Fe, etc.).

## 12.5 Conclusions

Cathodoluminescence investigation and trace element analyses revealed very complex properties of natural quartz. Obviously, the presence of impurities in the quartz lattice affects—at least partly—the luminescence emission. These interactions lead to a large variety of different luminescence colours in different quartz samples. The most important results can be summarised as follows:

- Luminescence spectra of quartz are composed of emission bands at 395, 450, 505, 570 and 650 nm. All emission bands have high width at half maximum resulting in a strong overlap of distinct bands. Deconvolution therefore is the first step for a reliable interpretation.
- The emission bands at 395, 505 and 650 nm are unstable and show decreasing intensity (395, 505 nm) or increasing intensity (650 nm) with ongoing electron irradiation.
- Impurity-related point defects are common in quartz, although the trace element content is generally low. Aluminium, Li and H are the most prominent elements incorporated in quartz which grew at low temperatures from an aqueous

solution, Ti is important in magmatic and high-grade metamorphic quartz (T > 400°C). Sodium, B, Ge, and Fe are commonly present at trace levels.
- Many impurity-related point defects can be modified by irradiation. Each defect may provide additional energy levels in the band gap but a confidential assignment of distinct defects to the emission bands is difficult. The UV-emission at 395 nm seems to be affected by Al-defects, water-related defects may be partly responsible for the increase of the 650 nm-emission.

**Acknowledgments** We wish to thank Thomas Pettke (Bern) and Jan Meijer (Bochum) for technical support with the LA-ICP-MS and the proton microprobe, respectively. Frank Preusser (Bern) kindly provided crystalline quartz samples from New Zealand. We gratefully acknowledge the financial support from the German Research foundation (DFG, Go 1089/3-1). We also thank an anonymous reviewer, Jens Götze (Freiberg), and Robert Möckel (Freiberg) for their suggestions on an earlier draft of the manuscript.

## References

Alonso PJ, Halliburton LE, Kohnke EE, Bossoli RB (1983) X-ray-induced luminescence in crystalline $SiO_2$. J Appl Phys 54(9):5369–5375

Bambauer HU (1961) Spurenelemente und $\gamma$-Farbzentren in Quarzen aus Zerrklüften der Schweizer Alpen. Schweiz Miner Petrogr Mitt 41:335–369

Bernstein LR (1985) Germanium geochemistry and mineralogy. Geochim Cosmochim Acta 49:2409–2422

Botis S, Nokhirn SM, Pan A, Xu Y, Bonli Th, Sopuck V (2005) Natural radiation-induced damage in quartz. I. correlations between cathodoluminescence colors and paramagnetic defects. Can Mineral 43:1565–1580

Demars C, Pagel M, Deloule E, Blanc P (1996) Cathodoluminescence of quartz from sandstones: Interpretation of the UV range by determination of trace element distributions and fluid-inclusion P-T-X properties in authigenic quartz. Am Mineral 81(7–8):891–901

Dersch O (2001) Wasseraufnahme von Quarz: Grundlage für eine Methode zur Datierung archäologischer Quarzartefakte. Ph.D. Thesis Goethe-University Frankfurt, 250 S, Frankfurt am Main, Germany

Glinka YD, Lin S-H, Chen Y-T (1999) The photoluminescence from hydrogen-related species in composites of $SiO_2$ nanoparticles. Appl Phys Lett 75(6):778–780

Glinka YD, Lin S-H, Hwang LP, Chen YT (2000) Photoluminescence from mesoporous silica: similarity of properties to porous silicon. Appl Phys Lett 77(24):3968–3970

Gorton NT, Walker G, Burley SD (1997) Experimental analysis of the composite blue cathodoluminescence emission in quartz. J Lumin 72–4:669–671

Götte T (2004) Petrographische und geochemische Untersuchungen zu den postvariszischen Mineralisationen im devonischen Massenkalk des nordwestlichen rechtsrheinischen Schiefergebirges unter besonderer Berücksichtigung der Kathodolumineszenz, Ph.D. Thesis, University of Bochum, Bochum, Germany, pp 186

Götte Th, Richter DK (2003) Late Palaeozoic and Early Mesozoic hydrothermal events in the northern Rhenish Massif: results from fluid inclusion analyses and cathodoluminescence investigations. J Geochem Explor 78–79:531–535

Götte Th, Richter DK (2006) Cathodoluminescence characterization of quartz particles in mature arenites. Sedimentology 53(6):1347–1359

Götte Th, Pettke Th, Ramseyer K, Koch-Müller M, Mullis J (2011) Cathodoluminescence properties and trace element signature of hydrothermal quartz: A fingerprint of growth dynamics. Am Mineral 96:802–813

Götze J (2001) Cathodoluminescence microscopy and spectroscopy in applied mineralogy. Freiberger Forschungshefte C 485:1–128

Götze J (2009) Chemistry, textures and physical properties of quartz—geological interpretation and technical application. Mineral Mag 73:645–671

Götze J, Plötze M, Fuchs H, Habermann D (1999) Defect structure and luminescence behaviour of agate—results of electron paramagnetic resonance (EPR) and cathodoluminescence (CL) studies. Mineral Mag 63(2):149–163

Götze J, Plötze M, Graupner T, Hallbauer DK, Bray CJ (2004) Trace element incorporation into quartz: a combined study by ICP-MS, electron spin resonance, cathodoluminescence, capillary ion analysis, and gas chromatography. Geochim Cosmochim Acta 68:3741–3759

Götze J, Plötze M, Trautmann T (2005) Structure and luminescence characteristics of quartz from pegmatites. Am Mineral 90:13–21

Ismail-Beigi S, Louie SG (2005) Self-trapped excitons in silicon dioxide: Mechanism and properties. Phys Rev Lett 95:156401-1-4

Itoh C, Tanimura K, Itoh N (1988) Optical studies of self-trapped excitons in $SiO_2$. J Phys C-Solid State Phys 21(26):4693–4702

Kempe U, Götze J, Dandar S, Habermann D (1999) Magmatic and metasomatic processes during formation of the Nb-Zr-REE deposits Khaldzan Buregte and Tsakhir (Mongolian Altai): indications from a combined CL-SEM study. Mineral Mag 63(2):165–177

King GE, Finch AA, Robinson RAJ, Hole DE (2010) The problem of dating quartz 1: spectroscopic ionoluminescence of dose dependence. Radiat meas 46:1–9

Komuro K, Horikawa Y, Toyoda S (2002) Development of radiation damage halos in low-quartz: cathodoluminescence measurement after $He^+$ ion implantation. Mineral Petrol 76:261–266

Krickl R, Nasdala L, Götze J, Grambole D, Wirth R (2008) Alpha-irradiation effects in $SiO_2$. Eur J Mineral 20:517–522

Kuzuu N, Matsumoto Y, Murahara M (1993) Characteristics of ArF-eximer-laser-induced 1.9 eV emission bands in type-III and soot-remelted silicas. Phys Rev B 48(19):6952–6956

Landtwing MR, Pettke T (2005) Relationships between SEM-cathodoluminescence response and trace-element composition of hydrothermal vein quartz. Am Mineral 90(1):122–131

Larsen RB, Henderson I, Ihlen PM, Jacamon F (2004) Distribution and petrogenetic behaviour of trace elements in granitic pegmatite quartz from South Norway. Contrib Mineral Petrol 147:615–628

Larsen RB, Jacamon F, Sørensen B (2008) Petrogenetic significance of trace elements in igneous quartz. 33rd International Geological Congress, Oslo, 2008

Larsen RB, Jacamon F, Kronz A (2009) Trace element chemistry and textures of quartz during the magmatic hydrothermal transition of Oslo Rift granites. Mineral Mag 73(4):691–707

Lehmann K, Pettke T, Ramseyer K (2011) Significance of trace elements in syntaxial quartz cement, Haushi Group sandstones, Sultanate of Oman. Chem Geol 280:47–57

Luff BJ, Townsend PD (1990) Cathodoluminescence of synthetic quartz. J Phys-Condens Matter 2(40):8089–8097

Meijer J, Stephan A, Adamczewski J, Bukow HH, Rolfs C, Pickart T, Bruhn F, Veizer J (1994) PIXE microprobe for geoscience applications. Nucl Instrum Methods B89:229–232

McKeever SWS, Chen CY, Halliburton LE (1985) Point-defects and the predose effect in natural quartz. Nucl Tracks Radiat Meas 10(4–6):489–495

Monecke T, Kempe U, Götze J (2002) Genetic significance of the trace element content in metamorphic and hydrothermal quartz: a reconnaissance study. Earth Planet Sci Lett 202:709–724

Müller A (2000) Cathodoluminescence of defect structures in quartz with applications to the study of granitic rocks. Ph.D Thesis, University Göttingen, Göttingen, p 229

Müller A, Koch-Müller M (2009) Hydrogen speciation and trace element contents of igneous, hydrothermal and metamorphic quartz from Norway. Mineral Mag 73(4):569–583

Neuser RD, Bruhn F, Götze J, Habermann D, Richter DK (1996) Kathodolumineszenz: methodik und anwendung. Zentralblatt für Geologie und Paläontologie, Teil I 1995:287–306
Perny B, Eberhardt P, Ramseyer K, Mullis J, Pankrath R (1992) Microdistribution of Al, Li, and Na in α-quartz—possible causes and correlation with short-lived cathodoluminescence. Am Mineral 77(5–6):534–544
Preusser F, Ramseyer K, Schlüchter Ch (2006) Characterisation of low OSL intensity quartz from the New Zealand Alps. Radiat Meas 41:871–877
Ramseyer K, Baumann J, Matter A, Mullis J (1988) Cathodoluminescence colors of quartz. Mineral Mag 52(368):669–677
Ramseyer K, Mullis J (1990) Factors influencing short-lived blue cathodoluminescence of α - quartz. Am Mineral 75(7–8):791–800
Rosa AL, El-Barbary AA, Heggi MI, Briddon PR (2005) Structural and thermodynamic properties of water related defects in quartz. Phys Chem Mineral 32:323–331
Rossman GR (1994) Colored varieties of the Silica minerals. In: Heaney PJ, Prewitt CT, Gibbs GV (eds) Silica—physical behaviour, Geochemistry and materials applications. Reviews in Mineralogy, vol 29, pp 433–367
Schilles T, Poolton NRJ, Bulur R, Bøtter-Jensen L, Murray AS, Smith GM, Riedi PC, Wagner GA (2001) A multi-spectroscopic study of luminescence sensitivity changes in natural quartz induced by high-temperature annealing. J Phys D-Appl Phys 34(5):722–731
Skuja L (1998) Optically active oxygen-deficiency related centers in amorphous silicon dioxide. J Non-Cryst Solids 239:16–48
Skuja L, Naber A (1997) Laser-induced luminescence in glassy $SiO_2$ and neutron-irradiated alpha-quartz: three types of non-bridging oxygen hole centres. Mater Sci Forum 239–241:25–28
Stenina NG (2004) Water related defects in quartz. Bulletin Geosci 79:251–268
Stevens-Kalceff MAS, Phillips MR (1995) Cathodoluminescence microcharacterization of the defect structure of quartz. Phys Rev B 52(5):3122–3134
Stevens-Kalceff MAS (2009) Cathodoluminescence microcharacterization of point defects in quartz. Mineral Mag 73:585–605
Tanimura K, Halliburton LE (1986) Polarization of the x-ray-induced blue luminescence in quartz. Phys Rev B 34(4):2933–2935
Wark DA, Watson EB (2006) TitaniQ: a titanium-in-quartz geothermometer. Contrib Mineral Petrol 152:743–754
Woda C, Schilles T, Rieser U, Mangini A, Wagner GA (2002) Point defects and the blue emission in fired quartz at high dose: a comparative luminescence and EPR study. Radiat Prot Dosim 100(1–4):261–264
Xu Y, Ching WY (1991) Electronic and optical properties for all polymorphic forms of silicon dioxide. Phys Rev B 44:11048–11059
Yang XH, McKeever SWS (1990) Point-defects an the predose effect in quartz. Radiat Prot Dosim 33(1–4):27–30

# Chapter 13
# Mineralogy, Geochemistry and Cathodoluminescence of Authigenic Quartz from Different Sedimentary Rocks

**Jens Götze**

**Abstract** Authigenic quartz is present in different sedimentary rocks of North-Eastern Germany. Single crystals of euhedral quartz were detected in the Permian (Zechstein) salt deposit of Roßleben, in quartz nodules within Triassic sandstone layers (*Chirotherien* sandstone, Bunter) from Jena, and Tertiary lignite deposits in the Leipzig region (Zwenkau, Cospuden). Mineralogical and geochemical investigations revealed that the authigenic quartz crystals from the different geological units differ in morphology (habit), characteristic inclusions, trace-element geochemistry and cathodoluminescence properties. Accordingly, the results allow not only to clearly distinguish between authigenic and detrital quartz, but also between authigenic quartz from different sedimentary environments. Authigenic quartz from Zechstein salt deposits shows characteristic euhedral forms dominated by rhombohedral faces or a combination of rhombohedral and prism faces, and mineral inclusions (halite or anhydrite) in dependence on the saliniferous facies. The crystals exhibit a blue luminescence, which can be related to a broad emission band at 450 nm. The authigenic quartz crystals from the Bunter sandstone are often intergrown, forming aggregates of several mm up to cm in size. At least three growth zones can be distinguished: spherulithic growth starting from calcite inclusions, quartz with complex internal CL structure, and a homogeneous outer zone with no visible luminescence. The second zone exhibits a cathodoluminescence pattern similar to that of agate with three emission bands at 650, 580 and 450 nm. Authigenic quartz from Tertiary lignites is characterized by doubly terminated crystals with prism and rhombohedral faces. Intergrowth of two or more crystals was observed. The CL is dominated by a transient emission band at 650 nm, which increases in intensity during electron irradiation. The crustal

J. Götze (✉)
Institute of Mineralogy, TU Bergakademie Freiberg,
Brennhausgasse 14, 09596 Freiberg, Germany
e-mail: goetze@mineral.tu-freiberg.de

J. Götze and R. Möckel (eds.), *Quartz: Deposits, Mineralogy and Analytics*, 287
Springer Geology, DOI: 10.1007/978-3-642-22161-3_13,

signature of all quartz REE distribution patterns and high contents of Al and Fe indicate the origin of the silica-bearing fluids from weathering solutions and do not show any influence of hydrothermal fluids. On the other hand, elevated concentrations of Na, K, Mg, Ca, and B can probably be related to the influence of saliniferous fluids during quartz precipitation. Although the specific physicochemical conditions may have been different for the various occurrences, the data suggest a formation of the authigenic quartz crystals during early diagenesis.

## 13.1 Introduction

Quartz is one of the most important minerals in the earth's crust occurring in large amounts in sedimentary rocks. Macrocrystalline quartz, chalcedony as well as poorly and non-crystalline silica (opal-A, opal-CT) play a central role in the composition and diagenesis of sediments. Therefore, investigations of quartz are extensively used for the evaluation of provenance in arenites (e.g., Zinkernagel 1978; Zuffa 1985; Seyedolali et al. 1997; Bahlburg and Floyd 1999; Götze and Zimmerle 2000; Boggs et al. 2002; Richter et al. 2003; Bernet and Basset 2005). On the other hand, secondary quartz neophormism in form of quartz overgrowth cements, fracture fillings and silicification of fossil remains is a common feature in sedimentary and diagenetic environments.

In particular authigenic quartz cement plays a major role in controlling the hydraulic properties and quality of reservoir sandstones (e.g., McBride 1989; Bjørlykke and Egeberg 1993; Wordan and Morad 2000). The application of cathodoluminescence (CL) to sandstone petrology by Sippel (1968) and Zinkernagel (1978) first revealed the spectacular difference between detrital quartz and authigenic quartz cements. More recent studies showed that the amount of diagenetic quartz may be quantified by image analysis (Evans et al. 1994). The combination of CL with other analytical methods can provide important information about polyphase quartz cementation and the burial history of sandstones (e.g., Sippel 1968; Houseknecht 1991; Hartmann et al. 2000). For instance, CL microscopy with fluid inclusion studies is ideal to correlate quartz cements with the formation of other authigenic minerals or dissolution events, and to gain information concerning temperature and salinity of fluids during precipitation (e.g., Burley et al. 1989; Walderhaug 1990, 1994). On the other hand, timing of cementation and evaluation of temperature and precipitation mechanisms is supported by high precision in situ $\delta^{18}O$ analysis (e.g., Lyon et al. 2000; Hiatt et al. 2007; Kelly et al. 2007). Authigenic quartz plays also an important role in more intensively tectonized sandstones with crack-seal structures during processes of brittle deformation and healing (e.g., Milliken and Laubach 2000; Markowitz and Milliken 2003).

In contrast to the widespread occurrence and significance of ordinary authigenic quartz overgrowths, isolated euhedral quartz crystals seem to be a relatively uncommon form of diagenetic silica formation in sediments. Authigenic quartz formation is described in different geological formations and from various sedimentary environments. A few publications document the occurrence of authigenic quartz crystals in soil (Dixon and Weed 1989), limestones and carbonates (Black 1949; Richter 1971; Molenaar and deJong 1987; Mišik 1995; Chavetz and Zhang 1998; Liu et al. 2004; Evans and Elmore 2006), sulphate rocks (Richter 1971; Friedman and Shukla 1980), salt deposits (Grimm 1962; Nachsel 1969; Sedletskyi 1971; Fabricius 1987; Fruth and Blankenburg 1992), bituminous coal and lignite (Baker 1946; Hoehne 1954; Leskevich 1959; Ruppert et al. 1985; Soong and Blattner 1986; Botz et al. 1986; Fruth and Blankenburg 1992) or oil deposits and bitumen veins (Füchtbauer 1961; Parnell et al. 1996). In most of these investigations, inclusions and/or isotope data have been used for the reconstruction of diagenetic conditions.

Considering these results, the present study focused on the comparison of authigenic quartz crystals from contrasting sedimentary environments. Therefore, the investigation includes authigenic quartz from Zechstein salt deposits, nodular quartz from Triassic sandstones, and single authigenic quartz crystals from Tertiary lignite seams. The different sedimentary quartz occurrences were analyzed and compared concerning their characteristic mineralogical and geochemical properties, and the conditions of formation.

## 13.2 Materials and Methods

Authigenic quartz from different geological formations was selected for the present study (see Fig. 13.1). The first sample set originates from the Zechstein salt deposit Roßleben south of the Harz Mountains (Heynke and Zänker 1970). Authigenic quartz crystals occur within the main potassium salt seam *K2 Staßfurt* in hard salt layers, which are distributed between the primary carnallitite and the impoverishment (Fig. 13.1). According to Nachsel (1969) they form various morphologies in dependence on the salt mineralization. Therefore, two types of quartz crystals were selected from the kieseritic and anhydritic hard salt, respectively. The quartz crystals were hand picked under a binocular microscope from the insoluble residues of the salt minerals.

A second sample set was taken from Triassic sandstones of the Middle Bunter. Several small outcrops of these sandstones were available during building activities in the city of Jena (Thuringia). Within the so called *Chirotherien sandstone* (Solling formation; Langbein 1974—Fig. 13.1), isolated nodules of clear authigenic quartz crystals occur. The nodules are some mm up to cm in size and partially associated with concretions of reddish chalcedony (carneol) and carbonate. Such nodular quartz is highly unusual in siliciclastic rocks and in contrast to the common homoaxial authigenic quartz overgrowths.

The quartz samples of Tertiary age were taken from lignite seams in the open pits of Zwenkau and Cospuden within the brown coal region south of Leipzig (Fig. 13.1). Within the *Böhlener Oberflöz* (Lower Oligocene), several layers of

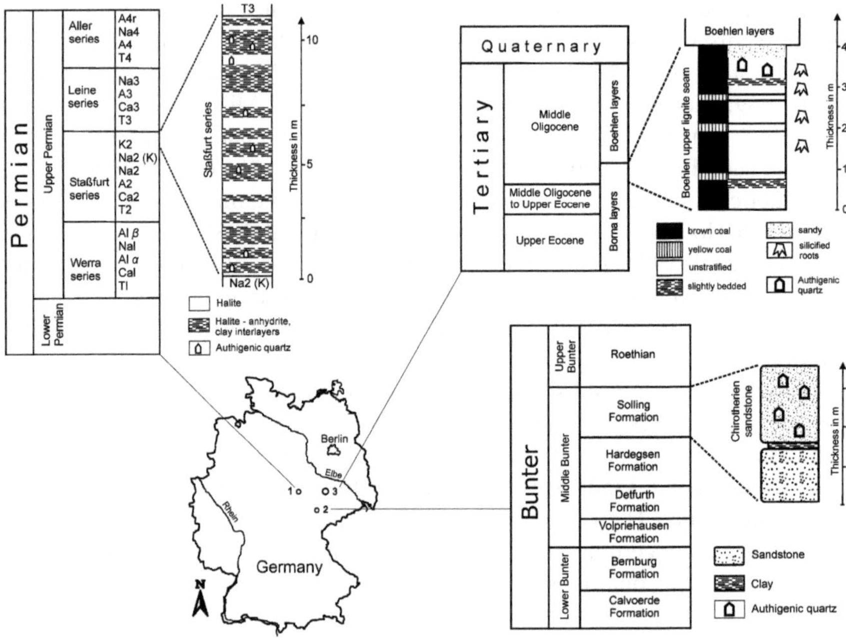


**Fig. 13.1** Location of investigated quartz samples in different sedimentary units of North-eastern Germany and schematic profiles of the relevant sediment units (stratigraphic classification according to Heynke and Zänker 1970; Langbein 1974; Bellmann 1986). *1* Zechstein salt, Roßleben, *2* Bunter sandstone, Jena, *3* Tertiary lignites, Zwenkau and Cospuden

silicified lenses and roots occur, which consist of small euhedral quartz crystals (Bellmann 1986). These crystals are also distributed within the surrounding coal. After careful crushing, washing and sieving of the quartz containing lignite samples, the authigenic quartz crystals were hand picked under a binocular microscope.

The selected quartz crystals of all samples were cleaned by ultrasonic agitation in distilled water and prepared for the different analytical methods. For SEM studies, single crystals were arranged on a sample holder. Polished thin sections were made for polarizing and CL microscopy.

The analytical procedure started with polarizing microscopy using a ZEISS Axio Imager A1 m microscope. The material was additionally studied by SEM (JEOL 6400 with EDX detector) to detect variations in grain size and morphology of the quartz crystals and identify microinclusions. CL measurements were done on carbon-coated, polished thin sections using a "hot cathode" CL microscope HC1-LM (Neuser et al. 1995). The system was operated at 14 kV accelerating voltage and a current density of about 10 $\mu A/mm^2$. Luminescence images were captured "on-line" during CL operations using a peltier-cooled digital video-camera (KAPPA 961-1138 CF 20 DXC). CL spectra in the wavelength range 380–900 nm were recorded with an Acton Research SP-2356 digital triple-grating spectrograph with a Princeton Spec-10 CCD detector that was attached to the CL microscope by a silica-glass fibre guide. CL spectra were measured under standardized conditions (wavelength calibration by a Hg-halogen lamp, spot width 30 μm, measuring time 10 s).

Separated aliquots of the quartz samples were investigated for trace-element composition by ICP-MS analysis. The sample material was carefully crushed and again separated by hand picking under a binocular microscope. The separated fractions were treated with distilled water to remove adhering particles and then air dried. 400–500 mg milled sample material ($\leq$30 μm grain size) were digested in a glassy carbon vessel with 5 ml concentrated HF and 3 ml concentrated $HNO_3$ at 50°C and analyzed using a Perkin Elmer Sciex Elan 5000 quadrupole instrument with a cross-flow nebuliser and a rhyton spray chamber. Details of the preparation and analytical conditions are summarized in Monecke et al. (2000).

Investigations by XRD and Micro-Raman were carried out to check the samples concerning other silica modifications than $\alpha$-quartz. X-ray powder diffraction analysis was carried out by means of an XRD 7 diffractometer using Cu-K$\alpha$ radiation and a secondary graphite monochromator. Samples were scanned with $2\theta$ step sizes at step times of 10 s per step. The Rietveld algorithm BGMN was used for refinement. Raman spectra were obtained by means of a JOBIN–YVON T 64000 Raman spectrometer with OLYMPUS microscope and macro-sample chamber. Spectra were excited by the 514.5 nm line of an $Ar^+$ laser with 0.5–1.5 mW beam power.

## 13.3 Results and Discussion

### *13.3.1 Mineralogy*

Comparison of authigenic quartz from the different sedimentary environments revealed significant differences in size and morphology of the crystals. The insoluble residues from the Zechstein salt contain a number of different minerals mainly including quartz, clay minerals, sulphates, boracite, and pyrite (Fruth and Blankenburg 1992). The size of the quartz crystals varies from some hundreds μm up to about 2 mm. The isolated single crystals are mainly of euhedral shape without visible twinning or intergrowth (Fig. 13.2a, b). The morphology of the authigenic quartz is variable, dominated by the rhombohedra and prism faces. The dependence of the morphology on the parent salt type was already reported by Grimm (1962) and Nachsel (1969). In the kieseritic hard salt, rombohedral faces dominate (resembling the habit of high-temperature volcanic quartz), whereas the quartz crystals from the anhydritic hard salt additionally show well developed prism faces (Fig. 13.2a, b). Sometimes, the quartz crystals may show incomplete crystal faces and/or features of corrosion (Fig. 13.3a).

Investigations by SEM, polarizing and CL microscopy revealed that the authigenic quartz crystals contain no identifiable nucleus. On the other hand, the different

**Fig. 13.2** SEM micrographs of typical morphologies of the investigated authigenic quartz samples. **a** Isolated euhedral quartz crystal from the kieseritic hard salt of the Zechstein salt deposit Roßleben (Germany); the crystal has dominant rhobohedral faces and almost no prism faces; **b** Authigenic quartz crystal from the anhydritic hard salt of the Zechstein salt deposit Roßleben (Germany) showing a combination of rhombohedral and prism faces; **c/d** Aggregate of authigenic quartz from a temporary outcrop of the *Chirotherien* sandstone (*Middle Bunter*) in Jena (Thuringia, Germany); the close-up (**d**) of the outer part shows well developed crystal faces; **e** Intergrowth of two isolated euhedral quartz crystals from the Tertiary lignite deposit Zwenkau (Germany); **f** Authigenic quartz crystals from the Tertiary lignite deposit Cospuden (Germany) with typical intergrowth (twinning?) of individual crystals

saliniferous environments are reflected in the mineral inclusions within the authigenic quartz crystals. Whereas the quartz from the kieseritic hard salt only contains halite inclusions, the quartz crystals from the anhydritic hard salt exclusively inherit microinclusions of calcium sulphate.

The authigenic quartz from the Bunter sandstone has a completely different appearance. In contrast to common diagenetic quartz overgrowths in sandstones, the crystals of nodular quartz may reach several mm or even cm in size, and isolated euhedral crystals are more or less absent. Instead, the crystals are intergrown and form aggregates (Fig. 13.2c, d). Polarizing microscopy revealed that the quartz shows zonal growth (Fig. 13.3c). Crystallization started with spherulitic growth and the formation of chalcedonic quartz, later changing into well developed crystals. A similar sequence has been described for quartz in silicified evaporites (e.g., Milliken 1979; Maliva 1987). The different stages are clearly detectable by CL microscopy. Minute inclusions of calcite are included in the central part of the aggregates. The quartz sample was checked by XRD and micro-Raman analyses, and only α-quartz was found as silica modification. This is in contrast to the associated reddish chalcedony (carneol), where moganite could be detected as a second silica phase besides α-quartz.

The authigenic quartz from the Tertiary lignite seams mainly consists of doubly terminated euhedra, varying from some 100 μm to some mm in size. Within the coal, the crystals are small and isolated, whereas also thin lenses and layers of clusters and aggregates exist. Again, the crystals contain no detectable nucleus and the morphology is dominated by prism and rhombohedral faces. The majority of crystals shows elongated forms with well developed prism faces (Fig. 13.2e, f), although few crystals with subordinate prism faces were also found. Moreover, twinning and intergrowth of euhedral crystals is common. Investigations in transmitted light revealed that inclusions of well preserved organic material are common (Fig. 13.3d, e) and microinclusions of calcite sometimes occur.

### *13.3.2 Cathodoluminescence*

Investigations by CL microscopy and spectroscopy showed that the quartz crystals from the different sedimentary environments show significant differences in their luminescence behaviour. Walker and Burley (1991) concluded that different luminescence characteristics of authigenic quartz in sediments may reflect different conditions of formation.

The authigenic quartz crystals from the saline environment exclusively exhibit a dark blue luminescence colour (Fig. 13.3a, b). Inclusions of halite (kieseritic hard salt) and anhydrite (anhydritic hard salt), respectively, are clearly detectable due to their brighter luminescence. Spectral CL measurements revealed that the CL emission consists of a single broad band in the blue centred at ca. 450 nm (2.75 eV—Fig. 13.4a). This luminescence emission can be related to the twofold coordinated silicon on an oxygen vacancy (Fitting et al. 2001). Other typical CL emission bands of quartz are lacking.

In contrast, the luminescence of the nodular quartz from the Bunter sandstone is more complex. At least three different growth zones are detectable with CL. The core of the quartz aggregates exhibits a yellow-orange CL with inclusions of

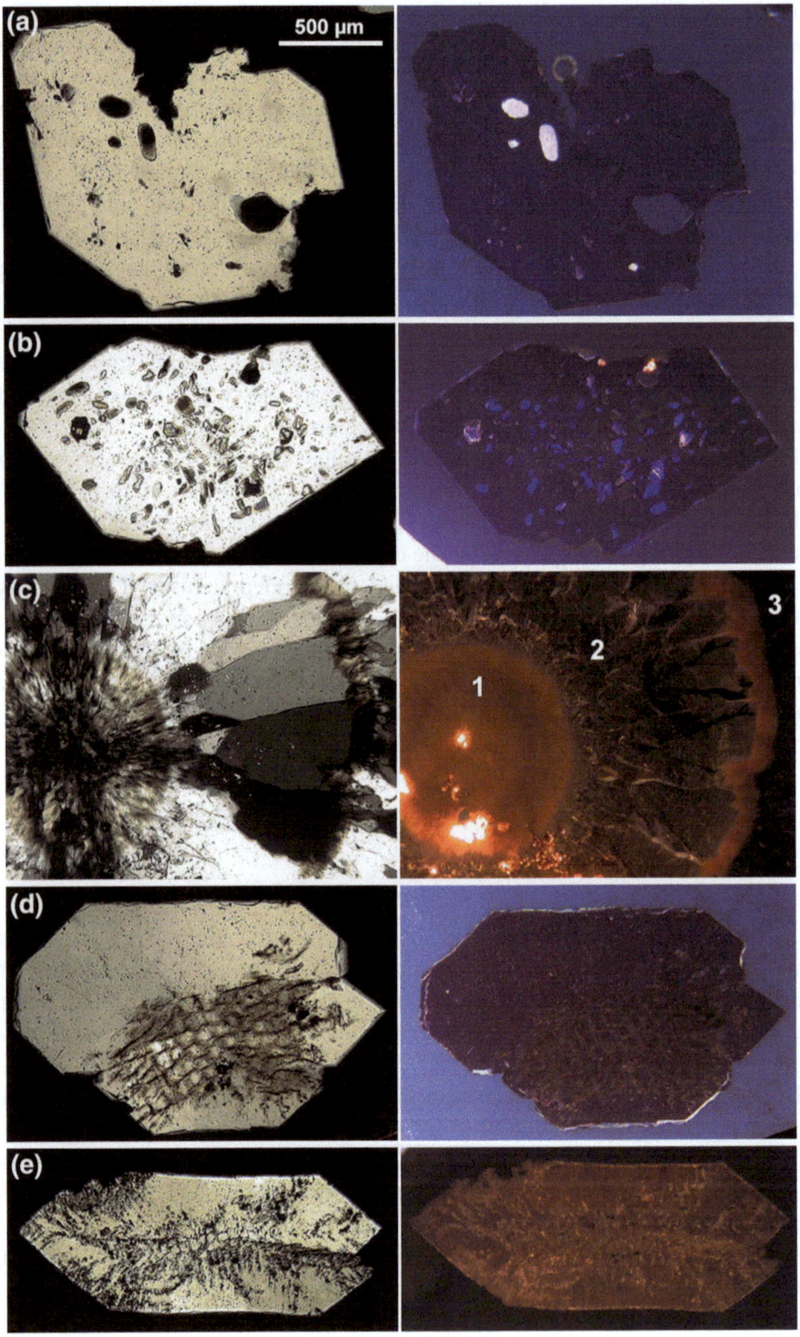
(a)
500 µm
(b)
(c)
1
2
3
(d)
(e)

◀ **Fig. 13.3** Pairs of micrographs from polarizing and cathodoluminescence microscopy of the authigenic quartz samples (uniform scale bar for all Figures). **a** Quartz crystal from the kieseritic hard salt of the Zechstein deposit Roßleben; the quartz has *dark blue* CL and brightly luminescing microinclusions of halite; the euhedral shape of the crystal is slightly corroded. **b** Authigenic quartz from the anhydritic hard salt of the Zechstein deposit Roßleben; the quartz shows *dark blue* CL; inclusions of anhydrite are detectable by their brighter luminescence. **c** Quartz aggregate from the Middle Bunter sandstone of Jena; at least three growth zones are detectable: (*1*) a central part with spherulitic growth and microinclusions of calcite, (*2*) a second zone with *yellow-brownish* CL and a complex internal structure, and (*3*) an outer zone of well crystallized quartz without visible luminescence. **d** Intergrowth of two authigenic quartz crystals from the Tertiary lignite deposit Zwenkau; the quartz exhibits an initial CL colour of *brownish-violet*; residuals of organic matter with preserved cell structures are visible; **e** Authigenic quartz from the Tertiary lignites of Cospuden with brownish luminescence after ca. 1 min of electron irradiation

brightly luminescent calcite (Fig. 13.3c). The microinclusion of calcite in the centre of the spherulitic quartz core may have acted as nucleus for the quartz crystallization. The second zone with a yellowish CL shows a complex internal structure of the quartz crystals. The outer zone has only very dull CL indicating low defect concentration (slow growth).

CL spectra were only available from the second zone. The CL of the core was influenced by the brightly luminescing calcite and in the outer zone the CL intensity was too low for spectral measurements. The associated CL spectra of the second zone are very complex and consist of at least three main emission bands (Fig. 13.4b), which is similar to CL spectra of agate (Götze et al. 1999). The spectrum is dominated by an emission band at ca. 650 nm and a second one with a maximum at ca. 580 nm. The 450 nm band in the blue is relatively weak.

The 650 nm (1.91 eV) CL emission band can be related to the non-bridging oxygen hole centre (NBOHC; Siegel and Marrone 1981). Possible precursors of this defect centre are Si–OH silanol groups from the mineralizing fluid. The yellow luminescence band at 580 nm (2.14 eV) can probably be associated with an E′ centre (electron defect on an oxygen vacancy—Götze et al. 1999). The common occurrence of the 580 nm emission band in quartz can possibly be related to high concentrations of defect centres due to rapid growth. The predominance of these defects in the authigenic quartz indicates a rapid crystallization from a silica-oversaturated solution.

The authigenic quartz samples from Tertiary lignite deposits show a different CL behaviour compared to the other occurrences. The quartz grains show a transient CL with initial violet colours changing into browns during electron irradiation (Fig. 13.3d, e). The spectra consist of a dominant 650 nm emission and a weak emission band in the blue (Fig. 13.4c, d). Time-resolved spectral CL measurements reveal a typical increase of the 650 nm emission, which is probably caused by the conversion of precursors (e.g., Si–OH silanol groups) into the non-bridging oxygen centres due to electron bombardment. This is a common phenomenon of quartz crystallized from aqueous solutions (Götze et al. 2001a).

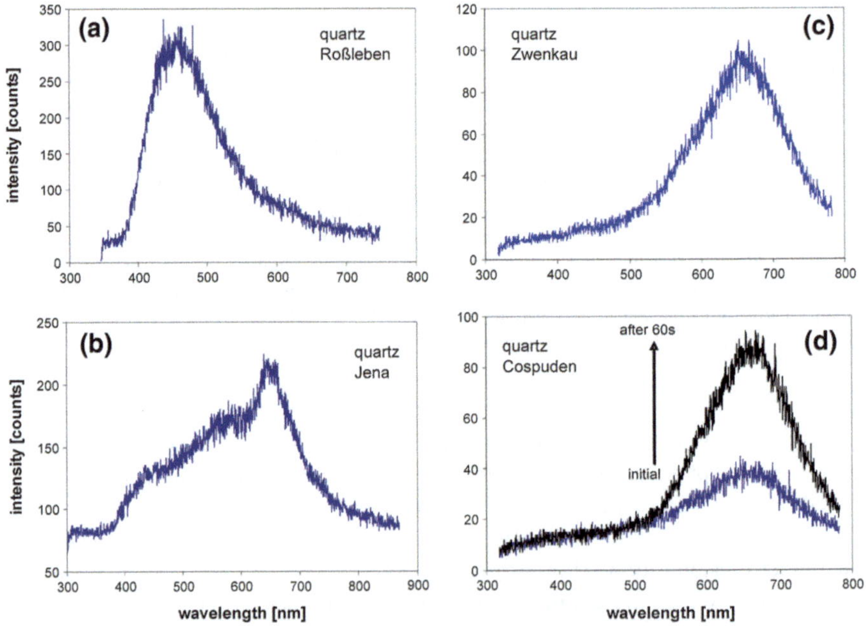


**Fig. 13.4** Cathodoluminescence spectra of investigated authigenic quartz samples. **a** The quartz from the Zechstein salt has a characteristic blue emission caused by a broad emission band at 450 nm. **b** The quartz from Jena (Bunter sandstone) shows a complex CL spectrum with at least three emission bands: a main band at 650 nm and two additional bands at 580 and 450 nm. **c/d** The quartz crystals from Tertiary lignite deposits (Zwenkau, Cospuden) have a dominant emission band at 650 nm. Time-resolved CL measurements (**d**) revealed that the 650 nm luminescence emission increases during electron irradiation

## *13.3.3 Geochemistry*

The trace-element data of the investigated quartz samples are summarized in Table 13.1. Because of the limited sample material of quartz from the salt deposits, analyses could only be realized for the quartz from the kieseritic hard salt.

In general, trace elements in quartz are either incorporated into the crystal lattice or contained in fluid or mineral inclusions. Because of the small ionic radius and the high charge of the $Si^{4+}$ ion, only a few elements are substitutionally incorporated into quartz. The most common ion is $Al^{3+}$; the structural incorporation of Ge, Ti, Ga, Fe or P (partially with charge compensating cations) was also proved (Weil 1984). Other elements are preferentially bound on microinclusions (Götze et al. 2004).

The contents of most trace elements in the investigated authigenic quartz samples are low. Some trace elements (e.g., Ga, Pb, Sn, V, Co, Cr, Mn, Nb, Ni, Zr) were not detectable in all analyzed quartz samples. Only some specific elements are enriched, which reflect both the specific physico-chemical environment of

**Table 13.1** Trace element contents (ppm) of authigenic quartz from different occurrences

| | QR | QCh | QZ | QCo |
|---|---|---|---|---|
| Al | 770 | 21.9 | 104 | 79.4 |
| Au | 0.043 | – | – | |
| B | 10 | 68 | 17 | nd |
| Ba | <7.7 | 2.83 | 7.33 | 6.13 |
| Ca | 146 | 351 | 580 | 28.1 |
| Cs | 0.108 | – | – | |
| Fe | 105 | 103 | 45.7 | 8.8 |
| Ge | nd | 2.10 | 0.04 | – |
| K | 1240 | 26.5 | 40.7 | 53.1 |
| Li | 60 | 0.97 | 0.80 | 3.57 |
| Mg | 640 | 6.09 | 15.1 | 8.21 |
| Na | 7360 | 53.0 | 88.2 | 43.1 |
| Rb | 1.11 | 0.05 | 0.11 | 0.06 |
| Sr | nd | 0.86 | 0.89 | 0.26 |
| Ti | nd | 7.24 | 1.88 | 9.26 |
| Th | 0.075 | | 0.08 | – |
| U | <0.08 | 1.32 | 0.02 | |
| Y | nd | 0.0448 | 0.1103 | 0.0490 |
| La | 1.020 | 0.0226 | 1.8929 | 0.3711 |
| Ce | <0.29 | 0.0695 | 2.6064 | 0.4364 |
| Pr | – | 0.0051 | 0.2412 | 0.0406 |
| Nd | – | 0.0163 | 0.2894 | 0.0401 |
| Sm | 0.038 | 0.0041 | 0.0435 | 0.0046 |
| Eu | <0.013 | 0.0013 | 0.0046 | 0.0009 |
| Gd | – | 0.0056 | 0.0259 | 0.0048 |
| Tb | 0.006 | 0.0012 | 0.0044 | 0.0010 |
| Dy | – | 0.0059 | 0.0211 | 0.0061 |
| Ho | – | 0.0016 | 0.0048 | 0.0018 |
| Er | – | 0.0051 | 0.0139 | 0.0064 |
| Tm | – | 0.0009 | 0.0018 | 0.0011 |
| Yb | 0.034 | 0.0053 | 0.0099 | 0.0070 |
| Lu | <0.009 | 0.0011 | 0.0015 | 0.0013 |

(–) = below detection limit; (nd) = not determined
QR—Zechstein salt deposit Roßleben; QCh—Bunter sandstone Jena
QZ—Tertiary lignite Zwenkau; QCo—Tertiary lignite Cospuden

formation and typical microinclusions, respectively. Al and Fe are more or less frequent in all quartz types. Assuming that the silica-rich solutions leading to the quartz formation derive from the weathering and alteration of silicate minerals, these elements may have also been enriched and trapped in the quartz crystals.

The quartz from the Zechstein salt (kieseritic hard salt) shows, in particular, elevated concentrations of the alkali and alkali earth elements K, Li, Na, Ca and Mg. This is in accordance with the observed fluid and mineral microinclusions. Fruth and Blankenburg (1992) reported a dominance of $MgCl_2/CaCl_2$-solutions

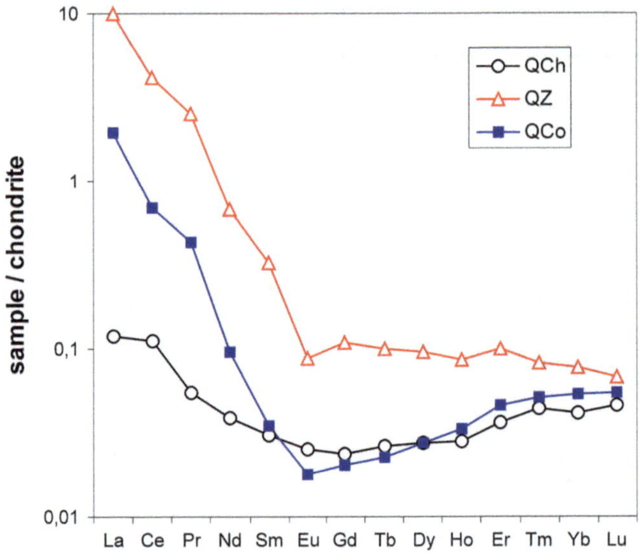


**Fig. 13.5** Chondrite-normalized REE distribution patterns of authigenic quartz samples from different sediments (normalization according to data of Mason 1979). QCh—Bunter sandstone, Jena; QZ—Tertiary lignites, Zwenkau; QCo—Tertiary lignites Cospuden

within the primary fluid-inclusions of quartz from the Zechstein salt deposits. Therefore, these elements were entrapped during the authigenic quartz formation within the saline environment. This conclusion is also supported by the elevated content of B, since authigenic boracite was found together with quartz in the insoluble residues of the salts. Additionally, soluble metal salts (e.g., Fe, Al, Au) could be transported and precipitated simultaneously with Si.

Looking at the trace-element composition of the authigenic quartz from nodules in the Bunter sandstone, at first the elements Al, Ca, Na and K are contained in elevated concentrations. Assuming a silica source from $SiO_2$-saturated pore fluids, these elements could probably originate from the silica supply during alteration and dissolution of silicate minerals (e.g., feldspar minerals). After release and migration within the silica bearing solutions, these elements could be incorporated into the authigenically forming quartz. The contents of some specific trace elements (e.g., B, Ge, U) are also remarkable. These elements show a similar geochemical behaviour like Si and thus, can be transported in the fluid together with silicon. This is interesting in so far as these trace elements are also characteristic elements in the composition of agates (Götze et al. 2001b).

The trace-element contents in the two samples from Tertiary lignites are in general low, except the already mentioned enrichment in Al and Fe. Elevated contents of Ca can be related to the occurrence of minor inclusions of calcite. Additionally, the elements Mg and Na are preferentially concentrated in fluid inclusions. Highly saline $MgCl_2$(NaCl)-solutions were detected in authigenic quartz crystals from Tertiary lignite (Fruth and Blankenburg 1992).

Figure 13.5 shows the chondrite-normalized REE distribution patterns of the quartz from Bunter sandstone and Tertiary lignites. Unfortunately, the REE contents in the saliniferous quartz were too low to obtain a REE distribution pattern.

The quartz from the sandstone has very low concentrations of REE. The normalized distribution is slightly enriched in light REE (LREE) but also shows a tendency of increasing heavy REE (HREE). The distribution can be explained by a crustal signature overprinted by an enrichment of HREE. Assuming that $HCO_3^-$ bearing fluids are present, the preferred complexation and transport of released HREE, and the following trapping during crystallization of quartz can explain this signature (Wood 1990). The existence of carbonate inclusions confirms the possibility of such interpretation. With that, the REE distribution pattern is different from that of marine silica (e.g., chert), which shows a typical sea water signature with a distinct negative Ce anomaly (Götze 1998). A marine origin of the silica can therefore be excluded.

The chondrite normalized REE distribution patterns of the two lignite quartzes show a similar shape, and differ only in the absolute REE concentrations. They both have a typical crustal signature with enriched LREE and a slight negative Eu anomaly, indicating an origin from terrestrial weathering solutions. The slight enrichment of the HREE in the sample from Cospuden may be explained by the same processes of preferred complexation by $HCO_3^-$ bearing solutions like in the sandstone sample. This idea can be supported by minute calcite inclusions in some of the quartz crystals from this locality.

### *13.3.4 Sources of Silica and Conditions of Quartz Formation*

Although the specific conditions of formation may have been different for the various quartz types, there are also similar features. In general, the formation should have taken place within the sediment in an early stage of diagenesis, when the permeability is high enough to enable the silica-bearing solutions to circulate within the pore space or in fissures within the sediment. The silica in the pore fluids may derive from the weathering and decomposition of silicate minerals or siliceous organic material. There is no indication (e.g., REE data) that silica-bearing hydrothermal fluids are involved in the formation of quartz in the investigated rocks. The formation of macrocrystalline quartz requires low $SiO_2$ concentrations (orthosilicic acid—$H_4SiO_4$), since otherwise chalcedony or opal will be precipitated.

#### 13.3.4.1 Zechstein Salt

The average concentration of silicic acid in ocean water is about 0.5–3 mg/l. During evaporation, the concentration of $SiO_2$ in solution can rise. An inflow of silica-bearing weathering solutions and the supply of detrital $SiO_2$ and silicates into the brine can further increase the silica concentration. Under the alkaline and anaerobic conditions, quartz and alumosilicates are dissolved or altered into Mg-rich sheet silicates, respectively, resulting in a release of silica. The solubility of $SiO_2$ in ocean water, which is normally in the range between 100 and 120 mg/l can significantly be increased in Ca–Na mineralized solutions up to >500 mg/l $SiO_2$ (Sedletskiy 1971). The release of $SiO_2$ can proceed until early diagenesis due to the dissolution of biogenic opal or volcanic glass just below the sediment/water interface.

The silica-bearing solutions migrate via pores and fissures over long distances until they reach saturation and crystallize. The observation of euhedral clear quartz crystals without any chalcedony or opal indicates a direct crystallization of the quartz crystals from the fluids with relatively low silica concentration. The preferred existence of orthosilicic acid is favoured by the high pH ($\geq$8) in the brines and the fact that $Cl^-$ ions can depolymerize the silica bearing solutions, leading to a stabilization of $H_4SiO_4$ (Sedletskyi 1971).

The precipitation of quartz can be initiated by oversaturation due to evaporation of the mother fluid or lowering of the pH from alkaline conditions to neutral values. Furthermore, ions of Mg or Al can decrease the solubility of silica (Sedletskyi 1971). Grimm (1962) additionally suggested the role of evaporite minerals (especially sulphates) in the nucleation of quartz crystallization. The frequent occurrence of inclusions of salt minerals within the authigenic quartz crystals indicates that processes of replacement of parent salt minerals by quartz can play a role in the formation process. Results of CL microscopy showed that the internal structure of the quartz crystals is homogeneous. Therefore, the crystals result from a continuous crystallization process from a mother fluid with relatively low silica content. The detection of oxygen vacancies clearly points to the anoxic conditions in the brine. The elevated concentrations of metals such as Fe or Au within the authigenic quartz can be related to the simultaneous precipitation from the saliniferous fluids.

The theory of a syngenetic or early diagenetic silica accumulation in evaporites is supported by the temperatures of quartz formation. Fruth and Blankenburg (1992) detected average homogenization temperatures of fluid inclusions of 76°C in authigenic quartz of kieseritic hard salt and 61°C within the anhydritic hard salt. These temperatures can be interpreted as primary temperatures of crystallization, because temperatures between 60 and 70°C were observed in recent salt lakes (Sedletskiy 1971) or may occur as post-depositional sub-surface temperatures.

#### 13.3.4.2 Bunter Sandstone

To interpret the possible silica sources of the quartz formation within the Bunter sandstone, the geological situation has to be considered. The *Chirotherien* sandstone was deposited in the transition zone between a sandy and marine clayey facies (Langbein 1974). The quartz and carneol-bearing horizon is characterized by the typical occurrence of carbonate concretions, which were formed under a semi-arid to arid climate. The formation of red chalcedony (carneol) is associated with the replacement of carbonate nodules by silica during the highly saline Roethian transgression (Langbein 1974). The mixing of saline water with silica-bearing pore fluids resulted in the dissolution/replacement of carbonate and precipitation of silica. The trace element data indicate that the $SiO_2$ in the pore solutions derives from the alteration and decomposition of silicate material in the sediment; no hydrothermal silica input could be detected.

Whereas most of the silica rapidly crystallized as chalcedony (carneol) and chalcedonic pore cement, lowering of the $SiO_2$ concentration favoured the crystallization of euhedral quartz crystals in cavities. The microscopic investigations showed that the formation of quartz took place in a multi-step process. Starting on a nucleus of carbonate, spherulithic quartz crystallized first. The formation of chalcedony is promoted both by the replacement of pre-existing carbonate and probably high silica concentrations. The following phanerocrystalline quartz zone seems to be the prosecution of chalcedony "fibres". The high defect density (characteristic 580 nm CL emission band) and the occurrence of sector zoning detected by CL may result from rapid crystallization under non-equilibrium conditions. Only the outer part of the quartz aggregates probably crystallized directly from an undersaturated fluid with respect to silica.

Milliken (1979) discussed similar crystallization sequences of quartz from silicified evaporates. She concluded that microcrystalline quartz formed earliest in fluids related to sea water, whereas megaquartz crystallized later from meteoric fluids. Applied to the quartz nodules from the Bunter sandstone, the changing quartz microtexture could be an indication for a changing ratio of pore fluids of saline and meteoric origin, and decreasing silica content.

In conclusion, specific microchemical conditions were necessary for the formation of the euhedral authigenic quartz crystals.

#### 13.3.4.3 Tertiary Lignite

In general, the authigenic quartz crystals in lignite seams should have formed by precipitation and crystallization from silica-rich solutions during chemical and/or biochemical processes. Earlier investigations of Walther and Götze (1994) showed a negative correlation between the ash content of Tertiary lignites and the amount of authigenic quartz. They concluded that the precipitation of silica is favoured during the time of decreased clastic input and intense chemical weathering in the surrounding area of the brown coal swamps. Moreover, the formation, abundance, and mineralogical features of authigenic quartz are influenced by the hydrophilic or hydrophobic properties of organic matter and the complexing by organic compounds (e.g., Spiro and Rozeson 1982; Bennett and Siegel 1987; Bennett 1991). Whereas certain organic acids can accelerate the dissolution and increase the solubility of quartz in water, a hydrophilic behaviour should support the quartz crystallization in solution. This is in accordance with observations of Schneider (1986), who found an early stage of authigenic quartz formation within the residues of plant material from Tertiary lignites.

The perfect euhedral shape of most authigenic quartz crystals in lignites, the absence of any detrital nucleus and the inclusions of organic matter and calcite support the theory of an authigenic formation within the coal seams. They may have formed by crystallization from silica-rich solutions migrating in the pore space and along fissures during an early stage of diagenesis. The presence of organic matter can increase the silica solubility (e.g., stabilization by protecting colloids). The diagenesis of the organic matter accompanied by $CO_2$ production lowers the pH and results in the precipitation of quartz (Zajic 1969). On the other hand, the inflow of saliniferous pore fluids during the transgression of the sea may have promoted the precipitation of quartz. This idea is supported by the detected high salinity of the fluid inclusions in the authigenic quartz crystals (Fruth and Blankenburg 1992).

The temperatures of quartz formation can be estimated from results of fluid inclusion and oxygen isotope studies, respectively. Oxygen isotope analysis of separated authigenic quartz crystals from a subbituminous coal seam (Charlston, New Zealand) gave a result of 26.5 ± 0.5 ‰ $\delta^{18}O$ (SMOW), which the authors interpreted with a maximum temperature of formation of about 65°C (Soong and Blattner 1986). This is in accordance with homogenization temperatures of quartz from Tertiary lignites in Germany, which gave minimum temperatures of crystallization between 39 and 60°C (Fruth and Blankenburg 1992). These temperatures are consistent with burial to only shallow depths given normal geothermal gradients.

## 13.4 Conclusions

The results of the present study illustrate that quartz authigenesis is common in different sedimentary environments. The formation of quartz crystals often went under similar conditions but, nevertheless, resulted in characteristic properties of authigenic quartz in dependence on the specific physico-chemical conditions and the sedimentary environment, respectively.

Most authigenic quartz crystals are suspended within the rock matrix. Microscopic studies (SEM, CL) show that they have a perfect euhedral shape and no identifiable nucleus. Therefore, nucleation of authigenic quartz on detrital quartz grains, as it is dominant for secondary silica overgrowths in sandstones, does not play any role in the formation of isolated authigenic quartz crystals. The morphology of the quartz crystals (dominating crystal faces) strongly depends on the environment of formation. Moreover, the microinclusions and also the trace-element composition of authigenic quartz reflect the specific geochemical environment during quartz growth.

CL microscopy and spectroscopy revealed that the quartz crystals from different sediments show completely different luminescence behaviour. Quartz from the salt deposit is characterized by a deep blue CL with an emission band at 450 nm. Authigenic nodular quartz from the Bunter sandstones shows zonal growth with variable luminescence, which is similar to the CL of agate (main emission bands at 650 and 580 nm). In contrast, the quartz from Tertiary lignite has a transient violet to brown CL with a main emission at 650 nm. These different CL characteristics reflect different defects in the quartz structure and, therefore, can be related to specific conditions of quartz formation. Oxygen vacancies in the quartz from salt indicate anoxic environment, whereas the defects and microtexture of the nodular quartz from the Bunter sandstone point to rapid growth under non-equilibrium conditions with varying silica saturation. The CL emission of the authigenic quartz from lignite reflects the existence of silanol groups due to the crystallization from aqueous solutions.

Conclusions concerning the sources of silica and the conditions of quartz formation could be drawn from geochemical data of the quartz. The crustal signature of the REE patterns excludes the influence of hydrothermal fluids and points to an origin of $SiO_2$ in pore fluids from the alteration and decomposition of silica and silicate minerals. The migration of these fluids through pores and fissures in the host rocks and the crystallization of authigenic quartz probably took place in an early stage of diagenesis. Mineral and fluid inclusions show that in all occurrences saliniferous fluids took part in the crystallization process of authigenic quartz. The nucleation was promoted by minerals or organic matter from the sediment matrix.

With the detected mineralogical and geochemical characteristics, the authigenic quartz from the various sedimentary environments clearly differs from quartz of crystalline rocks (Götze et al. 2004). Furthermore, the specific properties enable the certain recognition of their authigenic origin and a distinction between quartz from different sedimentary environments.

**Acknowledgment** M. Fruth is gratefully acknowledged for providing samples from the Zechstein salt deposits. U. Kempe and L. Nasdala kindly helped during the analytical work. The preparation of the drawings was supported by St. Kubel. Constructive reviews of the manuscript by Kitty L. Milliken and J. Kelly significantly improved the quality of the paper.

## References

Bahlburg H, Floyd PA (1999) Advanced techniques in provenance analysis of sedimentary rocks. Sedimentary Geology, vol 124. Elsevier, Amsterdam

Baker G (1946) Microscopic quartz crystals in brown coal, Victoria. Am Mineral 31:22–30

Bellmann HJ (1986) Zur Genese der verkieselten Hölzer und Braunkohlenquarzite im Raum Leipzig. Zeitschrift geologische Wissenschaften 13:699–704

Bennett P (1991) Quartz dissolution in organic-rich aqueous systems. Geochim Cosm Acta 55:1781–1797

Bennett P, Siegel DI (1987) Increased solubility of quartz in water due to complexing by organic compounds. Nat 326:684–686

Bernet M, Bassett K (2005) Provenance analysis by single-quartz-grain SEM-CL/optical microscopy. J Sed Res 75:492–500

Bjørlykke K, Egeberg PK (1993) Quartz cementation in sedimentary basins. Am Assoc Petrol Geol Bull 77:1538–1548

Black WW (1949) An occurrence of authigenic feldspar and quartz in Yoredale limestones. Geol Mag 86:129

Boogs S Jr, Kwon Y-I, Goles GG, Rusk BG, Krinsley D, Seyedolali A (2002) Is quartz cathodoluminescence a reliable provenance tool? A quantitative examination. J Sed Res 72:408–415

Botz RW, Hunt JW, Smith JW (1986) Isotope geochemistry of minerals in Australian bituminous coal. J Sed Petrol 56:99–111

Burley SD, Mullis J, Matter A (1989) Timing diagenesis in the Tartan Reservoir (UK North Sea): constraints from combined cathodoluminescence microscopy and fluid inclusion studies. Marine Petrol Geol 6:98–120

Chavetz HS, Zhang J (1998) Authigenic euhedral megaquartz crystals in a Quaternary dolomite. J Sediment Res 68:994–1000

Dixon JB, Weed SB (1989) Minerals in soil environments. Soil Science Society of America, Madison

Evans MA, Elmore RD (2006) Fluid control of localized mineral domains in limestone pressure solution structures. J Struct Geol 28:284–301

Evans J, Hogg AJC, Hopkins MS, Howarth RJ (1994) Quantification of quartz cements using combined SEM, CL, and image analysis. J Sediment Petrol A64:334–338

Fabricius J (1987) Natural Na-K-Mg-Cl solutions and solid derivatives trapped in euhedral quartz from Danish Zechstein salt. Chem Geol 61:95–112

Fitting H-J, Barfels T, Trukhin AN, Schmidt B (2001) Cathodoluminescence of crystalline and amorphous $SiO_2$ and $GeO_2$. J Non-Cryst Solids 279:51–59

Friedman GM, Shukla V (1980) Significance of authigenic quartz euhedra after sulphates; example from the Lockport formation (Middle Silurian) from New York. J Sediment Res 50:1299–1304

Fruth M, Blankenburg H-J (1992) Charakterisierung von authigenen idiomorphen Kohle- und Salinarquarzen durch Einschlussuntersuchungen. Neues Jahrbuch Mineralogie, Abhandlungen 165:53–64

Füchtbauer H (1961) Zur Quarzneubildung in Erdöllagerstätten. Erdöl und Kohle 14:169–173

Götze J (1998) Geochemistry and provenance of the Altendorf feldspathic sandstone in the Middle Bunter of the Thuringian basin (Germany). Chem Geol 150:43–61

Götze J, Zimmerle W (2000) Quartz and silica as guide to provenance in sediments and sedimentary rocks. Contributions to Sedimentary Geology, vol 21E. Schweizerbart'sche Verlagsbuchhandlung, Stuttgart, p 91

Götze J, Plötze M, Fuchs H, Habermann D (1999) Defect structure and luminescence behavior of agate-results of electron paramagnetic resonance (EPR) and cathodoluminescence (CL) studies. Mineral Mag 63:149–163

Götze J, Plötze M, Habermann D (2001a) Origin, spectral characteristics and practical applications of the cathodoluminescence (CL) of quartz: a review. Miner Petrol 71:225–250

Götze J, Tichomirowa M, Fuchs H, Pilot J, Sharp Z (2001b) Geochemistry of agates: a trace element and stable isotope study. Chem Geol 175:523–541

Götze J, Plötze M, Graupner T, Hallbauer DK, Bray C (2004) Trace element incorporation into quartz: a combined study by ICP-MS, electron spin resonance, cathodoluminescence, capillary ion analysis and gas chromatography. Geochim Cosmochim Acta 68:3741–3759

Grimm W-D (1962) Idiomorphe Quarze als Leitmineralien für salinare Fazies. Erdöl und Kohle 15:880–887

Hartmann BH, Juhász-Bodnár K, Ramseyer K, Matter A (2000) Polyphased quartz cementation and its sources: a case study from the Upper Paleocoic Haishi Group sandstones; Sultanate of Oman. IAS Special Publications 29:253–270

Heynke A, Zänker G (1970) Zur Ausbildung und Leitbankgliederung des Staßfurtsteinsalzes im Südharz-Kalirevier. Zeitschrift angewandte Geologie 16:344–356

Hiatt EE, Kyser TK, Fayek M, Polito P, Holk GJ, Riciputi LR (2007) Early quartz cements and evolution of paleohydraulic properties of basal sandstones in three Paleoproterocoic continental basins: evidence from in situ $\delta^{18}O$ analysis of quartz cements. Chem Geol 238:19–37

Hoehne K (1954) Zur Neubildung von Quarz in Kohlenflözen. Neues Jahrbuch Geologie Paläonthologie, Abhandlungen 99:209–220

Houseknecht DW (1991) Use of cathodoluminescence petrography for understanding compaction, quartz cementation, and porosity in sandstones. In: Baker CE, Kopp OC (eds) Luminescence microscopy: quantitative and qualitative aspects. SEPM, Dallas, pp 59–66

Kelly JL, Fu B, Kita NT, Valley JW (2007) Optically continuous silcrete quartz cements of the St. Peter sanstone: high precision oxygen isotope analysis by ion microprobe. Geochim Cosmochim Acta 71:3812–3832

Langbein R (1974) Zur Petrologie der Karneole des thüringischen Chirotherien Sandsteins (Solling-Folge). Chemie der Erde 33:301–325

Leskevich LE (1959) Quartz crystals in coal (in Russian). Doklady Akademii Nauk SSSR 124(3):575–577

Liu X, Wang S, Zhang F (2004) Fission track dating of authigenic quartz in red weathering crusts of carbonate rocks in Guizhou province (Chinese with English Abstract). Acta Geol Sinica 78:1136–1142

Lyon IC, Burley SD, McKeever PJ, Saxton PJ, Macaulay JM (2000) Oxygen isotope analysis of authigenic quartz in sandstones: a comparison of ion microprobe and conventional analytical techniques. In: Worden RH, Morad S (eds) Quartz cementation in sandstones. Blackwell Science, Oxford, pp 299–316

Maliva RG (1987) Quartz geodes: Early diagenetic silicified anhydrite nodules related to dolomitization. J Sediment Res 57:1054–1059

Markowitz A, Milliken KL (2003) Quantification of brittle deformation in burial compaction, Frio and Mount Simon formation sandstones. J Sediment Res 73:1007–1021

Mason B (1979) Cosmochemistry, Part I. Meteorites. In: Fleischer M (ed) Data of geochemistry, U.S. Geological Survey professional papers, 440-B1, 132 p

McBride EF (1989) Quartz cement in sandstones: a review. Earth Sci Rev 26:69–112

Milliken KL (1979) The silicified evaporate syndrome—two aspects of silicification history of former evaporate nodules from Southern Kentucky and Northern Tennessee. J Sediment Petrol 49:245–256

Milliken KL, Laubach SE (2000) Brittle deformation in sandstone diagenesis as revealed by scanned cathodoluminescence imaging with application to characterization of fractured reservoirs. In: Pagel M, Barbin V, Blanc P, Ohnenstetter D (eds) Cathodoluminescence in geosciences. Springer, Berlin, pp 225–243

Mišik M (1995) Authigenic quartz crystals in the Mesozoic and Paleogene carbonate rocks of the Western Carpathians. Geologica Carphatica 46:227–239

Molenaar N, deJong AFM (1987) Authigenic quartz and albite in Devonian limestones: origin and significance. Sedimentology 34:623–640

Monecke T, Bombach G, Klemm W, Kempe U, Götze J, Wolf D (2000) Determination of trace elements in quartz standard UNS-SpS and in natural quartz by ICP-MS. Geostand Newslett 24(1):73–81

Nachsel G (1969) Idiomorphe Quarze und Vertaubungen im Kaliflöz "Staßfurt" des Südharz-Kalireviers. Z Angew Geol 15:420–425

Neuser RD, Bruhn F, Götze J, Habermann D, Richter DK (1995) Kathodolumineszenz: Methodik und Anwendung. Zentralblatt für Geologie und Paläontologie Teil I H 1/2:287–306

Parnell J, Carey PF, Monson B (1996) Fluid inclusion constraints on temperatures of petroleum migration from authigenic quartz in bitumen veins. Chem Geol 129:217–226

Richter DK (1971) Fazies- und Diagenesehinweise durch Einschlüsse in authigenen Quarzen. Neues Jahrbuch Geologie Paläonthologie, Monatshefte H 10:604–622

Richter DK, Götte Th, Götze J, Neuser RD (2003) Progress in application of cathodoluminescence (CL) in sedimentary petrology. Miner Petrol 79:127–166

Ruppert LF, Cecil CB, Stanton RW, Christian RP (1985) Authigenic quartz in the Upper Freeport coal bed, west-central Pennsylvania. J Sediment Res 55:334–339

Schneider W (1986) Phytogene Verkieselungen in der miozänen Braunkohle und deren Aussagen für Stratigraphie, Fazies und Flözgenese. Zeitschrift geologische Wissenschaften 14:153–162

Sedletskiy VI (1971) Some features of authigenic quartz formation in evaporate basins (in Russian). Geologika Geofizika 5:72–77

Seyedolali A, Krinsley DH, Boggs S Jr, O'Hara PF, Dyavik H, Goles GG (1997) Provenance interpretation of quartz by scanning electron microscope-cathodoluminescence fabric analysis. Geology 25:787–790

Siegel GH, Marrone MJ (1981) Photoluminescence in as-drawn and irradiated silica optical fibers: An assessment of the role of nonbridging oxygen defect centers. J Non-Cryst Solids 45:235–247

Sippel RF (1968) Sandstone petrology, evidence from luminescence petrography. J Sediment Petrol 38:530–554

Soong R, Blattner P (1986) Biterminal authigenic $^{18}$O-enriched quartz in a subbituminous coal seam, Charleston, New Zealand. N Z J Geol Geophys 29:141–145

Spiro B, Rozenson I (1982) Formation and properties of authigenic minerals in bituminous calcareous shales, Ghareb Formation, Israel. Can Mineral 20:29–39

Walderhaug O (1990) A fluid inclusion study of quartz-cemented sandstones from offshore mid-Norway—possible evidence for continued quartz cementation during oil emplacement. J Sediment Petrol 60:203–210

Walderhaug O (1994) Temperature of quartz cementation in Jurassic sandstones from the Norwegian continental shelf—evidence from fluid inclusions. J Sediment Res A64:311–323

Walker G, Burley S (1991) Luminescence petrography and spectroscopic studies of diagenetic minerals. In: Barker CE, Kopp OC (eds.) Luminescence microscopy and spectroscopy: Qualitative and quantitative applications. SEPM, Tulsa, pp 83–96

Walther H, Götze J (1994) Zur Bildung von Quarziten und authigenen Quarzen in Braunkohlen. Eur J Mineral Beiheft 1(6):301

Weil JA (1984) A review of electron spin spectroscopy and its application to the study of paramagnetic defects in crystalline quartz. Phys Chem Miner 10:149–165

Wood SA (1990) The aqueous geochemistry of the rare-earth elements and yttrium. Chem Geol 88:99–125

Wordan RH, Morad S (2000) Quartz cementation in oil field sandstones: a review of the key controversies. In: Worden RH, Morad S (eds.) Quartz cementation in sandstones. Blackwell Science, Oxford, pp 1–20

Zajic JM (1969) Microbial biogeochemistry. Academic Press, New York

Zinkernagel U (1978) Cathodoluminescence of quartz and its application to sandstone petrology. Contrib Sedimentol 8:1–69

Zuffa GG (1985) Provenance of arenites. NATO ASI series C 148, Reidel Publ. Co., Boston, 393 p

# Chapter 14
# Cathodoluminescent Textures and Trace Elements in Hydrothermal Quartz

**Brian Rusk**

**Abstract** When viewed with scanning electron microscope-cathodoluminescence (SEM-CL), hydrothermal vein quartz displays textures that are unobservable using other techniques. These textures provide unique insights into the sequence of quartz precipitation and dissolution events during hydrothermal vein formation. Such textures relate specific quartz generations to specific mineralization events or fluid inclusion populations and may also relate quartz isotopic or trace element data to specific hydrothermal events. The most commonly observed CL textures in hydrothermal quartz include: (1) euhedral growth zones of oscillating CL intensity; (2) chalcedonic, coliform, and spheroidal textures; (3) mosaic textures; (4) CL-dark bands; (5) spider and cobweb texture; (6) rounded cores with overgrowths; (7) microbrecciation; (8) rounded or wavy concentric zonation; and (9) homogeneous (or slightly mottled) texture. These textures are present to varying degrees in quartz from different types of hydrothermal ore deposits depending on the pressure, temperature, or composition of hydrothermal fluids, and the rates and magnitude of fluctuations in these variables. In samples, where the geologic setting of quartz is not clear, CL textures distinguish among quartz derived from epithermal, porphyry-type, and orogenic Au deposits. CL textures result from defects in the quartz lattice, including those caused by trace element concentration variations. Like CL textures, trace element abundance and distribution result from variations in the physical and chemical conditions of quartz precipitation, followed by any subsequent solid-state changes in quartz chemistry. Here we show that CL textures, CL spectra, and trace element concentration vary systematically between quartz from various types of hydrothermal ore deposits. The information derived

B. Rusk (✉)
Economic Geology Research Unit, School of Earth and Environmental Sciences, James Cook University, Townsville 4811, Australia
e-mail: brian.rusk@jcu.edu.au

J. Götze and R. Möckel (eds.), *Quartz: Deposits, Mineralogy and Analytics*, Springer Geology, DOI: 10.1007/978-3-642-22161-3_14,

from quartz analysis can therefore be used to fingerprint the origin of quartz and make some inferences about the pressure, temperature, and fluid compositional changes that accompany hydrothermal quartz precipitation.

## 14.1 Introduction

Quartz is one of the most abundant minerals in the crust and is the dominant gangue mineral in many hydrothermal ore deposits, where it forms under a wide range of temperature and pressure conditions from fluids of diverse origins and compositions. Genetic models of ore deposit formation rely heavily on information obtained from fluid inclusions trapped in and isotopic data derived from vein quartz. The interpretation of such data relies upon our ability to understand the timing relationships between the analyzed quartz and the other minerals of interest. Whereas optical petrography offers some insight into these textural relationships (Adams 1920; Spurr 1926; Dowling and Morrison 1989; Vearncombe 1993; Dong et al. 1995), scanning electron microscope–cathodoluminescence (SEM-CL) reveals textures in quartz that are not observable by optical petrography (Sippel 1968, Smith and Stenstrom 1965). These unique textures illustrate the sequential history of quartz precipitation, dissolution, fracturing, recrystallization, and/or subsequent quartz growth. When deciphered, such textures have important implications for unraveling the physical and chemical history of a hydrothermal system.

Cathodoluminescent textural variations result from defects in the crystal lattice, many of which are related to the incorporation of trace elements (Marshall 1988; Ramseyer et al. 1988; Stevens-Kalceff and Phillips 1995; Götze et al. 1999, 2001; Walker 2000; Stevens-Kalceff et al. 2000; Götze 2009). The trace element variations and the resultant CL textures reflect the rate of crystallization, fluid/magma composition, pressure, or temperature, as well as any subsequent metamorphic and/or deformational history after quartz formation (Perny et al. 1992; Mullis et al. 1994; Lowenstern and Sinclair 1996; Ramseyer and Mullis 2000; Götze et al. 2001; Monecke et al. 2002; Larsen et al. 2004; Breiter and Müller 2009). Currently little is known about the specific conditions that are reflected by quartz trace elements, however an increasing number of studies of natural and synthetic quartz are aimed at quantifying the effect of crystal growth rate, temperature, pressure, and fluid composition on the incorporation of trace elements in quartz (Wark and Watson 2006; Rusk et al. 2008; Thomas et al. 2010; Götte et al. 2011). Because quartz precipitates throughout much of the history of hydrothermal systems, the interpretation of CL textures and quartz trace elements greatly enhances our ability to interpret the evolving physical and chemical conditions of hydrothermal fluid flow in the crust.

This paper provides an overview of CL textures observed and trace elements measured in quartz from a variety of hydrothermal environments. We compare CL textures, spectra, and trace elements among hydrothermal ore deposits and show that these data can be used to interpret the physical and chemical conditions of vein formation, to correlate hydrothermal events across a broad region, to relate fluid inclusion and isotopic data to specific mineralization events, and to fingerprint ore deposit types. This overview is based on existing literature as well as on trace element and SEM-CL studies of ~300 vein samples from ~50 hydrothermal ore deposits, with a focus on porphyry-Cu (-Mo-Au) deposits, epithermal deposits, and orogenic Au deposits. Comparison of textures and trace elements among quartz veins formed under different pressure and temperature conditions helps to constrain the most significant factors that affect CL textures and trace element incorporation. In addition to understanding geologic processes, this work has implications for understanding the formation of high purity quartz, which is required for the manufacture of many high-tech products such as silicon wafers used in computers and photovoltaic solar cells.

## 14.2 CL Textures in Quartz

Cathodoluminescence is the emission of light from a mineral when bombarded by an electron beam (Marshall 1988). The intensity and spectrum of the CL emission is dependant on the composition of the mineral and the abundance and distribution of trace impurities (Stevens-Kalceff 2009). Cathodoluminescence has been applied to many minerals, including quartz, carbonates, apatite, feldspars, zircon, monazite, anhydrite, garnets, and others, in order to understand a range of geologic phenomena (Machel and Burton 1991; Pagel et al. 2000; Götze et al. 2001). The main advantage of CL imaging is that the textures revealed are typically not observable by any other method of observation. As such, the textures offer unique information about the formation history of a mineral. CL images of quartz reveal otherwise unobservable textures that illustrate the growth history of the quartz and allow one to distinguish between quartz grown at different times, under different conditions, with different mineral assemblages (Fig. 14.1). The information derived from these images provides unique insights into the geologic history of the sample. Because quartz is ubiquitous in so many geologic environments, advances in understanding geologic histories based on quartz textures enjoys broad application across the geosciences.

Cathodoluminescent characteristics in quartz from volcanic, plutonic, metamorphic, sedimentary, and hydrothermal environments differ from one another, illustrating that quartz textures reflect the environment of quartz growth (Seyedolali et al. 1997; Boggs et al. 2002; Richter et al. 2003; Bernet and Bassett 2005; Boggs and Krinsley 2006) (Fig. 14.2). Figure 14.2 shows typical CL textures from plutonic, volcanic, intrusive (porphyritic), and metamorphic environments. CL growth zones are typical of phenocrysts in volcanic tuffs. These growth zones represent changes in the physical or chemical conditions of crystal growth within a magma chamber (Peppard et al. 2001; Wark et al. 2007). Due to relatively rapid cooling, growth textures in volcanic quartz phenocrysts are typically well

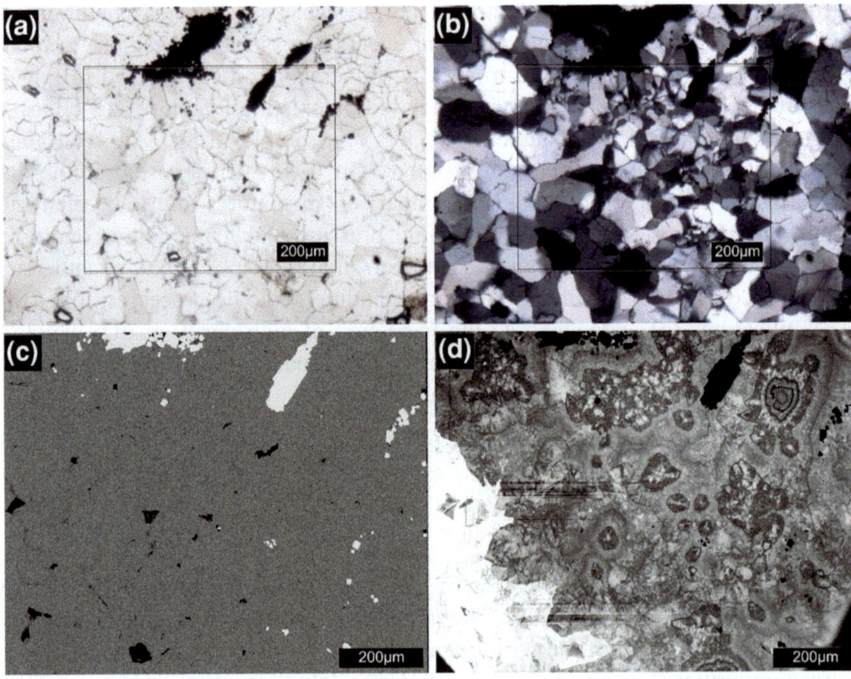


**Fig. 14.1** **a** Plane-polarized light; **b** Cross-polarized light; **c** Back-scattered electron; and **d** SEM-CL image of a quartz vein from the epithermal Au–Ag deposit in McLaughlin, CA. Plane- and cross-polarized light show typical mosaics of quartz with little obvious textural information to differentiate multiple generations of quartz growth or to relate sulfides to specific mineralization events. SEM-CL reveals micro-textures that are not visible using any other analytical technique. SEM-CL textures are key for deciphering the history of vein formation. Pyrite precipitated last in the banded microcrystalline quartz, but both pyrite and microcrystalline quartz are overgrown by CL-bright euhedral quartz (*lower left*)

preserved. Quartz-bearing porphyryitic rocks are common, especially near many magmatic hydrothermal ore deposits. Quartz phenocrysts from porphyrytic rocks also typically show growth zones, but growth textures are more complex and quartz phenocrysts commonly show signs of dissolution and healed fractures (Chang and Meinert 2004). Granitic quartz is characterized by either slight growth zonations or no growth zones and typically displays textures indicating hydrothermal fracturing, recrystallization, and/or dissolution (Kanaori 1986; Müller et al. 2005; Larsen et al. 2009). Metamorphic quartz shows the least CL textural variation of all quartz types. Metamorphic quartz is typically CL-homogenous with no growth zonation and little variation in CL intensity throughout the quartz grains (Sprunt et al. 1978; Spear and Wark 2009). The lack of CL texture results from the redistribution and homogenization of lattice defects during metamorphic processes.

Hydrothermal quartz displays a far wider range of CL textures than plutonic, intrusive, volcanic, or metamorphic quartz reflecting the wide range of conditions under which hydrothermal quartz crystallizes. CL textures reveal that many

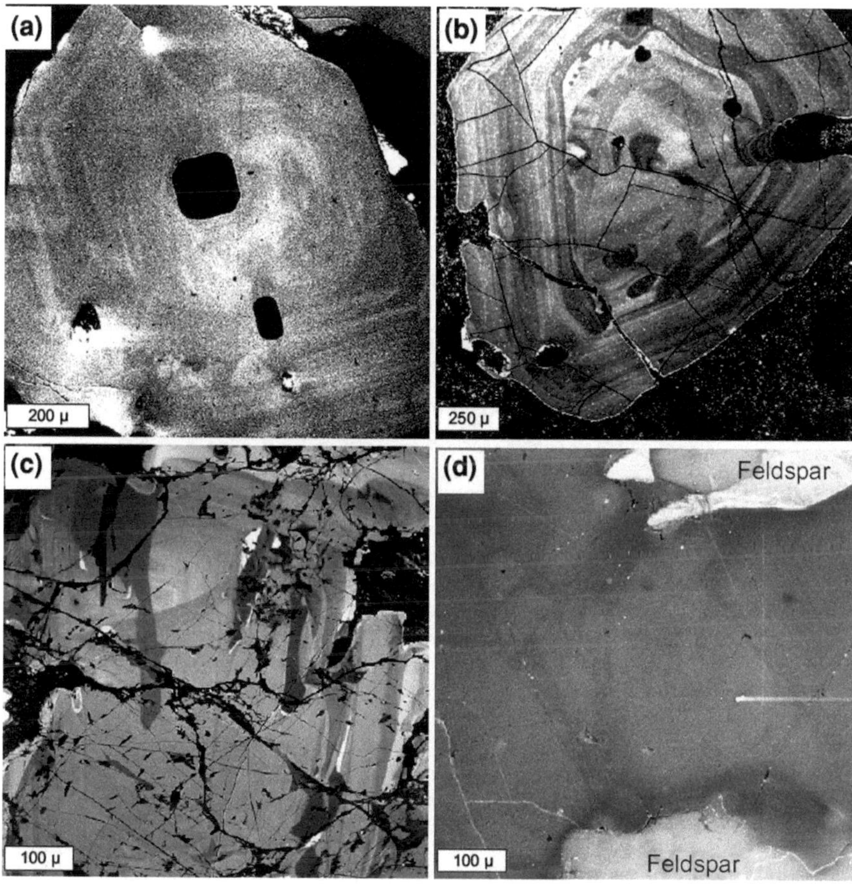


**Fig. 14.2** **a** SEM-CL image of a zoned phenocryst form the Tamagawa Tuff. Euhedral growth zones of oscillating CL intensity are typical of volcanic quartz phenocrysts; **b** Phenocryst from an intrusive porphyrytic rock from Summitville Colorado. Growth zones are slightly more complex than in typical volcanic rocks and embayments are common; **c** Quartz from intrusive igneous rocks, may or may not display euhdral growth zonations (Nimrod granite, OR). CL-dark fractures and spiders are common in plutonic quartz; **d** Metamorphic quartz typically displays nearly homogenous or slightly mottled texture, indicating annealing of original CL textures. This sample is from the Packsaddle schist, Texas

hydrothermal veins have complex histories that involve multiple stages of fluid flow, quartz precipitation, dissolution, fracturing, and/or recrystallization. These textures provide critical petrographic evidence that can be used to (1) Deconvolute the chronologic sequence of vein forming events; (2) To relate specific generations of quartz to specific hydrothermal events or fluid inclusion populations or (3) To infer distal and cryptic expressions of hydrothermal systems, (Boiron et al. 1992; Valley and Graham 1996; Wilkinson and Johnston 1996; King et al. 1997;

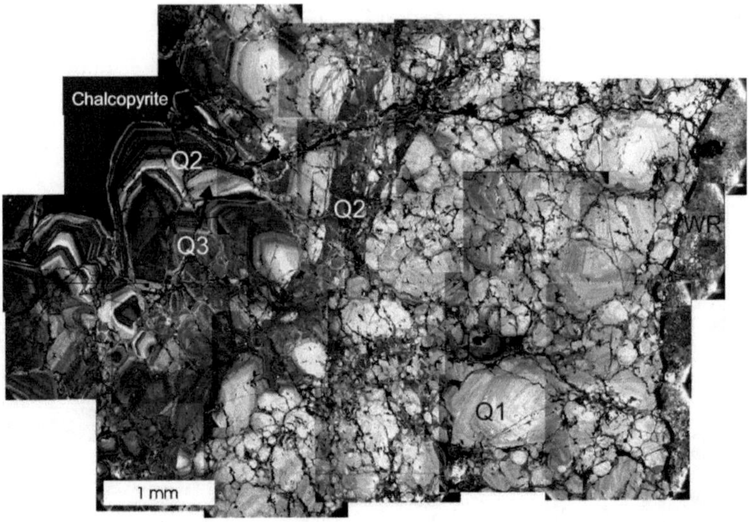


**Fig. 14.3** SEM-CL photomosaic of a chalcopyrite-bearing quartz vein from the porphyry-Cu-Mo deposit in Butte, Montana. CL image shows an early CL-bright generation of quartz that is cut by later CL-darker quartz. The later CL-dark quartz both cuts and overgrows the earlier CL-bright quartz. Where the later quartz grew into open space, it developed euhedral growth zones and filled a void along with chalcopyrite (*black* mineral in upper *left* corner). The later euhedral crystals and the earlier CL-bright crystals are all cut by a later generation of CL-dark fractures, which correspond to trails of fluid inclusions

Wilkinson et al. 1999; Graupner et al. 2000; Penniston-Dorland 2001; Fischer et al. 2003; Ioannou et al. 2004; Sekine 2003; Götte and Richter (2003); Bignall et al. 2004; Redmond et al. 2004). For example, the CL image in Fig. 14.3 shows that this apparently simple quartz vein actually consists of at least three successive quartz generations (Q1, Q2, and Q3), where Q2 appears to immediately precede or coincide with chalcopyrite precipitation. Primary fluid inclusions, in CL-bright Q1, pre-date chalcopyrite precipitation, whereas secondary fluid inclusions along the CL-dark healed fractures (Q3) either coincide with or post-date chalcopyrite precipitation. In some cases, CL textures serve not only as petrographic tools, but can be interpreted in light of the changing temperature, pressure, and fluid compositional history of a hydrothermal system. For example quartz dissolution textures, observed in numerous porphyry copper veins, have been interpreted to result from the cooling of hydrothermal fluids through the zone of retrograde quartz solubility (in the range of 400–600°C at pressures below ~800 bars) (Rusk and Reed 2002; Redmond et al. 2004; Müller et al. 2010).

A wide variety of CL textures are observed in hydrothermal quartz. Although many are complex and difficult to categorize, the most common textures can be summarized as follows: (1) euhedral growth zones of oscillating CL intensity; (2) chalcedonic, colliform, and spheroidal textures; (3) mosaic textures;

(4) CL-dark bands; (5) spider and cobweb texture; (6) rounded cores with overgrowths; (7) microbrecciation; (8) rounded or wavy concentric zonation; and (9) homogeneous (or slightly mottled) texture (Fig. 14.4). The above CL textures result from the initial (primary) conditions of quartz crystallization followed by secondary processes such as dissolution, fracturing, overgrowth, and recrystallization that modify the original textures. The presence of numerous primary and secondary CL textures in many veins from many hydrothermal ore deposits indicates that the veins have a complex history involving multiple fluid incursions under varying pressure and temperature conditions. Interpretation of the conditions under which these textures form leads to improved models of hydrothermal systems and hydrothermal fluid flow in the Earth's crust.

### *14.2.1 Primary Textures*

Primary textures include oscillatory zoned euhedral crystals, chalcedonic and spheroidal textures, and mosaic quartz textures. The most common texture observed in the hydrothermal quartz veins is oscillatory euhedral growth zones that vary in thickness, shape, and CL intensity. There is no obvious connection between temperature or deposit type and growth zone intensity, abundance, shape, thickness, or other morphological aspects; however, euhedral crystals are more common in epithermal and other low temperature quartz (Götze et al. 1999; Lehmann et al. 2009), where they are the dominant texture, than in high temperature deposits (orogenic Au, porphyry-type, iron oxide-Cu-Au, intrusion-related Au), where they are subordinate to other textures. Their common presence in hydrothermal quartz that formed between ~100 and ~>700°C indicates that euhedral growth zones form under a wide range of hydrothermal conditions. Sector zoning in euhedral quartz is commonly observed in epithermal and other low temperature deposits. Mosaics of fine-grained interlocking crystals of quartz may or may not have internal growth zonations. Mosaic texture is common in veins from porphyry copper deposits that display "sugary" texture in hand sample. Rusk and Reed (2002) suggest that mosaic texture results from rapid quartz precipitation upon pressure drop in porphyry-type hydrothermal systems. While this texture may be a primary texture, similar textures could be derived from recrystallization or mineral replacement.

### *14.2.2 Secondary Textures*

Secondary textures include those that modify or overprint previously existing primary quartz. Multiple secondary textures are common in many hydrothermal quartz veins. CL-dark bands, reflecting fracturing and healing is the most common secondary texture observed in hydrothermal quartz and commonly coincide with

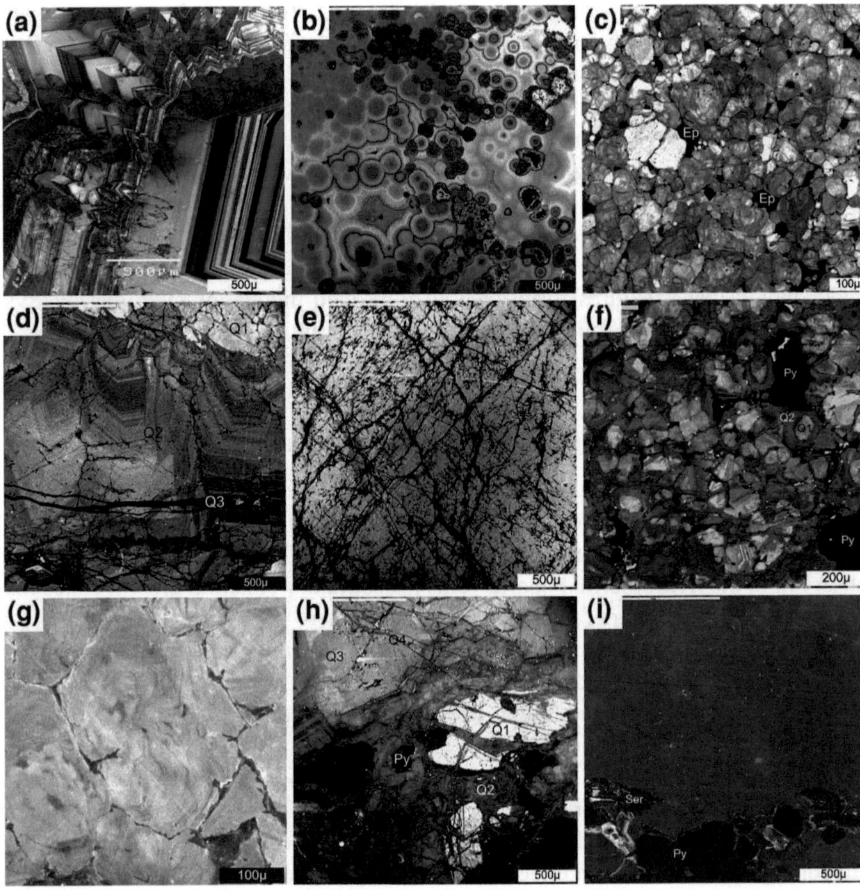


**Fig. 14.4** SEM-CL textures in hydrothermal quartz are much more variable and complex in hydrothermal quartz than in any other quartz type. **a** Euhedral growth zones are the most common texture observed in hydrothermal vein quartz (Tokatea epithermal deposit, NZ); **b** Spherical texture, only observed in epithermal veins (McLaughlin epithermal deposit, CA); **c** Mosaics of fine grained equant crystals of vein quartz from Refugio porphyry Au deposit, Peru. Grains within the quartz mosaic show internal wavy growth zonation; **d** Early CL-bright quartz cut by later CL-darker quartz with euhedral growth zones, cut by CL-dark fractures (Bata Hijau, Indonesia porphyry Cu deposit); **e** Spider and cobweb texture, typical of porphyry type deposits (Henderson porphyry MO deposit, CO); **f** CL-bright cores with CL-darker overgrowths. This texture has been recognized in multiple porphyry-type deposits and is attributed to dissolution, possibly caused by fluid cooling through the zone of retrograde quartz solubility (Gaby Sur, Chile); **g** Wavy concentric banding, seen only in quartz from porphyry-type deposits, and likely indicating some post-crystallization elemental diffusion (Butte, Montana); **h** Microbreccia texture showing early CL-bright quartz (Q1) and pyrite fractured and infilled by CL-darker Q2, which is subsequently overgrown by slightly brighter euhedral Q3. All of these quartz generations are then fractured by late CL-dark fractures (Q4) (Cordilleran base metal lodes, Butte, MT); **I** Homogenous CL texture, typical of orogenic Au veins (Valdez Creek, Alaska). Cp = chalcopyrite, Ep = epoxy, Ab = albite, WR = wallrock, Py = pyrite

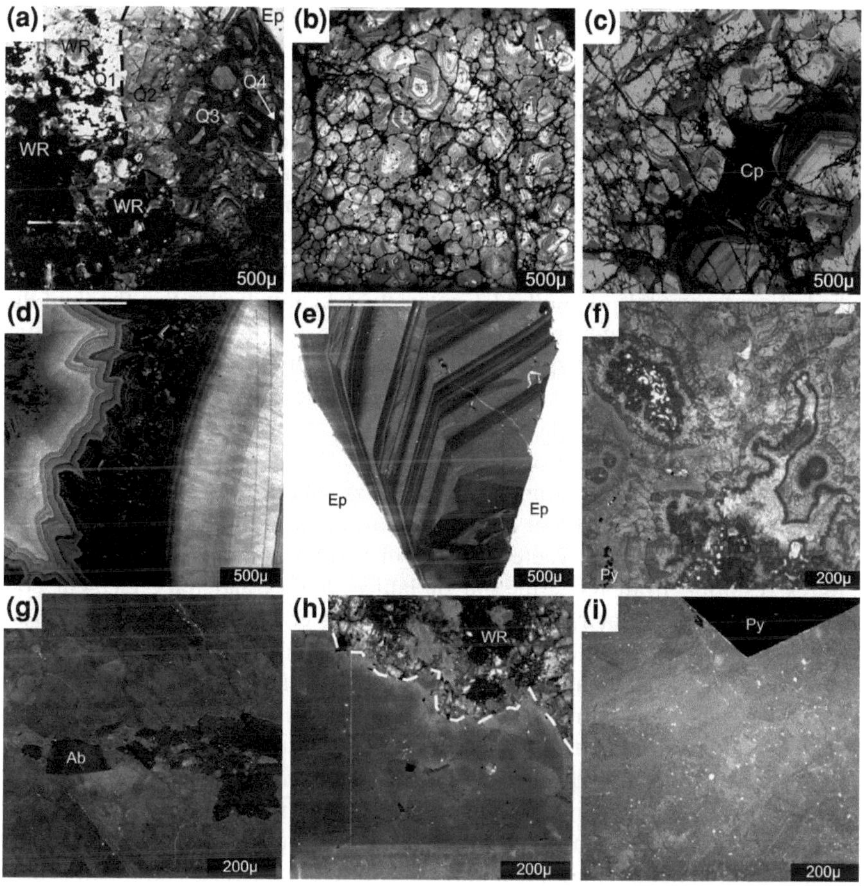


**Fig. 14.5** Textures typical of porphyry copper deposits (**a**, **b**, **c**), epithermal deposits (**d**, **e**, **f**), and orogenic Au deposits (**g**, **h**, **i**). The consistent suite of textures in each deposit type makes CL imaging useful for fingerprinting the origin of quartz in samples where the origin is not known (stream samples, soil samples, etc.). **a** Four generations of quartz (labeled Q1 to Q4) demonstrating superposition of multiple hydrothermal events. CL-brightest Q1 is igneous quartz in the wall rock. Q1 is overgrown by CL-bright mosaic quartz of Q2, which is dissolved and overgrown by CL-darker quartz of Q3. Q2 and Q3 are cut by a late CL-dark fracture of Q4 (Los Pelambres, Chile); **b** Mosaics of euhedrally zoned quartz (Grasberg, PNG); **c** Quartz displaying many features typical of porphyry copper deposits including inward oriented euhdral growth projecting into chalcopyrite, splatter and cobweb texture, and CL-dark bands (healed fractures) (El Teniente, Chile); **d** Chalcedonic bands intergrown with bands that are slightly more crystalline (Creede, CO, USA); **e** A fairly simple inward-oriented euhedral quartz crystal (Comstock Lode, NV, USA); **f** Microcrystalline quartz with irregular concentric and spherical patterns (McLaughlin, CA, USA); **g** Homogenous texture in quartz with some feldspar in the vein (Valdez Creek, AK, USA), **h** Mostly homogenous CL dark quartz (Norseman, Aus); **i** Slightly mottled to homogenous texture of quartz with pyrite (Carson Hill, CA, USA). Cp = chalcopyrite, Ep = epoxy, Ab = albite, WR = wallrock, Py = pyrite

trails of secondary fluid inclusions. Where CL-dark bands coincide with trails of secondary fluid inclusions, the inclusions are primary with respect to the dark quartz in which they lie. Thus if the event with which the fractures are associated can be determined through CL imaging, then the inclusions may be considered primary with respect to that event. In a few samples multiple generations of fracturing are obvious based on cross-cutting relations of CL-dark fractures of varying CL intensities. Some CL-dark bands are filled with inward–oriented euhedral quartz crystals indicating fracturing, then dilation, fluid infiltration, and quartz precipitation. In some examples CL-dark bands end in, or grade into euhedral quartz crystals of similar CL intensity, which indicates that the fluid that flowed through the fracture creating the band is the same fluid from which crystals precipitated into fluid-filled space (Fig. 14.3 for example).

Recrystallization of quartz is reflected by homogenous or mottled CL texture and low contrast rounded "ghosty" growth zones. However, in some cases, it is difficult to infer whether the quartz initially precipitated with little or no CL texture or whether a previously existing CL-texture was subdued by annealing. Homogenous texture is characteristic of quartz from metamorphic rocks (Seyedolali et al. 1997; Boggs et al. 2002), where the lack of texture is due to annealing of CL textures resulting from redistribution of lattice defects and trace element concentrations under metamorphic conditions (Sprunt et al. 1978; Spear and Wark 2009). Alternatively, mottled and homogenous textures may be a primary precipitation feature, indicating little internal lattice variation in quartz grains that grew under stable conditions at relatively high temperature and pressure, as suggested for pegmatitic quartz (Götze et al. 2005). Homogenous and mottled CL textures are common in orogenic Au samples and in high temperature veins with potassic alteration from some porphyry copper deposits.

Mineral replacement is another possible secondary texture, but the CL textures resulting from mineral replacement may vary widely. Mineral replacement is most obvious in the case of pseudomorphs, where the original crystal shape is preserved, such as bladed calcite crystals replaced by quartz, which is common in some epithermal ore deposits. Leach et al. (2005) conclude that interlocking euhedral quartz crystals with multiple crystallographic orientations formed from the replacement of barite by quartz in the Red Dog Pb–Zn deposit of AK.

Evidence for dissolution includes rounded or embayed CL-bright quartz cores with euhedrally zoned CL-dark overgrowths. This texture forms where early CL-bright quartz is fractured, pervaded, and dissolved by hydrothermal fluids, followed by precipitation of CL-dark quartz into the pore-space created by fracturing and dissolution. Dissolution is also inferred where early quartz has a rounded contact against younger quartz, which is almost always CL-darker. In some instances growth zones within early quartz are truncated by growth zones in the later overgrowth quartz, which have different zonation patterns. Evidence for dissolution is most convincing when the CL-bright cores are rounded or embayed against the later CL-darker quartz, and when the late quartz displays euhedral growth zones indicating growth into open space.

### 14.2.3 CL Textures and Environment of Formation

Several studies show that specific CL textures and intensities are related to specific mineralization events at a deposit or even district-wide scale (Wilkinson et al. 1999; Lubben 2004; Landtwing et al. 2005; Rusk et al. 2006). This, along with the observation that different types of ore deposits are characterized by different CL textures indicates that many CL textural variations reflect deposit-scale physical and chemical fluctuations rather than local fluctuations. Thus the presence or absence of specific textures in a given sample can be used to constrain the geologic history of that sample and fingerprint the type of deposit from which the sample was derived.

Porphyry copper deposits are characterized by superposition of multiple generations of quartz growth, commonly punctuated by dissolution. Successively later generations of quartz growth are typically CL-darker than earlier generations. Evidence for quartz dissolution is common as is abundant fracturing and recrystallization (Fig. 14.5). Epithermal quartz, on the other hand, is dominated by growth textures including repeated layers of chalcedonic textures and euhedrally zoned crystals. Some late fracturing and brecciation occurs, but evidence for dissolution is not common. Orogenic Au quartz is different from both of the above quartz types and shows the least variation in CL textures. Vein quartz from orogenic-Au deposits is dominated by homogenous and mottled textures that likely result from annealing of CL textures during metamorphism.

## 14.3 Trace Elements in Hydrothermal Quartz

### 14.3.1 Analytical Techniques

CL intensity variations result from structural defects in the quartz lattice many of which are directly or indirectly caused by the incorporation of trace elements. Numerous studies show the variations in trace element concentrations and their relations to CL textures, using both spot analyses and trace element maps. Such studies have become far more common in the past 10 years, owing to advances in micro-analytical techniques such as electron probe micro-analysis (EPMA), secondary ion mass spectrometry (SIMS), laser ablation-inductively coupled plasma-mass spectrometry (LA-ICP-MS), and micro-Fourier transfer infrared (FTIR) analysis that allow for the in situ analysis of trace elements in quartz with detection limits as low as a few ppb for some elements. These analytical techniques have various benefits and drawbacks relative to one another and they offer most value when integrated with one another and with CL textural analysis. Whereas trace element maps have the advantage of showing, more clearly than spots, the correlations among trace element concentrations and between CL textures and trace element distribution, spot analyses generally offer both higher precision and spatial resolution.

Electron microprobe offers the highest spatial resolution and smallest sampling volume of any of the above analytical techniques. This is critical in many samples where CL features, such as cross cutting quartz generations or euhedral growth zones, are only a few microns or tens of microns in width. However detection limits, typically in the range of several tens of ppm using long counting times and other techniques (see Donovan et al. 2011), are generally too high to detect all but the most abundant quartz trace elements. Typically the only elements quantifiable by microprobe are Al, K, and sometimes Ti and Fe. Important quartz trace elements Li, K, Na, P, and H cannot typically be analyzed via electron microprobe.

The main benefit of LA-ICP-MS is the ability to quantify greater than 30 elements simultaneously and achieve detection limits in the range of tens of ppb to a few ppm for many elements. However, in order to obtain such low detection limits, the volume of consumed sample material using laser ablation is much larger than by microprobe. Spot sizes >35 μ in diameter are required to obtain low detection limits for multiple elements. Because the interaction volume is so large with a laser, the accidental analysis of fluid, mineral, and melt inclusions is always a risk, especially in hydrothermal quartz, which can be rich in mineral and fluid inclusions. The accidental ablation of inclusions may be obvious if they are large, however it may be very difficult to determine whether an analyzed element is present in the quartz structure or as micro- or nano-inclusions. Like with EPMA, H cannot be analyzed using LA-ICP-MS, however most other elements of interest in quartz can.

Only a few studies exist that show the usefulness of trace element analysis in quartz using SIMS. The spatial resolution of SIMS is intermediate between EPMA and LA-ICP-MS, with a typical spot size of ~10–15 μ. Currently the main drawback of this technique is that no standards exist that can be used to quantify quartz trace element data. SIMS analyses can still be useful in showing element distribution in the form of qualitative trace element maps (Lehmann et al. 2009). The main advantage of SIMS is that it is the only instrument that can be used to map H distribution. If appropriate standards were developed or characterized, quantitative SIMS analyses of quartz would have the combined advantages of low detection limits and high spatial resolution (c.f. Behr et al. 2011).

Of the above techniques, Micro-FTIR is the only one able to quantify H concentration and speciation. However other trace elements cannot be simultaneously quantified and there are a number of complications in data quantification (Miyoshi et al. 2005; Müller and Koch-Müller 2009; Thomas et al. 2009). The resolution of this technique is on the order of 50 μ and it requires special sample preparation and inclusion-free quartz. The main advantage of this technique in studies of quartz trace elements is that it is the only in situ technique available for the quantification of H in quartz.

### *14.3.2 Trace Element Concentrations in Hydrothermal Quartz*

Quartz has a simple chemical composition, and is generally relatively pure, containing less than a few ppm of most trace elements (Götze 2009; Müller and

Wanvik 2012). Even though it is pure relative to most other minerals, hydrothermal quartz contains several important elements. As with CL textures, concentrations of trace elements in hydrothermal quartz vary more widely than do quartz trace elements concentrations in other geologic environments, reflecting the wide range of physical and chemical conditions under which hydrothermal quartz forms.

Of the trace elements that are commonly identified in quartz, Al is typically the most abundant, and concentrations vary widely from a few ppm up to ~5,000 ppm. Al is the only trace element that is ever present in concentrations greater than a thousand ppm. Other trace elements present in concentrations between a few tens and a few 100 ppm in quartz may include Ti, Li, K, Sb, Fe, Ca, Na, P. In addition to these, H, B, Ge, Ga, Sn, Ba, Cs, and As, may be present in the concentration range of a few 100 ppb to a few ppm in quartz from various environments (Penniston-Dorland 2001; Monecke et al. 2002; Flem et al. 2002; Breiter and Müller 2009; Müller et al. 2003; Larsen et al. 2004; Allan and Yardley 2007; Rusk et al. 2008). In addition, some studies indicate that Cl and $H_2O$ may be structurally bound in quartz (Allan and Yardley 2007; Stenina 2004).

Lithium varies from a few ppm to a few 100 ppm in hydrothermal quartz and is in the range of ~300 ppm in low temperature quartz where Al is in the range of 3,000 ppm. Numerous studies show that the most common correlation among quartz trace elements is the correlation between Al and Li. Although typically slightly less abundant, K and Na are also common in hydrothermal quartz and range from a few ppm up to a few 100 ppm and tend to correlate positively with Al concentration (for example Perny et al. 1992). However, in quantifying concentrations of Na and K in hydrothermal quartz, one must be cautious of the presence of fluid inclusions, which may contain several orders of magnitude more Na and K than quartz. The accidental analysis of volumetrically minor quantities of fluid inclusions will have significant impacts on the measured concentrations of these elements. The common observation of positive correlations between Al and these monovalent cations, provides evidence for a coupled a substitution in quartz where $Al^{3+}$ and $Li^{+}$, $K^{+}$, or $Na^{+}$ substitute for $Si^{4+}$, thus maintaining charge balance. In most quartz however, the molar ratio of Al/(Na + K + Li) is significantly greater than one and $H^{+}$ is most commonly suggested to make up the remainder of the charge balance, as it has been shown to be abundant in some quartz (Miyoshi et al. 2005, Lehmann et al. 2009). Considering all of the potentially significant trace elements in quartz, to maintain charge neutrality, the total of the trivalent cations Fe and Al should be equivalent to the total of monovalent cations plus $P^{5+}$. However in the few cases where $H^{+}$ is measured along with Al and the other cations, the total of H, Li, Na, K, and P are greater than Al and Fe by a factor of 1.5–2 (Müller and Koch-Müller 2009). The correlation between Al and the monovalent cations is displayed most strongly by low temperature hydrothermal quartz, such as that from detrital overgrowths on sedimentary rocks, alpine fissure quartz, and quartz from Mississippi Valley Type and epithermal ore bodies (Jourdan et al. 2009; Lehmann et al. 2009; Rusk et al. 2011; Götte et al. 2011) (Fig. 14.6). In higher temperature quartz, the correlation commonly exists but is weaker, likely owing to the relatively rapid and variable diffusion rates among Al and the monovalent cations in quartz (c.f. Cherniak 2010).

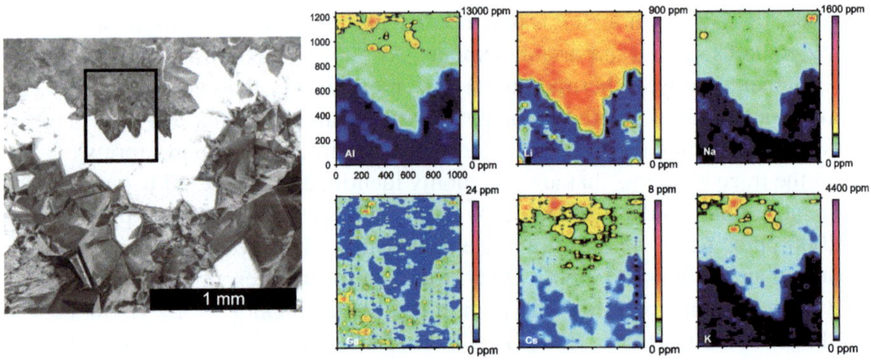


**Fig. 14.6** SEM-CL image and LA-ICP-MS trace element maps of quartz from the McLaughlin, CA epithermal deposit. Al concentrations are correlated with Li, Na, K, and Cs and are negatively correlated with Ga. Ti is below detection limit. Although the CL image that corresponds with the mapped area was lost in an unfortunate hardware malfunction, the CL image shown here is only millimeters away from the location of the original trace element map, and the CL textures shown here are nearly identical to those observed in the location of the trace element map. The boxed area in the CL image closely approximates the CL texture where the trace element maps were obtained

Titanium is also common in hydrothermal quartz, and ranges from a few ppb to a few 100 ppm. Titanium concentrations correlate with Al concentrations in some cases, but as a whole in hydrothermal quartz, there is not a consistent correlation between Ti and Al. Antimony is not typically measured or reported in quartz, however it is present in concentrations of several 100 ppm in some inclusion-free hydrothermal quartz from some low temperature (<300°C) ore deposits (Rusk et al. 2011). Phosphorous, B, Fe, Cl, and Ca are present in various quantities up to a few 100 ppm in some hydrothermal quartz, but no correlations among these elements or between them and any other elements have been recognized. Tin, Ge, Ga, and As are typically present in concentrations below ~5 ppm and also do not consistently correlate with each other or with any other element.

## *14.3.3 Correlations Between CL and Trace Elements in Hydrothermal Quartz*

Quartz monochromatic CL intensity variations almost always coincide with measurable trace element variations. However there is no consistent correlation between CL intensity and trace element abundance that applies to all quartz. This is because specific intrinsic and extrinsic defects in the quartz structure lead to luminescence at specific wavelengths. Monochromatic CL intensity, typically obtained from SEM-based CL imaging is not an absolute quantity, but represents the sum of the various CL emission peaks over the wavelengths of CL emissions (Tarashchan and Waychunas 1995; Stevens-Kalceff 2009; Götte et al. 2011).

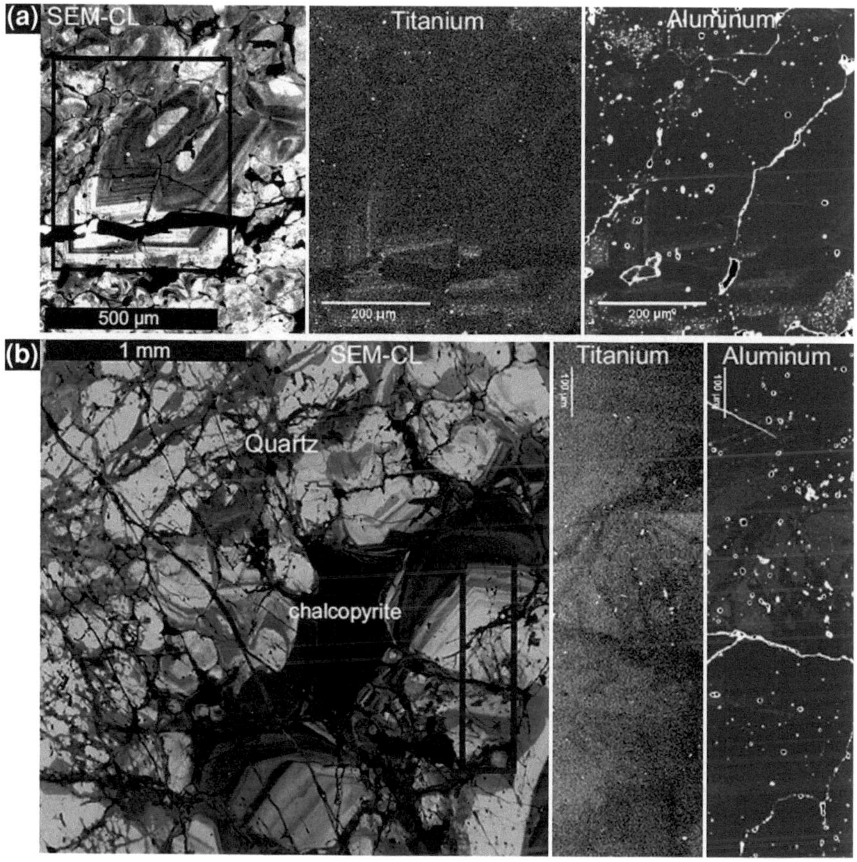


**Fig. 14.7** CL images of hydrothermal vein quartz and corresponding Al and Ti maps from **a** Los Pelambres; and **b** El Teniente, porphyry Cu deposits, Chile. Trace element maps were acquired by electron microprobe (cf. Rusk et al. 2008). Titanium concentrations correlate very closely with CL intensity. The correlation between Al and CL intensity is also close

In low temperature quartz ($<\sim 300°C$), where Ti concentrations are less than a few ppm, CL intensity variations commonly correlate with variations in the concentrations of Al and the correlated monovalent cations. In some quartz, the correlation between Al concentration and CL intensity is positive and sometimes the correlation is negative. On the other hand, quartz formed at temperatures above $\sim 400°C$, where Ti concentrations are greater than a few ppm, shows a consistent correlation between CL intensity and Ti concentration, such that increased Ti concentration is always reflected by increased CL intensity (Fig. 14.7). In this quartz, Al is also typically closely correlated with CL intensity, but the correlation between Al concentration and CL intensity is not as consistent as the correlation between Ti concentration and CL intensity. The correlation between CL intensity and Ti concentration has been shown in igneous quartz (Wark et al. 2007; Jacamon and Larsen 2009) and also in vein quartz from several porphyry copper deposits

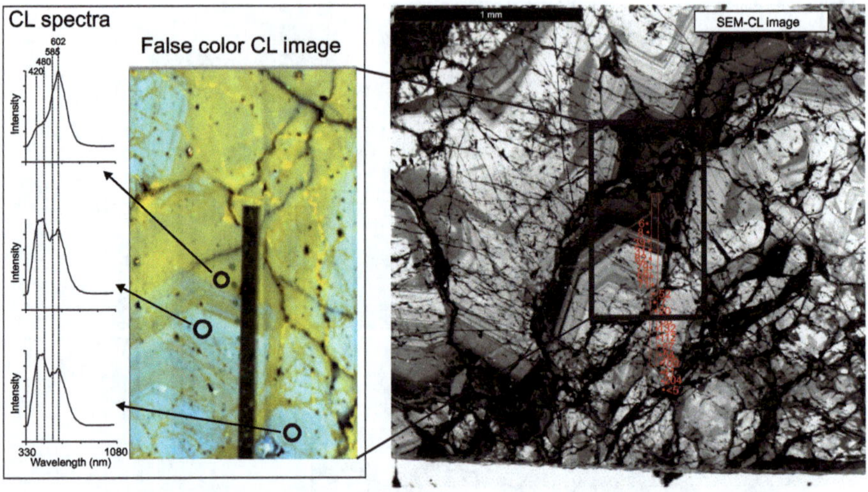


**Fig. 14.8** Monochromatic SEM-CL image with a false color spectral CL image. The wavelengths of the CL emissions at various points are shown. Ti concentrations measured by electron microprobe are also shown. There is a very close correlation between CL intensity and Ti concentration. CL-bright quartz is characterized by Ti concentrations in the range of ~80–160 ppm. Here the quartz shows a strong emission band in the range of 420–480 nm as well as another band in the range of 600 nm. The later, CL-darker quartz that is related to sulfide deposition (not shown) contains less than 10 ppm Ti. In this quartz, the CL emissions are dominated by the broad 600 nm peak and the 420–480 nm peak is barely visible. These spectral emissions and Ti concentrations are typical of vein quartz from porphyry copper deposits

including Butte, Bingham, Oyu Tolgoi, and El Slvador (Landtwing and Pettke 2005; Rusk et al. 2006; Müller et al. 2010; Donovan et al. 2011). Here the same relationship is shown for vein quartz from the Los Pelambres and El Teniente porphyry Cu deposits (Figs. 14.7 and 14.8). Other elements such as Fe, Ge, P, and Sb do not show systematic relationships to CL intensity variations.

The strong correlation between CL intensity and Ti concentration in hydrothermal quartz, where Ti concentrations are above several ppm, suggests that the substitution of $Ti^{4+}$ for $Si^{4+}$ in the quartz structure leads to increased CL intensity. This hypothesis is supported by comparisons between quartz trace element concentrations and spectral maps of quartz CL emissions from numerous hydrothermal ore deposits. The 450 nm (~2.7 eV) emission band dominates in Ti-bearing quartz (Fig. 14.8), such as from porphyry copper deposits, but it is small or absent in low temperature quartz (<~300°C), such as that from epithermal ore deposits (Fig. 14.9). An emission band in this spectral range has also been shown to be important in Ti-bearing igneous quartz (Müller et al. 2002; Wark and Spear (2005). Further, in vein quartz from porphyry type deposits, the intensity of the 450 nm emission band correlates with Ti concentration such that in CL-bright quartz with tens or hundreds of ppm Ti, the 450 nm peak dominates, but in CL-darker quartz, containing only a few ppm Ti, the 450 nm band is present but

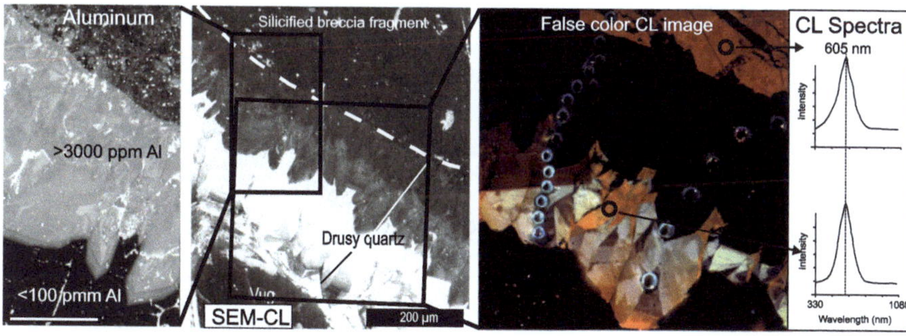


**Fig. 14.9** Monochromatic SEM-CL image, Al map, false color CL image, and CL spectra of hydrothermal breccia-infill quartz from the Jerritt Canyon Carlin-type Au deposit, NV. The CL-bright euhedral quartz contains <100 ppm Al, and <2 ppm Ti. This quartz is characterized by a broad peaks at 600 nm. CL-dark euhedral quartz is characterized by >3,000 ppm Al (and also <2 ppm Ti) and the spectral emission is weak, but has a broad peak around 605 nm (not shown). The CL-dark chert, which is a breccia fragment, shows similar luminescence characteristics to the late CL-bright euhedral quartz, with a peak ~605 nm

subordinate and a broad peak at 600 nm dominates. The broad 600 nm peak was observed in hydrothermal quartz of all origins, regardless of temperature of formation, and dominates in low temperature quartz, such as that found in epithermal deposits. This peak has not been recognized before and is interpreted here to be a composite of overlapping peaks at ~580 nm and ~620–650 nm, both of which are common in quartz (see Götte et al. 2011; Stevens-Kalceff 2009). A peak at 380 nm is common in Al-rich quartz and diminishes rapidly upon interaction with the electron beam (Götte et al. 2011; Stevens-Kalceff 2012). This peak was not observed here, probably resulting from CL imaging by SEM prior to obtaining spectral analyses, thus reducing the intensity of this peak.

### *14.3.4 Implications for Hydrothermal Systems and Ore Deposits*

Like CL textures, concentrations of trace elements vary systematically among ore deposit types and can be used to fingerprint the type of ore deposit based on trace element composition. Figure 14.10 shows Ti and Al concentrations in quartz from around 30 hydrothermal ore deposits including porphyry-type (Cu-Mo-Au) deposits, orogenic Au deposits, and epithermal deposits. Each of these ore types can be distinguished from one another based on Al and Ti concentrations alone (Fig. 14.10) (Rusk 2009). Results from both microprobe and LA-ICP-MS show that most porphyry type quartz contains between ~1 and 200 ppm Ti. In porphyry-type quartz, concentrations of <~15 ppm Ti are common in CL-dark quartz related to sulfide deposition (Figs. 14.7 and 14.8) (Landtwing and Pettke 2005; Rusk et al. 2006; Müller et al. 2010). Quartz from porphyry-type deposits typically contains between 50 and 500 ppm Al. A fairly linear correlation exists

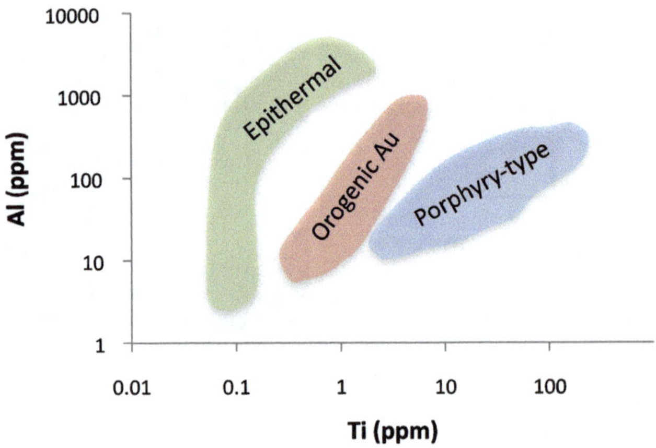


**Fig. 14.10** Titanium versus Aluminum concentrations in hydrothermal quartz from epithermal, orogenic Au, and porphyry-type deposits. The different deposit types can be distinguished from one another based on trace element concentrations alone. Quartz trace elements reflect the physio-chemical conditions of quartz growth and subsequent modifications

between these elements with an Al/Ti ratio between ~1 and 10. In distinct contrast, quartz from epithermal ore deposits always contains less than about 3 ppm Ti and Al concentrations range widely between ~20 and 4,000 ppm Al. Al/Ti ratios range widely between ~100 and ~10,000. Orogenic Au quartz is intermediate between these two with Al concentrations typically between 100 and 1,000 and Ti concentrations between 1 and 10 ppm. Al/Ti ratios vary from 10 to 100.

The systematic variation in trace element concentration among quartz crystallized under different physical and chemical conditions is evidence that the trace elements incorporated in quartz reflect geologic conditions of formation, and subsequent modifications. The relations between most quartz trace elements and the conditions of formation are not well constrained, but recent studies suggest that pressure, temperature (Thomas et al. 2010), fluid composition (Perny et al. 1992; Rusk et al. 2008; Lehmann et al. 2011), and crystallization rate (Lowenstern and Sinclair 1996) all influence quartz trace element composition. A major aim of recent studies has been to determine quantitative links between the conditions of quartz formation, and the resultant trace element endowment of the quartz. For example the experimentally calibrated Ti in quartz thermobarometer, TitaniQ (Wark and Watson 2006; Thomas et al. 2010), is commonly applied to understand pressure and temperature variations in metamorphic, igneous, and hydrothermal systems. In hydrothermal quartz, combined application of CL petrography and the Ti in quartz thermobarometer (Figs. 14.7 and 14.8), shows that in many mineralized veins from porphyry copper deposits, chalcopyrite is related to late CL-dark quartz that cuts earlier CL-bright quartz. Early CL bright quartz typically contains between about 50 and 200 ppm Ti, whereas later sulfide-related CL-dark quartz typically contains <10 ppm Ti. Rutile is common in early CL-bright quartz, but less common in volumetrically minor later CL-dark quartz. Application of the Ti in quartz thermobarometer suggests that chalcopyrite precipitation occurred after the hydrothermal system cooled from roughly 550–650 to ~375–425°C. Similar results to those shown here for El Teniente and Los Pelambres were obtained for the Bingham Canyon (Landtwing and Pettke 2005), Butte (Rusk et al. 2006), and the Oyu Tolgoi porphyry copper deposits (Müller et al. 2010).

Other studies have attempted to understand the controls on Al concentrations in quartz, and although little experimental data exists, Perny et al. (1992) and Rusk et al. (2008) suggest that quartz Al concentrations are affected by Al solubility in the hydrothermal fluid, which is strongly influenced by pH. Lehmann et al. (2011) also recently suggested a link between fluid composition and quartz Al concentrations, although $CO_2$ concentration in the fluid was the important controlling factor rather than pH. Still, the controls on Al concentrations remain to be deciphered. Understanding the controls on Al concentrations in quartz would be a great advance since Al is typically the most abundant and widely ranging trace element in quartz from any environment. Further studies of natural and synthetic quartz are required to quantify relationships between physical and chemical variables and quartz composition.

## 14.4 Conclusions

Scanning electron microscope-cathodoluminescence textures provide unique insights into the origin of hydrothermal quartz. These textures distinguish the chronologic relations among quartz generations and can be used to relate fluid inclusions, mineral precipitation, and isotopic data to specific mineralization events. The CL textures and related trace element fluctuations in quartz reflect the fluid pressure, temperature, and composition at the time of precipitation, as well as the rate of fluctuation of these variables. Together CL textures and quartz trace element compositions can be used to fingerprint the deposit type from which the quartz likely originated. Since quartz is ubiquitous in hydrothermal ores and resilient against physical and chemical weathering, this discrimination technique can readily be applied to exploration for hydrothermal ore deposits.

**Acknowledgments** I would like to thank Mark Reed, David Krinsley, Alan Koenig, Heather Lowers, Yi Hu, and Kevin Blake for stimulating discussions about quartz and quartz analysis. I also thank Thomas Götte and Jens Götze for constructive reviews, which improved the content and presentation of the manuscript.

## References

Adams SF (1920) A microscopic study of vein quartz. Econ Geol 15:623–664

Allan MM, Yardley BWD (2007) Tracking meteoric water infiltration into a magmatic hydrothermal system: A cathodoluminescence, oxygen isotope, and trace element study of quartz from Mt. Leyshon. Australia Chem Geol 240:343–360. doi:10.1016/j.chemgeo.2007.03.004

Behr W, Thomas J, Hervig R (2011) Calibrating Ti concentrations in quartz for SIMS determinations using NIST silicate glasses and application to the TitaniQ geothermobarometer. Am Mineral 96:1100–1106

Bernet M, Bassett K (2005) Provenance analysis by single-quartz grain SEM-CL/optical microscopy. J Sed Res 75:492–500

Bignall G, Sekine K, Tsuchiya N (2004) Fluid–rock interaction processes in the Te Kopia geothermal field (New Zealand) revealed by SEM-CL imaging. Geothermics 33:615–635

Boggs S Jr, Krinsley D (2006) Application of cathodoluminescence imaging to study of sedimentary rocks. Cambridge University Press, New York

Boggs S Jr, Kwon Y, Goles GG, Rusk BG, Krinsley D, Seyedolali A (2002) Is quartz cathodoluminescence color a reliable provenance tool? A quantitative examination. J Sed Res 72:408–415

Boiron M, Essarraj S, Sellier E, Cathelineau M, Lespinasse M, Poty B (1992) Identification of fluid inclusions in relation to their host microstructural domains in quartz by cathodoluminescence. Geochim et Cosmochim Acta 56:175–185

Breiter K, Müller A (2009) Evolution of rare-metal granitic magmas documented by quartz chemistry. Eur J Mineral 21:335–346

Chang Z, Meinert LD (2004) The magmatic hydrothermal-transition-evidence from quartz phenocryst textures and endoskarn abundance in Cu-Zn skarns at the Empire Mine, Idaho, USA. Chem Geol 210:149–171

Cherniak DJ (2010) Diffusion in quartz, melilite, silicate perovskite, and mullite. In: Zhang Y, Cherniak DJ (eds) Diffusion in minerals and melts, vol 72, p 735–756. Reviews in mineralogy and geochemistry, mineralogical society of america and the geochemical society, Chantilly, Virginia

Dong G, Morrison G, Jaireth S (1995) Quartz textures in epithermal veins, queensland-classification, origin, and implication. Econ Geol 90:1841–1856

Donovan J, Lowers H, Rusk B (2011) Improved electron probe microanalysis of trace elements in quartz. Am Mineral 96:274–282

Dowling K, Morrison G (1989) Application of quartz textures to the classification of gold deposits using north Queensland examples. Econ Geol Monogr 6:342–355

Fischer M, Roeller K, Kuester M, Stoeckhert B, McConnell VS (2003) Open fissure mineralization at 2600 m depth in long valley exploratory well (California); insight into the history of the hydrothermal system. J Volc Geotherm Res 127:347–363

Flem B, Larsen RB, Grimstvedt A, Mansfield J (2002) In situ analysis of trace elements in quartz by using laser ablation inductively coupled plasma mass spectrometry. Chem Geol 182:237–247

Götte T, Richter DK (2003) Late palaeozoic and early Mesozoic hydrothermal events in the northern Rhenish Massif; results from fluid inclusion analyses and cathodoluminescence investigations. J Geochem Explor 78–79:531–535

Götte T, Pettke T, Ramseyer K, Koch-Müller M, Mullis J (2011) Cathodoluminescence properties and trace element signature of hydrothermal quartz: a fingerprint of growth dynamics. Am Min 96:802–813

Götze J (2009) Chemistry, textures and physical properties of quartz- geological interpretation and technical application. Min Mag 73:645–671

Götze J, Plötze M, Fuchs H, Habermann D (1999) Defect structure and luminescence behavior of agate–results of electron paramagnetic resonance (EPR) and cathodoluminescence (CL) studies. Min Mag 63:149–163

Götze J, Plötze M, Habermann D (2001) Origin, spectral characteristics and practical applications of the cathodoluminescence of quartz—a review. Min Pet 71:225–250

Götze J, Plötze M, Trautmann T (2005) Structure and luminescence characteristics of quartz from pegmatite's. Am Min 90:13–21

Graupner T, Götze J, Kempe U, Wolf D (2000) CL for characterizing quartz and trapped fluid inclusions in mesothermal quartz veins: Muruntau Au ore deposit, Uzbekistan. Min Mag 64:1007–1016

Ioannou SE, Götze J, Weiershäuser L, Zubowski SM, Spooner ETC (2004) Cathodoluminescence characteristics of Archean VMS–related quartz: Noranda, Ben Nevis, and Matagami districs, Abitibi subprovince. Canada Geochem Geophys Geosys. doi:10.1029/2003GC000613

Jacamon F, Larsen RB (2009) Trace element evolution of quartz in the charnockitic Kleivan granite, SW Norway: the Ge/Ti ratio of quartz as an index of igneous differentiation. Lithos 107:281–191

Jourdan A, Vennemann TW, Mullis J, Ramseyer K (2009) Oxygen isotope sector zoning in natural hydrothermal quartz. Min Mag 73:615–632

Kanaori Y (1986) A SEM cathodoluminescence study of quartz in mildly deformed granite from the region of Atotsugawa fault, central Japan. Tectonophys 131:133–146

King EM, Barrie CT, Valley JW (1997) Hydrothermal alteration of oxygen isotope ratios in quartz phenocrysts, Kidd Creek mine, Ontario: magmatic values are preserved in zircon. Geol 25:1079–1082

Landtwing M, Pettke T (2005) Relationships between SEM–cathodoluminescence response and trace element composition of hydrothermal vein quartz. Am Min 90:122–131

Landtwing M, Pettke T, Halter WE, Heinrich CA, Redmond PB, Einaudi MT, Kunze K (2005) Copper deposition during quartz dissolution by cooling magmatic hydrothermal fluids: the Bingham porphyry. Earth Planet Sci Lett 235:229–243

Larsen RB, Henderson I, Ihlen PM, Jacamon F (2004) Distribution and petrogenetic behaviour of trace elements in granitic quartz from South Norway. Contrib Mineral Petrol 147:615–628

Larsen RB, Jacamon F, Krontz A (2009) Trace element chemistry and textures of quartz during the magmatic-hydrothermal transition of Oslo Rift granites. Min Mag 73:691–707

Leach DL, Marsh E, Emsbo P, Rombach CS, Kelley KD, Anthony M (2005) Nature of Hydrothermal Fluids at the Shale-Hosted Red Dog Zn-Pb-Ag Deposits, Brooks Range, Alaska. Econ Geol 99:1449–1480

Lehmann K, Berger A, Götte T, Ramseyer K, Wiedenbeck M (2009) Growth related zonations in authigenic and hydrothermal quartz characterized by SIMS-, EPMA-, SEM-CL- and SEM-CC-imaging. Min Mag 73:633–643

Lehmann K, Pettke T, Ramseyer K (2011) Significance of trace elements in syntaxial quartz cement, Haushi group sandstones, Sultanate of Oman. Chem Geol 280:47–57

Lowenstern JB, Sinclair WD (1996) Exsolved magmatic fluid and its role in the formation of comb–layered quartz at the Cretaceous Logtung W-Mo deposit, Yukon Territory, Canada. Trans Roy Soc Edinburgh Earth Sci 87:291–303

Lubben J (2004) Silicification across the Betze-Post carlin-type Au deposit; clues to ore fluid properties and sources, northern Carlin Trend, Nevada. Unpublished masters thesis, University of Nevada, Las Vegas

Machel HG, Burton EA (1991) Factors governing cathodoluminescence in calcite and dolomite, and their implications for studies of carbonate diagenesis. In: Barker CE, Kopp OC (eds) Luminescence microscopy and spectroscopy: qualitative and quantitative applications: SEPM short course, Tulsa, 25:37–57

Marshall DJ (1988) Cathodoluminescence of geological materials. Unwin Hyman, Boston

Miyoshi N, Yamaguchi Y, Makino K (2005) Successive zoning of Al and H in hydrothermal vein quartz. Am Mineral 90:310–315

Monecke T, Kempe U, Götze J (2002) Genetic significance of the trace element content in metamorphic and hydrothermal quartz: A reconnaissance study. Earth Planet Sci Let 202:709–724

Müller A, Koch-Müller M (2009) Hydrogen speciation and trace element contents of igneous, hydrothermal and metamorphic quartz from Norway. Min Mag 73:569–583

Müller A, Wanvik J (2012) Petrological and chemical characterisation of high-purity quartz deposits with examples from Norway, This volume

Müller A, Lennox P, Trzebski R (2002) Cathodoluminescence and micro-structural evidence for crystallization and deformation processes of granites in the Eastern Lachlan Fold Belt (SE Australia). Cont Min Pet 143:510–524

Müller A, Wiedenbeck M, Van den Kerkhof AM, Kronz A, Simon K (2003) Trace elements in quartz: a combined electron microprobe, secondary ion mass spectrometry, laser–ablation ICPMS, and cathodoluminescence study. Eur J Min 15:747–763

Müller A, Breiter K, Seltman R, Pecskay Z (2005) Quartz and feldspar zoning in the Eastern Erzgebirge pluton (Germany, Czech Republic): Evidence of multiple magma mixing. Lithos 80:201–207

Müller A, Herrington R, Armstrong R, Seltman R, Kirwin D, Stenina N, Kronz A (2010) Trace elements and cathodoluminescence of quartz in stockwork veins of Mongolian porphyry-style deposits. Mineralium Deposita 45:707–727

Mullis J, Dubessy J, Poty B, O'Neil J (1994) Fluid regimes during late stages of a continental collision: Physical, chemical, and stable isotope measurements of fluid inclusions in fissure quartz from a geotraverse through the Central Alps, Switzerland. Geochim Cosmochim Acta 58:2239–2267

Pagel M, Barbin V, Blanc P, Ohnenstetter D (eds) (2000) Cathodoluminescence in geoscience. Springer, Berlin, Heidelberg, New York, p 514

Penniston-Dorland SC (2001) Illumination of vein quartz textures in a porphyry copper ore deposit using scanned cathodoluminescence: grasberg igneous complex, Irian Jaya, Indonesia. Am Min 86:652–666

Peppard BT, Steele IM, Davis AM, Wallace PJ, Anderson AT (2001) Zoned quartz phenocrysts from the rhyolitic Bishop Tuff. Am Min 81:1034–1052

Perny B, Eberhardt P, Ramseyer K, Mullis J, Pankrath R (1992) Microdistribution of Al, Li, and Na in alpha quartz: possible causes and correlation with short-lived cathodoluminescence. Am Min 77:534–544

Ramseyer K, Mullis J (2000) Geologic application of cathodoluminescence of silicates. In: Pagel M, Barbin B, Blanc C, Ohnstetter (eds) Cathodoluminescence in geosciences. Springer, Berlin, pp 177–191

Ramseyer K, Baumann J, Matter A, Mullis J (1988) Cathodoluminescence colours of alpha-quartz. Mineral Mag 52:669–677

Redmond PB, Einaudi MT, Inan EE, Landtwing MR, Heinrich CA (2004) Copper deposition by fluid cooling in intrusion–centered systems: new insights from the Bingham porphyry ore deposit, Utah. Geol 32:217–220

Richter DK, Götte T, Götze J, Neuser RD (2003) Progress in application of cathodoluminescence in sedimentary petrology. Min Pet 79:127–166

Rusk B (2009) Insights into hydrothermal processes from cathodoluminescence and trace elements in quartz. In: Williams PJ, Rusk B, Oliver N (eds) Smart sciences for exploration and mining, proceedings of the 10th Biennial SGA meeting. Townsville, pp 749–751. https://www.e-sga.org/index.php?id=231&tx_commerce_pi1[showUid]=1650&tx_commerce_pi1[catUid]=43&cHash=3feeb527d1

Rusk BG, Reed MH (2002) Scanning electron microscope–cathodoluminescence of quartz reveals complex growth histories in veins from the Butte porphyry copper deposit, Montana. Geol 30:727–730

Rusk B, Reed M, Dilles J, Kent A (2006) Intensity of quartz cathodoluminescence and trace element content of quartz from the porphyry copper deposit in Butte, Montana. Am Min 91:1300–1312

Rusk B, Lowers H, Reed M (2008) Trace elements in hydrothermal quartz; relationships to cathodoluminescent textures and insights into hydrothermal processes. Geol 36:547–550

Rusk B, Koenig A, Lowers H (2011) Visualizing trace element distribution in quartz using cathodoluminescence, electron microprobe, and laser ablation inductively coupled plasma mass spectrometry. Am Mineral 96:703–708

Sekine K (2003) Development of fracture and fluid migration in granite during uplift and emplacement. Dissertation, Tohoku University, p 256

Seyedolali A, Krinsley DH, Boggs S, O'Hara PF, Dypvik H, Goles GG (1997) Provenance interpretation of quartz by scanning electron microscope-cathodoluminescence fabric analysis. Geol 25:783–786

Sippel RF (1968) Sandstone petrology, evidence from luminescence petrography. J Sed Pet 38: 530–554
Smith JV, Stenstrom RC (1965) Electron-excited luminescence as a petrological tool. J Geol 73:627–635
Spear FS, Wark DA (2009) Cathodoluminescence imaging and titanium thermometry in metamorphic quartz. J Metamorph Geol 27:187–205
Sprunt ES, Dengler LA, Sloan D (1978) Effects of metamorphism on quartz cathodoluminescence. Geol 6:305–308
Spurr JE (1926) Successive banding around rock fragments in veins. Econ Geol 21:519–537
Stenina NG (2004) Water-related defects in quartz. Bullet Geosci 79:251–268
Stevens-Kalceff M (2009) Cathodoluminescence microcharacterization of point defects in a-quartz. Min Mag 73:585–605
Stevens-Kalceff MA (2012) Cathodoluminescence microanalysis of the defect microstructures of bulk and nanoscale ultrapure silicondioxide polymorphs for device applications. (this volume)
Stevens-Kalceff MA, Phillips MR (1995) Cathodoluminescence microcharacterization of the defect structure of quartz. Phys Rev B 52:3122–3134
Stevens-Kalceff MA, Phillips MR, Moon AR, Kalceff W (2000) Cathodoluminescence microcharacterization of silicon dioxide polymorphs. In: Pagel M, Barbin B, Blanc C, Ohnstetter D (eds) Cathodoluminescence in geosciences, Springer, Berlin, pp 193–224
Tarashchan AN, Waychunas G (1995) Interpretation of luminescence spectra in terms of band theory and crystal field theory. Sensitization and quenching, photoluminescence, radioluminescence, and cathodoluminescence. In. Marfunmin AS (ed) Advanced mineralogy 2, Methods and instrumentations: results and recent developments. Springer, Berlin, pp 124–135
Thomas SM, Koch-Müller M, Reichart P, Rhede D, Thomas R, Wirth R (2009) IR calibrations for water determination in olivine, r-GeO2 and SiO2 polymorphs. Phys Chem Miner 36:489–509
Thomas JB, Watson EB, Spear FS, Shemella PT, Nayak SK, Lanzirotti A (2010) TitaniQ under pressure: the effect of pressure and temperature on the solubility of Ti in quartz. Contrib Mineral Petrol 160:743–759
Valley JW, Graham CM (1996) Ion microprobe analysis of oxygen isotope ratios in quartz from Skye granite: healed micro–cracks, fluid flow, and hydrothermal exchange. Contrib Min Pet 124:225–234
Vearncombe JR (1993) Quartz vein morphology and implications for formation depth and classification of Archaean gold-vein deposits. Ore Geol Rev 8:407–424
Walker G (2000) Physical parameters for the identification of luminescence centres in minerals. In: Pagel M, Barbin B, Blanc C, Ohnstetter D (eds) Cathodoluminescence in geosciences. Springer, Berlin, pp 23–40
Wark DA, Spear FS (2005) Titanium in quartz: cathodoluminescence and thermometry. Geochim Cosmochim Acta Suppl 69:A592
Wark DA, Watson BE (2006) TitaniQ: a titanium in quartz geothermometer. Contrib Min Pet 152:743–754. doi:10.1007/s00410-006-0132-1
Wark DA, Hildreth W, Spear FS, Cherniak DJ, Watson EB (2007) Pre-eruption recharge of the Bishop magma system. Geol 35:235–238
Wilkinson JJ, Johnston JD (1996) Pressure fluctuations, phase separation, and gold precipitation during seismic fracture propagation. Geol 24:395–398
Wilkinson JJ, Boyce AJ, Earls G, Fallick AE (1999) Gold remobilization by low–temperature brines: evidence from the Curraghinalt gold deposit, northern Ireland. Econ Geol 94:289–296

# Chapter 15
# Quartz Regeneration and its Use as a Repository of Genetic Information

**Ulf Kempe, Jens Götze, Enchbat Dombon, Thomas Monecke and Mikhail Poutivtsev**

**Abstract** Quartz represents one of the most widespread minerals and is widely used in geosciences to reconstruct physic-chemical conditions of rock and mineral formation. However, interpretation of analytical data may be limited by the ability of quartz to regenerate during secondary alteration processes occurring under metamorphic or hydrothermal conditions. This behaviour distinguishes quartz from most minerals commonly associated with. Primary genetic information is obliterated during quartz regeneration. This includes features related to the real structure of quartz, but also to fluid and mineral inclusions. The present contribution examines examples covering various fields of mineral research, namely the genetic interpretation of trace element content in quartz, quartz provenance analysis using cathodoluminescence (CL) colour imaging, and the analysis of mineral and fluid inclusions in quartz. It is demonstrated in all cases that care

U. Kempe (✉) · J. Götze · E. Dombon · T. Monecke · M. Poutivtsev
Institute of Mineralogy, TU Bergakademie Freiberg, Brennhausgasse 14, 09596 Freiberg, Germany
e-mail: kempe@mineral.tu.freiberg.de

*Present Address:*
E. Dombon
Department of Sciences, Technology, and Innovation, Mongolian University of Sciences and Technology, P.O. Box 520, Ulaanbatar 210646, Mongolia

*Present Address:*
T. Monecke
Department of Geology and Geological Engineering, Colorado School of Mines, 1516 Illinois Street, Golden, CO 80401, USA

*Present Address:*
M. Poutivtsev
Maier-Leibnitz-Laboratory, Faculty of Physics, TU Munich, James-Franck-Strasse, 85748 Garching, Germany

J. Götze and R. Möckel (eds.), *Quartz: Deposits, Mineralogy and Analytics*,
Springer Geology, DOI: 10.1007/978-3-642-22161-3_15,

needs to be taken when interpreting genetic information encoded. Distinction of features related to primary growth or secondary alteration is not simple and requires application of complementary analytical techniques.

## 15.1 Introduction

Quartz is one of the most common rock-forming minerals in the earth crust. Due to its widespread occurrence and formation under various geological conditions, the behaviour and real structure of quartz are an important focus of geoscientific research. Quartz properties are studied in petroleum exploration and exploitation, in mineral exploration, and in manufacturing industries using quartz as a raw material.

Despite the low variability in the main chemical and structural parameters, a wide range of analytical methods and approaches is now routinely used in investigations on quartz. These include trace element analysis (Bambauer 1961; Dennen 1966; Novgorodova et al. 1984; Watt et al. 1997; Monecke et al. 2002; Müller et al. 2003; Götze et al. 2004; Larsen et al. 2004; Landtwing and Pettke 2005; Rusk et al. 2006; Breiter and Müller 2009; Whiting et al. 2010; Agangi et al. 2011; Götte et al. 2011; Rusk et al. 2011), oxygen isotope analysis (Blatt 1987; Girard and Deynoux 1991; Vennemann et al. 1992; Barton et al. 1992; Onasch and Vennemann 1995; Williams et al. 1997; Jourdan et al. 2009), electron paramagnetic resonance (EPR) spectroscopy (Götze and Plötze 1997; Götze et al. 1999; Botis et al. 2005, 2006), cathodoluminescence (CL) imaging (Seyedolali et al. 1997; Watt et al. 1997; Götze 2000; Penniston-Dorland 2001; Rusk and Reed 2002; Larsen et al. 2004; Rusk et al. 2011), and CL spectroscopy (Claffy and Ginther 1959; Ramseyer and Mullis 1990; Perny et al. 1992; Götze 2000; Götze et al. 2001; Lehmann et al. 2009). Quartz is also used together with associated minerals in the context of fluid inclusion analysis (e.g., Roedder 1984; Samson et al. 2003) and melt inclusion studies (e.g., Frezzotti 2001; Webster 2006).

Genetic interpretation of data obtained by these techniques often relies on the assumption that quartz is a resistant repository for genetic information and that trace element content, CL characteristics, or melt and fluid inclusion inventory directly relate to the environment of quartz formation. This view is based, at least in part, on the relatively high hardness of quartz, the lack of a distinct cleavage, and its resistance to weathering (e.g., Füchtbauer et al. 1982; Zuffa 1985; Hallbauer 1992; Bodnar 2003b; Thomas et al. 2005).

However, growing evidence in the recent literature suggests that this assumption is not always valid. Under some circumstances quartz is more susceptible to secondary alteration than many associated minerals. Alteration often proceeds without significant destruction or modification of the quartz grains and aggregates. Quartz shows a conspicuous tendency to structural regeneration. For instance, recrystallization, ductile deformation, healing of cracks or continuation of crystal growth immediately after brittle deformation, as well as formation of secondary fluid inclusions are common phenomena observed in quartz in magmatic, metamorphic, and hydrothermal environments (e.g., Rusk and Reed 2002; Landtwing and Pettke 2005; Müller et al. 2010). Therefore, information obtained from the investigation of quartz cannot always be simply related to the processes of primary quartz growth. The behaviour of quartz under these conditions appears to differ from many other common minerals.

The present study provides an overview over advantages and pitfalls of some of the analytical techniques commonly used in the study of quartz. Four different case studies are presented that either focused directly on the investigation of quartz (trace element analysis and CL colour imaging) or considered the relationships of quartz to associated minerals (fluid inclusion analysis and analysis of mineral sequences in hydrothermal vein deposits). Although all four case studies had different objectives and employed different analytical techniques, they all demonstrate that quartz does not represent a stable repository of genetic information under metamorphic and hydrothermal conditions but can be used for reconstruction of the alteration history of the mineral.

## 15.2 Quartz with Metamorphic Micro-Fabrics and Hydrothermal Trace Element Signature from Wall Rocks of the Ehrenfriedersdorf Tin Deposit, Erzgebirge, Germany

Quartz is one of the minerals which commonly show low trace-element contents. Only few elements such as aluminium, titanium, iron, gallium, and germanium can substitute for the small and highly charged silicon atoms in the crystal structure (e.g., Bambauer 1961; Weil 1984, 1993; Plötze, 1995; Monecke et al. 2000a; Götze et al. 2004; Lehmann et al. 2011; Rusk et al. 2011). Lithium, sodium, aluminium, and potassium may also be incorporated in interlattice positions within the comparably large structural channels in the crystal structure (e.g., Ramseyer and Mullis 1990; Miyoshi et al. 2005). Most other elements including rubidium, strontium, yttrium, and the rare earth elements (REEs) likely do not fit into the quartz crystal structure and are probably mostly contained in fluid inclusions or adsorbed to internal and external surfaces of quartz grains (Novgorodova et al. 1984, Bottrell et al. 1988; Götze et al. 2004; Haßler et al. 2005). However, the exact mode of incorporation of many elements in quartz is still not well constrained.

Early studies introduced the aluminium content in quartz as a possible geothermometer (Dennen et al. 1970; Agel and Petrov 1990). However, problems may arise when bulk geochemical data are used which cannot distinguish between aluminium replacing silicon and aluminium occurring in some other mode within the quartz sample (Plötze 1995). A recent and more popular approach is the use of the titanium content in quartz for this purpose (Wark and Watson 2006; Wark et al. 2007; Rusk et al. 2008; Spear and Wark 2009). To enable a high spatial resolution, measurements of the Ti content of quartz are typically performed by electron microprobe analysis.

To measure the low-level contents of aluminium, titanium, or other trace elements, more sensitive methods are required such as instrumental neutron activation analysis (INAA) or inductively coupled-plasma mass spectrometry (ICP-MS). This way, quartz of different origin may be distinguished. For instance, a number of previous studies have shown that metamorphic and hydrothermal quartz display different levels of trace element contents (Monecke et al. 2002 and references therein). The uniformly low concentrations of lithium, aluminium, potassium, rubidium, strontium, and yttrium of metamorphic quartz result from the purification of the crystal structure during plastic deformation and recrystallisation under metamorphic conditions (Monecke et al. 2002; Sørensen and Larsen 2009). In addition, the fluid inclusion inventory changes during recrystallisation. As a result, trace element characteristics of quartz change during structural reorganisation and recrystallisation. In contrast to metamorphic quartz, hydrothermal quartz typically shows higher concentrations of the above mentioned trace elements, which is a reflection of the chemical composition of the hydrothermal fluids. In addition to the structural incorporation of impurities during crystal growth, entrapment of fluid inclusions results in the generally higher element concentrations of bulk quartz samples collected from hydrothermal veins (Monecke et al. 2002). Metamorphic and hydrothermal quartz can also be distinguished based on their chondrite-normalised REE patterns (Monecke et al. 2002). Interpretation is, however, complicated by the fact that hydrothermal alteration can also change the trace element concentrations in quartz as documented by quartz from the metamorphic wall rock of the Ehrenfriedersdorf tin deposit, Germany (Monecke et al. 2002).

The Ehrenfriedersdorf tin deposit formed in association with a multiphase Hercynian granite that intruded in a metamorphic host rock succession. Cassiterite mineralisation occurs in greisens bodies within the endocontact and in quartz veins in the endo- and exocontact of the granite (Hösel 1994). Monecke et al. (2002) and Haßler et al. (2005) have examined quartz from pegmatite, quartz—cassiterite veins, and the metamorphic wall rocks. Following a careful procedure of sample preparation to rule out the presence of mineral inclusions, hand-picked bulk samples of quartz were analysed by ICP-MS for their trace element contents (Monecke et al. 2002).

The sampled pegmatite is located in the endocontact of the hidden granite (Haßler et al. 2005). Vein quartz from both endo- and exocontact positions was sampled (Haßler et al. 2005). Two samples from the wall rocks were taken from typical metamorphic quartz lenses: one (ED-1) at a distance of approximately 300 m from the steeply sloping granite contact and the other (ED-520) 28 m from the shallow granite surface within the ore field (Monecke et al. 2002).

Haßler et al. (2005) showed that the pegmatite quartz as well as quartz from endo- and exocontact quartz—cassiterite veins are typified by high contents of lithium, aluminium, manganese, rubidium, and low concentrations of strontium (Table 15.1). Chondrite-normalised REE distribution patterns of these quartz samples show a relative enrichment of the heavy REEs, which is typical for

**Table 15.1** Trace element content of quartz lenses from the metamorphic host rocks of the Ehrenfriedersdorf tin deposit, Germany, in comparison to quartz from the cassiterite-bearing ore veins and pegmatite

| | Metamorphic quartz ED-1 (ppm) | Metamorphic quartz in the alteration halo ED-520 (ppm) | Vein and pegmatite quartz (ppm) |
|---|---|---|---|
| Li | 0.4 | 0.9 | 15–60 |
| Al | 14.5 | 121 | 150–640 |
| Sr | 0.22 | 0.19 | 0.06–0.09 |
| Rb | 0.05 | 0.29 | 0.8–1.2 |
| Mn | 0.4 | 0.5 | 1–2 |

Data from Monecke et al. (2002) and Haßler et al. (2005)

minerals and rocks from tin deposits (Monecke et al. 2011 and references therein). Samples from the endocontact have a strong negative europium anomaly while vein quartz from the exocontact displays a positive europium anomaly (Haßler et al. 2005).

The quartz sample from the metamorphic lens outside the alteration halo displays typical characteristics of metamorphic quartz with low contents of lithium, aluminium, manganese, strontium, and rubidium. The sample close to the granite contact has trace element characteristics that are intermediate between metamorphic quartz and quartz from the ore veins with elevated lithium, aluminium, and rubidium concentrations (Monecke et al. 2002; cf. Table 15.1). The two quartz samples collected from the wall rocks at different distances to the granite intrusion also have distinct REE patterns (Fig. 15.1a). The metamorphic quartz sample collected distal to the intrusion (ED-1) shows a distribution pattern typical of metamorphic quartz samples with an enrichment of the light REEs over heavy REEs, a positive cerium anomaly, and no significant anomaly of europium. In contrast, the sample collected close to the granite (ED-520) has a small positive cerium anomaly and a distinct positive europium anomaly. The heavy REEs are distinctly enriched.

The observed differences in the trace element abundances suggest that the metamorphic quartz lens close to the granite was overprinted by hydrothermal fluids related to the tin mineralisation without visible changes in the metamorphic fabrics. At the same time, the metamorphic quartz located at a larger distance was not notably affected by hydrothermal alteration. Both samples show similar microfabrics and mineral associations with the quartz being intergrown with biotite, muscovite, minor garnet, albite, apatite, ilmenite, and zircon. The only difference is the occurrence of rare inclusions of fluorite and of euhedral tourmaline in the altered quartz. Fluorite and tourmaline are also associated with the tin mineralisation in the ore zones.

The conclusion that the metamorphic sample collected close to the granite contact was affected by hydrothermal alteration is also supported by CL microscopy. The metamorphic quartz sampled away from the granite contact showed mostly stable brownish CL typical of metamorphic quartz, while the hydrothermally altered quartz has a transient bluish CL that is replaced by brownish CL during electron irradiation. Such behaviour is characteristic for some types of hydrothermal quartz (Fig. 15.2 see below).

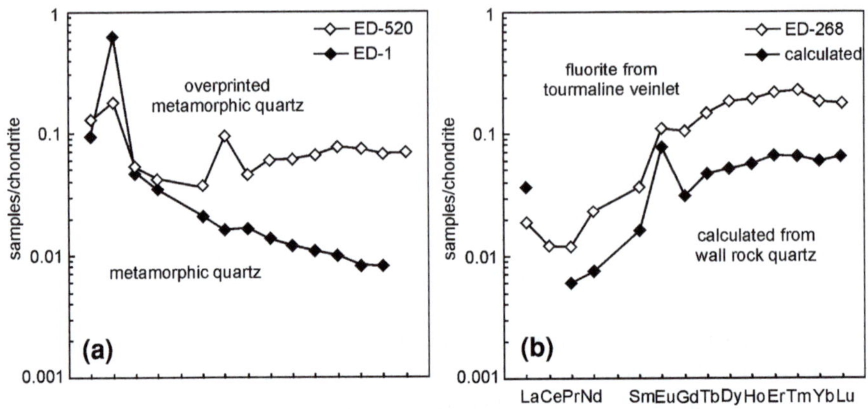


**Fig. 15.1** REE distribution patterns of quartz from lenses in the metamorphic host rocks and of fluorite from an early tourmaline veinlet (Ehrenfriedersdorf tin deposit, Erzgebirge, Germany). **a** Comparison of REE distribution patterns of quartz from metamorphic quartz lenses distal (ED-1) and proximal (ED-520) to the granite intrusion and ore zones. **b** REE distribution pattern for the hydrothermal fluid, calculated from the difference between the patterns of the distal and proximal quartz samples, in comparison to fluorite from an early tourmaline veinlet (ED-268). ICP-MS data of quartz from Monecke et al. (2002). ICP-MS data of fluorite from Dulski and Kempe (unpublished)

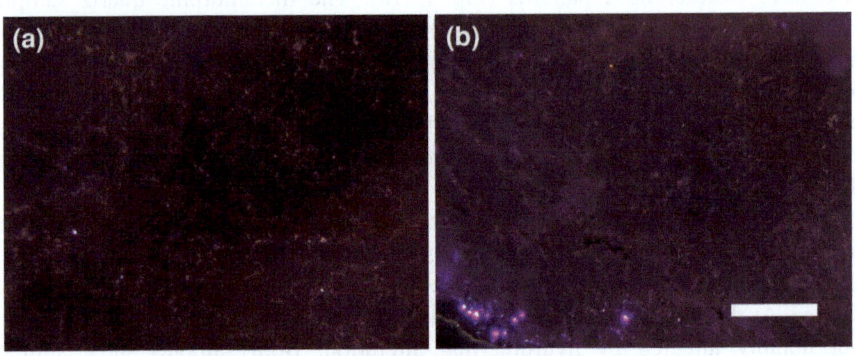


**Fig. 15.2** Optical microscope-based cathodoluminescence (CL) images of quartz from lenses hosted by the metamorphic wall rocks of the Ehrenfriedersdorf tin deposit, Germany. **a** Brownish and relatively homogeneous CL of quartz collected distal to the granite and tin ore zones (ED-1). **b** Remnants of transient bluish CL of quartz sample collected close to the granite (ED-520). Note the heterogeneous distribution of colour intensity. *Bright bluish spots* in the *lower left* corner of the image are small inclusions of fluorite. Both images were taken after 1 min of irradiation time. Initial strong bluish transient CL in the hydrothermally altered quartz (ED-520) could be observed, but not imaged due to the comparably long exposure time required. Scale bar is 0.3 mm

Assuming that the REE patterns of both metamorphic quartz lenses investigated were identical prior to hydrothermal alteration of the quartz lens located proximal to the granite, the differences between both patterns are a direct reflection of the

REE signature of the hydrothermal fluids. The REE pattern of the hydrothermal fluids reconstructed this way (except for cerium which is partly removed during alteration) shows a depletion of the light REEs, strong enrichment of the heavy REEs, and a positive europium anomaly (Fig. 15.1b). The calculated REE pattern of the hydrothermal fluid is similar to that of fluorite occurring in cassiterite-bearing tourmaline veinlets, which represent the earliest set of hydrothermal veins formed during ore formation (Monecke et al. 2000b, 2011). The REE pattern of the fluorite has been interpreted to record the trace element signature of the pristine ore-forming hydrothermal fluids prior to extensive interaction with the metamorphic host rocks (Gavrilenko et al. 1997; Monecke et al. 2000b, 2011).

It is important to note that the observed differences between the two quartz samples from the metamorphic wall rocks could not have been interpreted correctly without integration of observations made by optical microscopy, scanning electron microscopy, and cathodoluminescence microscopy. Hydrothermal alteration of quartz not only resulted in distinct changes in the trace element signature, but also modified the CL properties of the quartz. The occurrence of new accessory minerals provides additional evidence corroborating the conclusion that the metamorphic quartz lens close to the granite contact was affected by a process of hydrothermal alteration. The CL observations suggest that the entire volume of the metamorphic quartz lens was affected by this process. Although changes in the trace element contents such as those of the REEs may be assigned to changes in the fluid inclusion inventory, variations in the contents of trace elements commonly located in the crystal structure and changes in the CL characteristics suggest that alteration was not restricted to the formation of secondary fluids or the absorption of elements to internal and external surfaces. Hydrothermal alteration changed the quartz characteristics without visible deformation or destruction of the quartz lens.

## 15.3 Limitations to Provenance Analysis: Cathodoluminescence (CL) Colour Imaging of Quartz Pebbles from the Witwatersrand Basin, South Africa

CL imaging represents one of the most widely used techniques in quartz investigations. Especially, CL colour imaging is applied in provenance analysis in sedimentary petrology (Richter et al. 2003). In a pioneering work, Zinkernagel (1978) found that quartz grains of magmatic, metamorphic, and hydrothermal origins show distinct CL colours. Based on these results and evidence from subsequent studies, it is now widely accepted that (1) plutonic quartz, quartz phenocrysts, and quartz from rapidly cooled high grade metamorphic rocks show a stable blue to violet CL, (2) quartz in the groundmass of volcanic rocks often displays red luminescence, (3) quartz from many regional metamorphic rocks appears brownish under CL, but (4) authigenic quartz mostly exhibits weak or no visible CL but may luminesc in the UV range. Finally, hydrothermal quartz (5) often (but by far not always) shows transient greenish to blue and sometimes yellowish luminescence (Götze 2000; Götze and Zimmerle 2000; Götze et al. 2001 and references therein).

However, recent work indicates that such a rigid classification based on empirical evidence may be an oversimplification. The CL colours of quartz of different origins may show more variability, requiring either changes to the classification scheme or independent tests establishing the validity of the interpreted provenance (e.g., Boggs et al. 2002). In addition, post-depositional processes may affect CL characteristics. A study on quartz from the gold-bearing reefs of the Witwatersrand, South Africa, provides evidence that secondary processes can indeed modify the CL behaviour of quartz in sedimentary rocks.

Over the past decades, the genesis of the gold mineralisation in the reefs has been the focus of much debate. Traditionally, the gold deposits were regarded as conglomerate-hosted paleoplacers (e.g. Robb and Meyer 1990; Minter et al. 1993). However, abundant evidence points to the importance of hydrothermal activity (e.g., gold morphology and composition, formation of pyrophyllite and chlorite, association with sulphides and uraninite with high Th, Y, and REE contents; Barnicoat et al. 1997; Vollbrecht et al. 2002), implying that the gold is of hydrothermal origin (e.g., Phillips and Myers 1989; Barnicoat et al. 1997; Phillips and Law 2000; Law and Phillips 2005). Considering the detrital nature of some minerals and the fact that the host rocks are conglomerates, most of recent authors prefer a modified placer model that assumes a detrital genesis of quartz, pyrite, gold, and uranium with a late hydrothermal overprint and remobilisation of at least some of the gold (Robb and Meyer 1991; Frimmel and Gartz 1997; Gartz and Frimmel 1999; Vollbrecht et al. 2002; Frimmel et al. 2005; Schaefer et al. 2010 and references therein). Some authors proposed possible relationships between the gold and the Vredefort impact event (Kamo et al. 1996; Gartz and Frimmel 1999; Gibson and Reimold 1999).

To constrain the provenance of quartz from the reefs, Vollbrecht et al. (1996, 2002) used colour imaging by CL microscopy. These studies focused on the sand fraction (<5 mm) and cements in conglomerates from reefs in the ore fields. Vollbrecht et al. (1996, 2002) found that mono- or polycrystalline quartz displaying bluish CL colours and with rare growth zoning is the most abundant type. This prevailing quartz type was interpreted to be magmatic in origin. It may have been derived from plutons, pegmatites, and (rare) felsic porphyries located in the hinterland of the Witwatersrand basin. Brownish luminescent mono- and polycrystalline grains were less abundant, but still common, and were interpreted to have been derived from metamorphic rocks. Hydrothermal quartz identified by short-lived luminescence was found only in rare cases. Vollbrecht et al. (2002) also recognised the occurrence of secondary alteration features introduced during ductile and brittle deformation (undulate extinction, deformation lamella, pressure solution, fragmentation, and shearing) and hydrothermal alteration (replacement, alteration rims, and rims from radioactive irradiation).

A study on quartz pebbles from mineralised and barren parts of the Dominion, Vaal, Ventersdorp Contact, and Black Reefs was carried out by Poutivtsev (2001), Poutivtsev et al. (2001), and Kremenetsky et al. (2005). These studies employed the same CL techniques as used by Vollbrecht et al. (1996, 2002) but also included trace element analyses and analytical scanning electron microscopy. The trace element characteristics of quartz from the pebbles were found to be similar to those from mesothermal gold vein deposits. In addition, it was noted that cements in the mineralised parts of the reefs contain pyrrhotite, a sulphide commonly found in such gold deposits, and is absent in the barren parts of the reefs (Poutivtsev et al. 2001).

The CL imaging of the quartz pebbles yielded results that were generally comparable to the observations on the sand fraction made by Vollbrecht et al. (1996, 2002). The most common quartz CL was stable bluish to violet. Primary growth zoning was occasionally found. Quartz showing a brownish luminescence was less abundant, but still common. Transient bluish luminescence, with the CL colour changing to brown during continuous electron irradiation, was observed only in one quartz sample. In the gold-bearing conglomerates, orange luminescent rims on the pebbles were observed that formed as a result of radioactive irradiation of quartz. Rims with radiation-induced CL were less common in the unmineralised parts of the reefs.

Based on the observations by Vollbrecht et al. (1996, 2002), Poutivtsev (2001), Poutivtsev et al. (2001), and Kremenetsky et al. (2005), it can be concluded that quartz pebbles and quartz from the sand fraction are mainly magmatic in origin with subordinate but significant amounts derived from metamorphic rocks. However, more detailed inspection of the mineralised conglomerates from the Witwatersrand revealed that deformation and widespread hydrothermal alteration modified the CL behaviour of the quartz (Fig. 15.3). Larger quartz grains and pebbles showed evidence for the occurrence of processes of grain size reduction, in particular along their margins. Recrystallisation and development of subgrains (Fig. 15.3a, b) contributed to the overall grain size reduction. Bluish luminescent quartz grains commonly showed deformation lamellae that are clearly recognisable under CL as narrow, parallel zones of brownish CL (Fig. 15.3c). In addition, zones of microbreccia are observed (Fig. 15.3c, d). Brittle and plastic quartz deformations are thus clearly documented. The reduction in grain size and the formation of subgrains and quartz fragments was accompanied by a decrease in CL intensity and by a change in CL colour from bluish through dull violet to brownish. Evidence for changes in CL colour as a result of post-depositional fluid flow is provided by the occurrence of networks of secondary fluid inclusions trails that are surrounded by large halos and irregular trails of brownish CL (Fig. 15.3a, c, e). Embayed grain boundaries that formed by processes of quartz dissolution are often paralleled by rims of orange and zoned orange/bluish luminescence (Fig. 15.3c, e). As noted by previous workers, this change in CL colour is probably related to radioactive irradiation of the quartz during circulation of uranium-rich hydrothermal fluids. Similar zones of modified CL are also observed around inclusions of uranium-rich minerals.

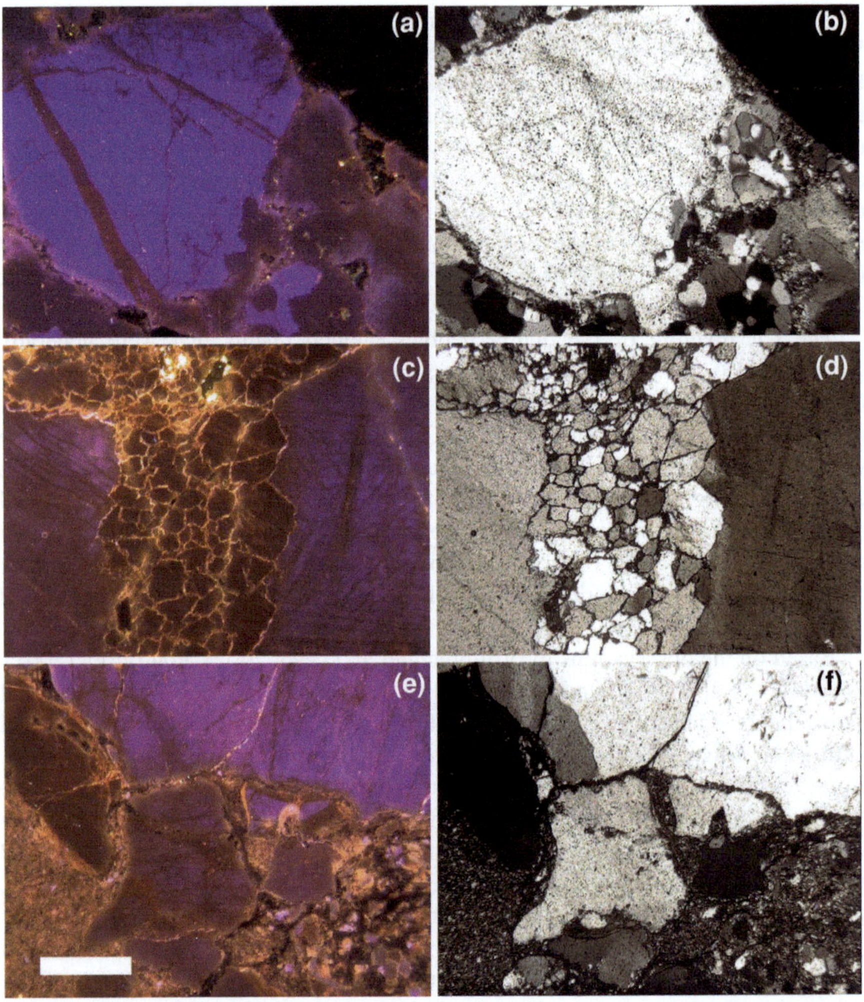


**Fig. 15.3** Optical microscope-based CL (*left*) and transmitted light crossed polar (*right*) images of quartz pebbles from reefs of the Witwatersrand, South Africa. **a, b** Fluid trails (*left side*, *brownish* CL) in a fragmented quartz pebble; recrystallisation and formation of smaller grains and subgrains (*right side*; *bluish*, *dull violet* to *brownish* CL); dissolution embayed grain boundaries, sometimes with *orange* luminescent rims from radioactive irradiation. Sample W63A. Gold-bearing conglomerate from the Ventersdorp Contact Reef, East Drifontein; **c, d** Deformation lamellae (*upper left* and *right*), fluid trails (*left*, *right*, *brownish* CL, *orange* CL around inclusions in the trails) and local brecciation (*middle*, *brownish* CL) in a quartz pebble; *orange* luminescent rims from radioactive irradiation. Sample 20-1a. Barren conglomerate from the Vaal Reef, Johannesburg, Vickers Road; **e, f** Fragmentation at the rim of a quartz pebble and fluid inclusion trails (*brownish* CL). Sample BR1/2. Barren conglomerate from the *Black Reef* near Klerkskraal. Figures 15.4c–f after Poutivtsev (2001). Scale bar is 0.3 mm

Obviously, the texture of the quartz pebbles, their CL properties, trace element contents, and fluid inclusion characteristics were changed under ductile and brittle deformation and during hydrothermal alteration. These observations may satisfactorily explain the hydrothermal trace element signature of the quartz pebbles observed by previous workers (Poutivtsev 2001; Poutivtsev et al. 2001; Kremenetsky et al. 2005). The identified ductile and brittle deformation features (Vollbrecht et al. 1996, 2002; this work) possibly correspond to thrust-related fracture networks observed by Jolley et al. (1999) and Gartz and Frimmel (1999) for the Ventersdorp Contact Reef. Together with these observations made at the macro- and microscales, new findings may open ways to a more comprehensive interpretation of the provenance of the conglomerates and their response to deformation by thrusting and/or a meteorite impact.

The textural relationships found for the quartz pebbles suggest that the brownish CL colour of quartz from the Witwatersrand is not primary but also does not result simply from low-grade regional metamorphism. A similar transformation from bluish/violet to brownish CL was also observed for quartz in gold-bearing brecciated quartz bodies from Muruntau (Uzbekistan). In analogy to samples from the Witwatersrand, the vein quartz at Muruntau was affected by ductile/brittle deformation and hydrothermal alteration at elevated temperatures (Graupner et al. 2000). Furthermore, observations by Sprunt et al. (1978) demonstrate that even the bluish CL of quartz is not always related to a magmatic origin of the quartz as high-grade metamorphism can also produce this CL colour. Additional evidence for the ambiguous nature of blue and brownish CL of quartz were presented by Boggs et al. (2002) and Spear and Wark (2009).

Summarising, the results of the CL study on quartz pebbles from the Witwatersrand basin show that provenance analysis is not possible using this technique in cases where quartz experienced regional or dynamic metamorphism and/or intense hydrothermal alteration. This conclusion is consistent with suggestions made by Boggs et al. (2002).

## 15.4 Fluid Inclusions in Quartz: Relationships to Fluids Involved in Wolframite Precipitation in Tungsten Deposits of the Mongolian Altai, Mongolia

Fluid and melt inclusion studies on quartz are routinely used to evaluate the temperature, pressure, and chemical conditions of quartz formation, especially in the context of igneous petrology and economic geology (Roedder 1984; Frezzotti 2001; Samson et al. 2003; Webster 2006, and references therein). In most cases, the investigation of fluid inclusions is one of a few direct approaches to reconstruct the physicochemical conditions of mineral growth. Quartz represents the most common transparent mineral in rocks or veins permitting the use of standard microscopic techniques.

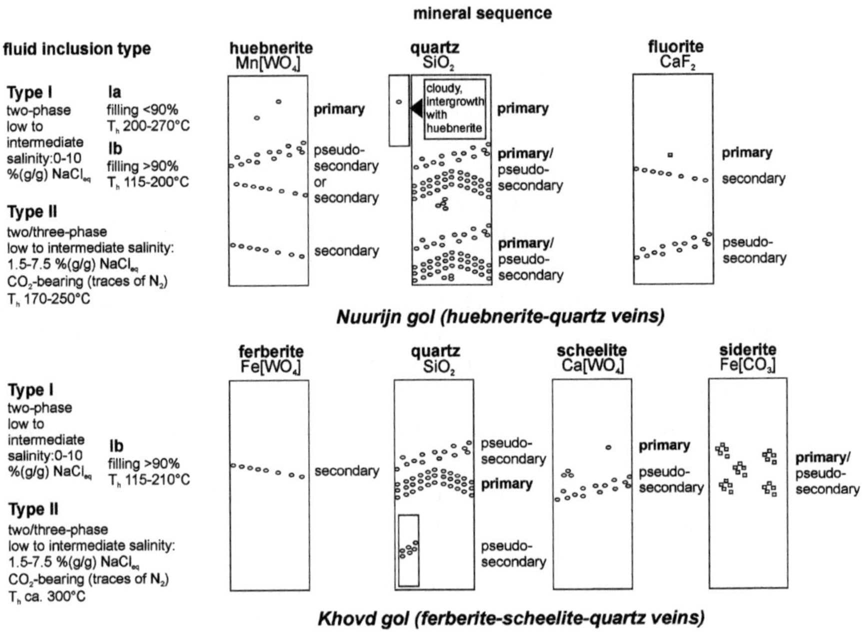


**Fig. 15.4** Summary of fluid inclusion investigations on minerals from the quartz-wolframite vein deposits Nuurijn gol and Khovd gol, Mongolian Altai. Several fluid inclusion types may be distinguished which occur as primary, pseudo-secondary, and secondary inclusions in wolframite, quartz, fluorite, scheelite, and siderite. Note the rare occurrence of Type Ia inclusions as primary inclusions in quartz from Nuurijn gol and of Type II inclusions in quartz from Khovd gol. See text for further explanation. Data from Enchbat (2007)

Based on textural criteria, fluid inclusions in minerals are conveniently classified as primary, pseudosecondary, and secondary inclusions (Roedder 1984; Van den Kerkhof and Hein 2001; Bodnar 2003a). As shown by these authors, a correct classification may be possible only for a small part of the fluid inclusion inventory as unequivocal textural evidence may be missing. Additional limitations are imposed by the fact that quartz often contains a large variability and high quantity of fluid inclusions assemblages. Interpretation of fluid inclusion data may also be limited due to the occurrence of reequilibration phenomena during hydrothermal alteration (e.g., Bakker 2009). A study on two tungsten deposits in the Mongolian Altai in the western Mongolia (Enchbat et al. 1999; Enchbat 2007) highlights some of the pitfalls and difficulties arising in the study of fluid inclusions in quartz.

Enchbat et al. (1999) and Enchbat (2007) studied the fluid inclusion inventory of various vein minerals occurring in the quartz-wolframite veins of the Nuurijn gol and Khovd gol deposits to better constrain the conditions of tungsten mineralisation in the Mongolian Altai. Based on field observations and textural evidence collected in thin section, the sequence of mineral formation in the veins was determined (Fig. 15.4). Two fluid inclusion types could be distinguished in both deposits: (1) low (to intermediate) saline, two-phase inclusions with NaCl being the dominant salt (Type I) and (2) low (to intermediate) saline, two- and three-phase $CO_2$-bearing inclusions with NaCl also being the dominant salt (Type II). Based on the degree of filling and the related homogenisation temperatures $T_h$, the first type may be further subdivided into Type Ia (filling below 90%) and Type Ib (filling above 90%; Fig. 15.4). Raman spectroscopy showed that Type II fluid inclusions sometimes contain traces of $N_2$ in the vapour phase (Enchbat 2007).

Comparison of vein quartz with the other vein minerals shows that quartz contains the highest total abundance of fluid inclusions. The quartz is also typified by more complex fluid inclusion characteristics (Fig. 15.4). It proved to be difficult to distinguish between inclusions in quartz formed during primary growth or secondary processes. Fluid inclusion assemblages of both $CO_2$-bearing and $CO_2$-barren types (Type I and Type II, respectively) appear to have been entrapped during primary quartz growth. This finding is somewhat unexpected as the general importance of $CO_2$ in wolframite precipitation is widely discussed in the literature (Higgins 1980, 1985; Campbell et al. 1988; Graupner et al. 1999). The role of both fluid types in ore formation is therefore fundamental for a correct understanding of hydrothermal processes in both deposits.

To explain the apparently primary occurrence of both inclusion types in quartz, several hypotheses were tested. There were no direct indications for heterogeneous fluid entrapment as all inclusions in one assemblage are of the same type. Alternatively, the occurrence of primary and pseudosecondary inclusions of both types could be the result of fluid evolution that paralleled quartz formation. However, investigation on a vein profile from Khovd gol failed to provide support for such a hypothesis (Enchbat 2007). It was further tested whether textural relationships were misinterpreted with a resulting incorrect classification of the inclusions to be primary or pseudosecondary. The nature of inclusions in quartz was evaluated by CL imaging. As shown by Roedder (1984), Boiron et al. (1992), Graupner et al. (2000), and Van den Kerkhof and Hein (2001) as well as by the study of quartz pebbles of the Witwatersrand discussed above, optical CL microscopy or scanning electron microscopy often reveals complex networks of secondary alteration trails in quartz to which fluid inclusions are confined and which may therefore be recognised as secondary in origin. No clear indications for a secondary origin of fluid inclusions were found by CL imaging for quartz from the Khovd gol deposit.

The occurrence of fluid inclusions in planes parallel to visible growth zones of a crystal is usually a reliable criterion for a primary origin (e.g., Roedder 1984; Fig. 15.4 and 15.5). However, as shown by Goldstein ( 2001) for carbonates, secondary alteration can sometimes proceed along certain growth zones resulting in secondary formation of fluid inclusions. A study on the Muruntau gold deposit in Uzbekistan demonstrated that similar situations may also occur in quartz (Graupner et al. 2000). However, no indication for a secondary formation of inclusions arranged along primary growth zones could be found in quartz from Nuurijn gol and Khovd gol (Fig. 15.5).

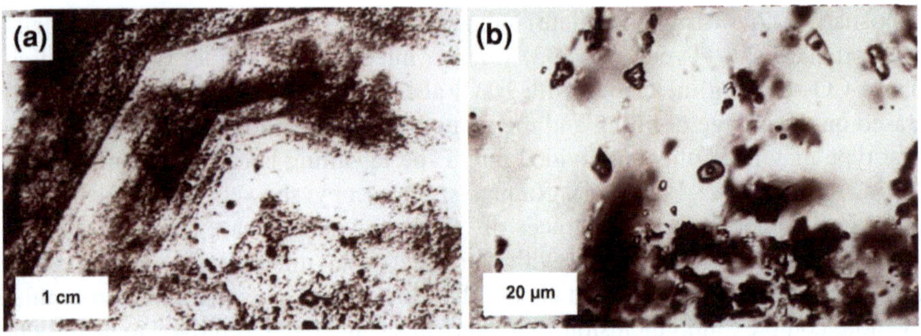


**Fig. 15.5** Microphotographs of fluid inclusion assemblages in quartz from Khovd gol, Mongolian Altai. **a** Typ I fluid inclusions arranged parallel to growth zones in quartz; **b** Close-up image of a group of fluid inclusions of Type I. Modified from Enchbat (2007)

Besides the occurrence of numerous types of fluid inclusions in vein quartz, it is important to note that some types of fluid inclusions are extremely rare. Just these inclusions may, however, be crucial for a correct understanding of the fluid evolution in the deposit. In the present example, this applies to Type Ia inclusions in cloudy quartz from Nuurijn gol that is intergrown with huebnerite and to the $CO_2$-bearing Type II inclusions in quartz from Khovd gol (Fig. 15.4). Due to their scarcity, the latter inclusions were not found during previous research on this deposit (Kempe et al. 1994).

Despite careful fluid inclusion petrography and the application of various microscopic techniques, the study on quartz from Nuurijn gol and Khovd gol yielded inconclusive results. It could not be shown unambiguously which fluid type relates to ore formation. Although fluorite from Nuurijn gol and scheelite and siderite from Khovd gol also contain primary and pseudosecondary fluid inclusion assemblages, these minerals are paragenetically even later and may not have been formed from the same hydrothermal fluids as wolframite. To constrain the nature of the mineralising fluids unequivocally, the fluid inclusion inventory of wolframite needs to be studied directly. However, with the exception of the manganese-endmember huebnerite, members of the wolframite solid solution are opaque and are not accessible to standard fluid inclusion techniques. (Enchbat 2007) investigated fluid inclusions in huebnerite from Nuurijn gol and ferberite from Khovd gol by conventional and IR microscopy, respectively. The results of this study showed that fluid inclusion assemblages observed in quartz are at variance with those occurring in wolframite (Fig. 15.4). In the wolframite from Nuurijn gol and Khovd gol, $CO_2$-bearing inclusions are not found or occur only as secondary inclusions. This suggests that $CO_2$ does not play a major role in wolframite precipitation, at least not in the case of the two deposits studied.

The investigations by Enchbat (2007) demonstrate that the study of vein quartz does not necessarily permit a reconstruction of the nature of the ore-forming hydrothermal fluids. In the case of the two tungsten deposits in the Mongolian Altai, the composition of the fluid inclusions entrapped in quartz would suggest that $CO_2$ was a major component of the fluids precipitating wolframite, which is not supported by direct investigations on wolframite. Results from similar studies on quartz and wolframite from deposits in Cornwall, Germany, Portugal, New Mexico, and Peru by Campbell and Robinson-Cook (1987), Campbell et al. (1988), Campbell and Panter (1990), and Lüders (1996) also demonstrated that inclusions in both minerals may be distinct. Differences between the fluid inclusions in quartz and ore minerals are not restricted to tungsten deposits as recently shown by Kouzmanov et al. ( 2010). These authors documented similar relationships for quartz and pyrite from the Rosia Poieni porphyry deposit, Romania.

In the context of the present study, the question arises why fluid inclusion assemblages in quartz can be so complex. Several important aspects need to be considered. Although quartz behaves brittle during deformation accompanying vein formation, vein quartz commonly shows features related to dissolution, recrystallisation, and repeated growth. This behaviour of quartz during vein formation is favourable for the repeated formation of fluid inclusions in the mineral by crack-healing. However, by far not all fluid inclusion trails observed in quartz may simply be related to brittle deformation. Large alteration zones containing fluid inclusions are often revealed by CL imaging (e.g., Rusk and Reed 2002; Fig. 15.3). It appears possible that the quartz structure may yield pathways for penetration of fluids without previous fragmentation or dissolution. Recent investigations on re-equilibration (e.g., stretching, leakage, decrepitation, necking down) of fluid inclusions in quartz by Bodnar (2003b) and Bakker (2009) support and necking down such a view.

## 15.5 Intergrowth Relationships Between Quartz and Ore Minerals: Evidence for South Crofty Tungsten Mine, Cornwall, United Kingdom

The study of textural relationships in rocks and ore veins provides the basis for the establishment of paragenetic relationships in geological systems. The identification of the relative timing of mineral formation is an important step in reconstruction of fluid evolution in ore deposits. As quartz is a common gangue mineral, the reconstruction of paragenetic relationships between quartz and ore minerals represents an integral part of most studies on ore deposits.

To establish paragenetic sequences of minerals in magmatic rocks and hydrothermal veins, the so-called inclusion principle is widely used. According to this principle, a (smaller) mineral grain included in another (larger) grain must have formed prior to, or at least contemporaneously with, the host mineral grain. The euhedral shape of inclusions is commonly cited as additional evidence supporting this conclusion (Taylor 1992; Charoy and Noronha 1996).

However, this textural relationship may not provide unequivocal constraints on timing in situations where secondary euhedral inclusions form (Kempe et al.

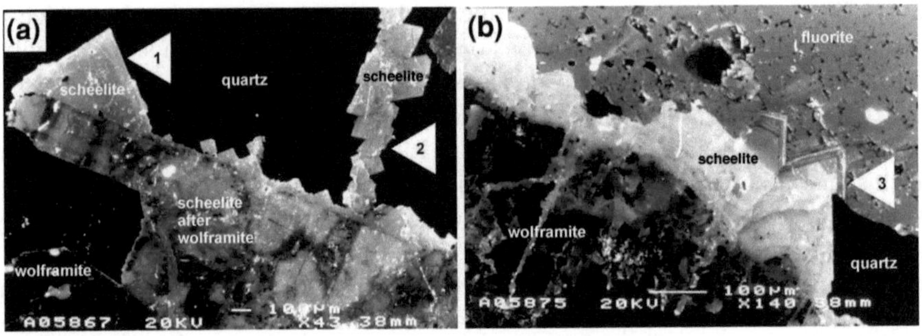


**Fig. 15.6** Scanning electron microscope CL images of pseudomorphs of scheelite after wolframite and of late scheelite overgrowth enclosed in quartz, images were taken using a Jeol JSM 6400 with a Mini CL detector operated at 20 kV and 600 pA. **a** Overgrowth of euhedral scheelite with primary growth zoning on pseudomorphs of scheelite after wolframite (*arrow* 1) and aggregate of euhedral scheelite crystals with primary growth zoning that infill a fracture in quartz (*arrow* 2). **b** Overgrowth of scheelite on pseudomorphs of scheelite after wolframite. Scheelite with oscillatory growth zoning replaces fluorite (*arrow* 3). Note the porous character of the fluorite aggregate (modified from Gavrilenko et al. 1998)

1999). For instance, Kempe et al. (1999) reported on the secondary formation of euhedral albite laths within K-feldspar and quartz grains in hydrothermally altered alkaline rocks. Similarly, euhedral tourmaline crystals in hydrothermally altered quartz occurring within the metamorphic host rocks from the Ehrenfriedersdorf tin deposit clearly postdate quartz formation (see above). An additional example of such a textural relationship between vein quartz and euhedral ore minerals is provided by the South Crofty mine near Pool, Cornwall (Gavrilenko et al. 1998).

The quartz-wolframite veins at South Crofty contain pseudomorphs of scheelite after wolframite, which are frequently overgrown by euhedral scheelite (Fig. 15.6a). Euhedral scheelite grains that are fully enclosed by quartz also occur abundantly in the veins. Crystal faces are always developed at the grain boundaries with quartz (Fig. 15.6a). The scheelite forming the euhedral crystals and the scheelite within the pseudomorphs display different CL characteristics and internal structures, suggesting variations in the mechanisms of crystal growth. The CL spectra of both scheelite types are distinct (Gavrilenko et al. 1998). Visible oscillatory growth zoning in the euhedral scheelite crystals forming the overgrowth are the result of dynamic processes common during primary crystal growth (cf. Shore and Fowler 1996 and references therein). These observations along with the occurrence of scheelite as inclusions in quartz may be taken as evidence for the following sequence of events: (1) crystallisation of wolframite, (2) brittle deformation and formation of pseudomorphs of scheelite after wolframite, (3) growth of euhedral scheelite as an overgrowth, (4) precipitation of quartz surrounding and enclosing the tungsten minerals.

Closer examination, however, suggests a different interpretation. Some scheelite crystals enclosed in quartz are arranged in trails. These trails commonly appear to have developed along fractures in the quartz (Fig. 15.6a) and must, therefore, postdate quartz precipitation. At the same time, individual crystals in these aggregates show euhedral grain shapes and oscillatory growth zoning similar to the euhedral scheelite overgrowing the pseudomorphs after wolframite. These relationships suggest that the vein quartz probably formed prior to the scheelite, with scheelite crystallising during or after brittle deformation of the vein. Similar textural relationships are also observed between scheelite and fluorite with scheelite forming crystals growing into the fluorite (Fig. 15.6b). In analogy to quartz, the scheelite crystals display oscillatory growth zoning and well developed crystal faces at the contact with the fluorite (Fig. 15.6b). The example of the quartz-wolframite veins from the South Crofty mine, Cornwall, highlights the fact that ore mineral inclusions in quartz may not necessarily predate, or are contemporaneous with, quartz formation and that this textural criterion may not be a reliable indicator of relative age relationships, at least not in environments where processes of metasomatic replacement occur (cf. Kempe et al. 1999).

## 15.6 Discussion and Conclusions

Quartz is commonly regarded to be a stable phase that is resistant to processes of secondary alteration. Based on this assumption, it is widely accepted that the trace element content of quartz, its CL properties, fluid inclusion inventory, and textural relationships with other minerals are related to the primary growth of the quartz and are not easily modified by secondary processes. However, the case studies discussed above, along with data scattered throughout the literature, demonstrate that quartz is more susceptible to processes of recrystallisation and structural readjustment accompanying hydrothermal alteration and deformation than previously recognised. In environments of high fluid activity, conventional interpretation of analytical data emphasising the fact that quartz is resistant to secondary processes may be misleading. It appears that more complicated relations may be revealed only by detailed studies integrating multiple analytical methods. Data evaluation should be strictly driven by observations as some of the interpretation may seem counterintuitive at first.

Over the past years, an increasing number of the trace element studies of quartz have been performed on quartz to constrain certain aspects of primary quartz growth such as the temperature of formation. The case study at the Ehrenfriedersdorf tin deposit, Germany, shows, however, that the trace element content of quartz can change during metamorphism and hydrothermal alteration postdating initial quartz growth. The formation of secondary fluid inclusions and adsorption of elements to the external and internal surfaces of quartz grains during hydrothermal alteration appear to be simple mechanisms that can modify the trace element content of quartz. However, hydrothermal alteration also affects the contents of elements that are most likely hosted by the quartz crystal structure itself (e.g., lithium, aluminium, and titanium). As compositional changes were accompanied by a modification of the CL colour, hydrothermal alteration of the metamorphic quartz must have affected its real structure through the formation of new ionic and electron defects which define the CL behaviour of quartz (Götze et al. 2001; Larsen et al. 2009).

The case studies at the Ehrenfriedersdorf tin deposit, Germany, and the Witwatersrand, South Africa, show that the CL colour characteristics of quartz can change in response to deformation and hydrothermal alteration. Due to potential post-depositional changes in CL colour, this property of quartz cannot always be used to unequivocally constrain provenance of quartz grains in sediments and sedimentary rocks. The results of CL investigations need to be tested against data obtained from complementary techniques. In the case of the conglomerates from the Witwatersrand basin, CL colour imaging of quartz from the large pebbles and the sand-sized fractions suggests that the quartz has been largely derived from magmatic and metamorphic rocks. However, more detailed CL studies, compounded by trace element analysis, suggest that the brownish CL interpreted to be indicative for a metamorphic origin of some quartz is in fact introduced during post-depositional deformation and hydrothermal alteration. The conclusion that brown CL of quartz can result from secondary processes is in accordance with earlier observations made on brecciated and altered gold quartz veins at the Muruntau gold deposit, Uzbekistan (Graupner et al. 2000). Furthermore, it appears possible that even the bluish CL, which is widely regarded to be indicative for a magmatic origin of the quartz, is not of primary origin. Although in a different geological environment, dark blue CL in quartz has been shown to develop in response to a high-grade metamorphic overprint (e.g., Sprunt et al. 1978; Boggs et al. 2002). Based on the evidence, it is suggested here that the CL characteristics of quartz from the Witwatersrand reefs are probably of secondary origin and should, therefore, not be used to interpret primary provenance of the conglomerate.

The formation of secondary fluid inclusions in quartz is a common phenomenon, providing evidence for fluid flow through quartz crystals after their formation (Van den Kerkhof and Hein 2001; Rusk and Reed 2002). As demonstrated in this study and in the literature, CL imaging can be used to identify pathways of fluid flow through quartz crystals (Van den Kerkhof and Hein 2001; Rusk and Reed 2002), including those that are not marked by secondary fluid inclusions (this study). Recent studies also suggest that primary fluid inclusions hosted by quartz may be affected by secondary processes through re-equilibration (Van den Kerkhof and Hein 2001; Bodnar 2003a, b; Bakker 2009). These changes occur without obvious quartz degeneration as brittle deformation, fragmentation, or dissolution (i.e. they occur when quartz remained in solid state). Fluid inclusion data can only be interpreted correctly if primary, pseudosecondary, and secondary fluid inclusion assemblages can be confidently distinguished, which is not always trivial or possible without the help of advanced analytical techniques. Moreover, establishment of paragenetic relationships between quartz and ore minerals is critical. The nature of the ore-forming fluid can only be constrained if it can be demonstrated that the quartz and ore minerals formed contemporaneously. The case study on the quartz-wolframite vein deposits in the Mongolian Altai showed that the results of fluid inclusion studies on quartz alone would have suggested that $CO_2$

plays a significant role in wolframite precipitation, which is in accordance with results from earlier studies on quartz from other wolframite deposits. However, fluid inclusion studies on wolframite itself indicate that $CO_2$ was not directly involved in wolframite crystallisation in these deposits. Although similar fluid inclusions lacking $CO_2$ were also found in quartz, their significance could not be identified unambiguously from the quartz. One factor contributing to this uncertainty is the presence of high quantities of fluid inclusions in the quartz and the occurrence of several distinct types of fluid inclusions, which contrasts other vein minerals.

The study of textural relationships between minerals occurring in veins is commonly used to establish paragenetic sequences in veins, allowing the reconstruction of the evolution of hydrothermal fluids during deposit formation. As demonstrated for the quartz-wolframite veins from the South Crofty mine, the occurrence of euhedral inclusions of ore minerals such as scheelite in quartz may suggest formation prior to quartz precipitation. However, more detailed investigations revealed that that formation of scheelite more likely occurred during or after brittle deformation of vein quartz. The inclusion principle should be used with caution in environments where fluid flow accompanied or followed vein or rock formation.

The examples discussed in this contribution all demonstrate that quartz is prone to alteration at elevated temperatures and pressure and/or fluid activity. Despite high hardness, the lack of distinct cleavage, and resistance against mechanical and chemical weathering, the trace element contents, CL properties, textural relationships to other minerals, and fluid and mineral inclusion inventory can change during secondary processes. These changes are not always accompanied by obvious recrystallisation, dissolution, and reprecipitation, implying that such transformations may possibly occur by solid state reactions.

It is assumed here that the reasons for the exceptional behaviour of quartz under metamorphic and hydrothermal conditions are related to the peculiarities of the quartz crystal structure. At elevated temperature and pressure conditions, enhanced diffusion of trace elements along the hexagonal channels that are parallel to the symmetry axis of the quartz structure may occur. Diffusion and electrodiffusion of $H^+$, $Li^+$, $Na^+$, $K^+$, and $Ca^{2+}$ in quartz are well documented in experimental studies (Ramseyer and Mullis 1990; Perny et al. 1992; Cherniak 2010 and references therein). At present, possible diffusion mechanisms for other elements, especially for those normally confined to atomic positions in the quartz structure, are not well constrained.

In addition to enhanced diffusion, the effect of hydrolytic weakening of quartz may be a key for the understanding of the quartz behaviour at elevated temperatures and pressures. The effect of hydrolytic weakening of quartz (i.e. a drastic drop in strength of synthetic quartz containing water at temperatures above 400°C) was discovered by Griggs and Blacic(1965). Investigations on this effect showed that quartz may incorporate and release water during deformation in the temperature range between 400 and 700°C (e.g., Griggs and Blacic 1965; McLaren et al.

1983; Doukhan and Trépied 1988). Formation of small water inclusions in the synthetic quartz was observed after the experiments (Doukhan and Trépied 1988).

Experimental investigations on diffusion and weakening using synthetic quartz provide critical insights into the mechanisms leading to changes in water content and mechanical behaviour of natural quartz, the formation of sub-grain boundaries, and the origin of fluid inclusions at elevated temperatures and/or pressure. One important point is that water is not only confined to the structural channels but can also hydrolyse silicon–oxygen bounds (Griggs and Blacic 1965). The conditions at which diffusion and hydrolytic weakening of quartz are observed in the experiments correspond to the temperature—pressure ranges at which transformation of quartz is observable in nature.

In summary, this contribution shows that quartz represents a mineral that is quite susceptible to recrystallisation and structural changes under metamorphic and hydrothermal conditions. This behaviour of quartz may be effectively used to identify and monitor secondary alteration processes. At the same time, care needs to be taken to ensure that primary and secondary characteristics are correctly identified and distinguished. Future research on quartz would undoubtedly benefit from a better integration of analytical data obtained by complementary techniques.

**Acknowledgments** We gratefully acknowledge reviews by Torsten Graupner and Brian Rusk, which helped us to significantly improve the manuscript. We thank Dieter Wolf for useful discussion on the behaviour of quartz. The trace element analyses would not have been possible without the analytical support by Gisela Bombach and Werner Klemm.

## References

Agangi A, McPhie J, Kamenetsky VS (2011) Magma chamber dynamics in a silicic LIP revealed by quartz: the Mesoproterozoic Gawler Range Volcanics. Lithos 126:68–83

Agel A, Petrov I (1990) Substitutional aluminium in the quartz lattice as indicator for the temperature of formation. Eur J Mineral 2 (Bh.1):144 (abstract, in German)

Bakker RJ (2009) Reequilibration of fluid inclusions: bulk diffusion. Lithos 112:277–288

Bambauer HU (1961) Spurenelementgehalte und $\gamma$-Farbzentren in Quarzen aus Zerrklüften der Schweizer Alpen. Schweiz Mineral Petrol Mitt 41:335–369

Barnicoat AC, Henderson IHC, Knipe RJ, Yardley BWD, Napier RW, Fox NPC, Kenyon AK, Mutingh DJ, Strydom D, Winkler KS, Lawrence SR, Cornford C (1997) Hydrothermal gold mineralization in the Witwatersrand basin. Nature 386:820–824

Barton JM, Wenner DB, Hallbauer DK (1992) Oxygen isotopic study of the nature and provenance of large quartz and chert clasts in gold-bearing conglomerates of South Africa. Geology 20:1123–1126

Blatt H (1987) Oxygen isotopes and the origin of quartz. J Sediment Petrol 57:373–377

Bodnar RJ (2003a) Introduction to fluid inclusions. In: Samson I, Anderson A, Marshall D (eds) Fluid inclusions: analysis and interpretation. Mineral Assoc Canada, Short Course vol 32, Quebec, pp 1–8

Bodnar RJ (2003b) Reequilibration of fluid inclusions. In: Samson I, Anderson A, Marshall D (eds) Fluid inclusions: analysis and interpretation. Mineral Assoc Canada, Short Course vol 32, Quebec, pp 213–230

Boggs S, Kwon YI, Goles GG, Rusk BG, Krinsley D, Seyedolali A (2002) Is quartz cathodoluminescence color a reliable provenance tool? A quantitative examination. J Sediment Res 72:408–415

Boiron MC, Essarraj S, Sellier E, Cathelineau M, Lespinasse M, Poty B (1992) Identification of fluid inclusions in relation to their host microstructural domains in quartz by cathodoluminescence. Geochim Cosmochim Acta 56:175–185

Botis S, Nokhrin SM, Pan Y, Xu Y, Bonli T (2005) Natural radiation-induced damage in quartz. I. Correlations between cathodoluminescence colors and paramagnetic defects. Can Mineral 43:1565–1580

Botis S, Pan Y, Bonli T, Xu Y, Zhang A, Nokhrin S, Sopuck V (2006) Natural radiation-induced damage in quartz. II. Distribution and implications for uranium mineralization in the Athabasca basin, Saskatchewan, Canada. Can Mineral 44:1387–1402

Bottrell SH, Yardley B, Buckley F (1988) A modified crush-leach method for the analysis of fluid inclusion electrolytes. Bull Minéral 111:279–290

Breiter K, Müller A (2009) Evolution of rare-metal granitic magmas documented by quartz chemistry. Eur J Mineral 21:335–346

Campbell AR, Panter KS (1990) Comparison of fluid inclusions in coexisting (cogenetic?) wolframite, cassiterite, and quartz from St. Michael's Mount and Cligga Head, Cornwall England. Geochim Cosmochim Acta 54:673–681

Campbell AR, Robinson-Cook S (1987) Infrared fluid inclusion microthermometry on coexisting wolframite and quartz. Econ Geol 82:1640–1645

Campbell AR, Robinson-Cook S, Amindays C (1988) Observation of fluid inclusions in wolframite from Panasqueira, Portugal. Bull Minéral 111:251–256

Charoy B, Noronha F (1996) Multistage growth of a rare-element, volatile-rich microgranite at Argemela (Portugal). J Petrol 37:73–94

Cherniak DJ (2010) Diffusion in quartz, melilite, silicate perovskite, and mullite. In: Zhang Y, Cherniak DJ (eds) Diffusion in minerals and melts. Rev Mineral Geochem 72:735–756

Claffy EW, Ginther RJ (1959) Red-luminescing quartz. Am Mineral 44:987–994

Dennen WH (1966) Stoichiometric substitution in natural quartz. Geochim Cosmochim Acta 30:1235–1241

Dennen WH, Blackburn WN, Quesada A (1970) Aluminium in quartz as a geothermometer. Contrib Mineral Petrol 27:332–342

Doukhan JC, Trépied L (1988) Plastic deformation of quartz single crystals. Bull Minéral 108:97–123

Enchbat D (2007) Die erzbildenden Fluide in den Au-W-Mineralisationen des Mongolischen Altai: Untersuchungen zu Fluideinschlüssen und Spurenelementchemismus von Erz- und Gangmineralen. Freiberg Forschungsh C520, TU Bergakademie Freiberg, 101 pp

Enchbat D, Kempe U, Dandar S, Wolf D (1999) Fluid inclusion characteristics and trace element chemistry of ore and vein minerals from the Au-W deposits in Altai tectonic zone of Mongolian Altai. In: Lüders V, Schmidt-Mumm A, Thomas R (eds) ECROFI XV: European current research on fluid inclusions. abstracts and program, vol 99(6). Alfred-Wegener-Stiftung, Potsdam, Terra nostra, pp 93–94

Frezzotti ML (2001) Silicate-melt inclusions in magmatic rocks: applications to petrology. Lithos 55:273–299

Frimmel HE, Gartz VH (1997) Witwatersrand gold particle chemistry matches model of metamorphosed, hydrothermally altered placer deposits. Mineral Deposita 32:523–530

Frimmel HE, Groves DI, Kirk J, Ruiz J, Chesley J, Minter WEL (2005) The formation and preservation of the Witwatersrand gold fields, the world's largest gold province. Econ Geol 100th Anniversary Volume:769–797

Füchtbauer H, Leggewie R, Gockeln C, Heinemann C, Schröder P (1982) Methoden der Quarzuntersuchung, angewandt auf mesozoische und pleistozäne Sandsteine und Sande. N Jb Geol Paläont Mh 193–210

Gartz VH, Frimmel HE (1999) Complex metasomatism of an Archean placer in the Witwatersrand basin, South Africa: the Ventersdorp Contact reef—a hydrothermal aquifer? Econ Geol 94:689–706

Gavrilenko VV, Kempe U, Gaidamako IM (1998) Formation of chemical heterogeneity in minerals during their metasomatic growth and replacement (on the example of scheelitization of wolframite). Zapiski Vserossijskogo Mineralogicheskogo Obshchestva 127(6):75–78 (in Russian)

Gavrilenko V, Morozov M, Kempe U, Smolenskiy V, Wolf D (1997) Unusual REE distribution patterns in fluorites from Sn-W deposits of the quartz-cassiterite and quartz-wolframite type. In: Novák M, Janoušek V, Košler J (eds) MAEGS—10, challenges to chemical geology. J Czech Geol Soc 42(3):36 (abstract)

Gibson RL, Reimold WU (1999) The significance of the Vredefort Dome for the thermal and structural evolution of the Witwatersrand basin, South Africa. Mineral Petrol 66:5–23

Girard JP, Deynoux M (1991) Oxygen isotope study of diagenetic quartz overgrowths from the upper Proterozoic quartzites of western Mali, Taoudeni basin: implications for conditions of quartz cementation. J Sediment Petrol 61:406–418

Goldstein RH (2001) Fluid inclusions in sedimentary and diagenetic systems. Lithos 55:159–193

Götte T, Pettke T, Ramseyer K, Koch-Müller M, Mullis J (2011) Cathodoluminescence properties and trace element signature of hydrothermal quartz: a fingerprint of growth dynamics. Am Mineral 96:802–813

Götze J (2000) Cathodoluminescence microscopy and spectroscopy in applied mineralogy. Freiberg Forschungsh C485, TU Bergakademie Freiberg, 128 pp

Götze J, Plötze M (1997) Investigation of trace-element distribution in detrital quartz by electron paramagnetic resonance (EPR). Eur J Mineral 9:529–537

Götze J, Zimmerle W (2000) Quartz and silica as guide to provenance in sediments and sedimentary rocks. Contrib Sediment Petrol 21, Schweizerbart'sche Verlagsbuchhandlung, Nägele & Obermiller, Stuttgart, 91 pp

Götze J, Plötze M, Fuchs H, Habermann D (1999) Defect structure and luminescence behaviour of agate—results of electron paramagnetic resonance (EPR) and cathodoluminescence (CL) studies. Mineral Mag 63(2):149–163

Götze J, Plötze M, Graupner T, Hallbauer DK, Bray CJ (2004) Trace element incorporation into quartz: a combined study by ICP-MS, electron spin resonance, cathodoluminescence, capillary ion analysis, and gas chromatography. Geochim Cosmochim Acta 68:3741–3759

Götze J, Plötze M, Habermann D (2001) Origin, spectral characteristics and practical applications of the cathodoluminescence (CL) of quartz—a review. Mineral Petrol 71:225–250

Graupner T, Kempe U, Dombon E, Pätzold O, Leeder O, Spooner ETC (1999) Fluid regime and ore formation in the tungsten(-yttrium) deposits of Kyzyltau (Mongolian Altai): evidence for fluid variability in tungsten-tin ore systems. Chem Geol 154:21–58

Graupner T, Götze J, Kempe U, Wolf D (2000) CL for characterizing quartz and trapped fluid inclusions in mesothermal quartz veins: Muruntau Au ore deposit Uzbekistan. Mineral Mag 64(6):1007–1016

Griggs DT, Blacic JD (1965) Quartz: anomalous weakness of synthetic crystals. Science 147:292–295

Hallbauer DK (1992) The use of selected trace elements in vein quartz and quartz pebbles in identifying processes of formation and source rocks. Proc Geol Soc South Africa, Bloemfontain, 157–159

Haßler S, Kempe U, Monecke T, Götze J (2005) Trace element content of quartz from the Ehrenfriedersdorf Sn-W deposit, Germany: results of an acid-wash procedure. In: Mao Y, Bierlein FP (eds) Mineral deposit research: meeting the global challenge. Springer, Berlin, pp 397–400

Higgins NC (1980) Fluid inclusion evidence for the transport of tungsten by carbonate complexes in hydrothermal solutions. Can J Earth Sci 17:823–830

Higgins NC (1985) Wolframite deposition in a hydrothermal vein system: the Gray River tungsten prospect, Newfoundland, Canada. Econ Geol 80:1297–1327

Hösel G (1994) Das Zinnerz-Lagerstättengebiet Ehrenfriedersdorf/Erzgebirge. Sächsisches Landesamt für Umwelt und Geologie, Radebeul und Sächsisches Oberbergamt, Freiberg, 196 pp

Jolley SJ, Henderson HC, Barnicoat AC, Fox PC (1999) Thrust-fracture network and hydrothermal gold mineralization: Witwatersrand Basin, South Africa. In: McCaffery KJW,

Lonergan L, Wilkinson JJ (eds) Fractures, fluid flow and mineralization. Geol Soc, vol 155. Spec Publ, London, pp 153–165

Jourdan AL, Vennemann TW, Mullis J, Ramseyer K (2009) Oxygen isotope sector zoning in natural hydrothermal quartz. Mineral Mag 73:615–632

Kamo SL, Reimold WU, Krogh TE, Colliston WP (1996) A 2.023 Ga age for the Vredefort impact event and a first report of shock metamorphosed zircons in pseudotachylitic breccias and granophyre. Earth Planet Sci Lett 144:369–387

Kempe U, Dandar S, Getmanskaya TI, Wolf D (1994) The tungsten-antimony mineralization (Focussed on new occurrences in the Mongolian Altai). In: Seltmann R, Kämpf H, Möller P (eds) Metallogeny of Collisional Orogens. Czech Geological Survey, Prague, pp 301–308

Kempe U, Götze J, Dandar S, Habermann D (1999) Magmatic and metasomatic processes during formation of the Nb-Zr-REE deposits Khaldzan Buregte and Tsakhir (Mongolian Altai): indications from a combined CL-SEM study. Mineral Mag 63(2):165–177

Kouzmanov K, Pettke T, Heinrich CA (2010) Direct analysis of ore-precipitating fluids: combined IR microscopy and LA-ICP-MS study of fluid inclusions in opaque ore minerals. Econ Geol 105:351–373

Kremenetsky AA, Maksimyuk IE, Yushko NA, Kempe U, Poutivtsev M (2005) Trace elements in quartz from conglomerates in the Witwatersrand Basin (South African Republic) and its role in the understanding of the deposit formation. In: Burenko EK, Kremenetsky AA (eds) Prikladnaya Geokhimiya. IMGRE, Moscow, vyp. 7, 1 87–100 (in Russian)

Landtwing MR, Pettke T (2005) Relationships between SEM-cathodoluminescence response and trace-element composition of hydrothermal vein quartz. Am Mineral 90:122–131

Larsen RB, Henderson I, Ihlen PM, Jacamon F (2004) Distribution and petrogenetic behaviour of trace elements in granitic pegmatite quartz from South Norway. Contrib Mineral Petrol 147:615–628

Larsen RB, Jacamon F, Kronz A (2009) Trace element chemistry and textures of quartz during the magmatic hydrothermal transition of Oslo Rift granites. Mineral Mag 73:691–707

Law JDM, Phillips GN (2005) Hydrothermal replacement model for Witwatersrand gold. Econ Geol 100th Anniversary Volume:799–811

Lehmann K, Berger A, Götte T, Ramseyer K, Wiedenbeck M (2009) Growth related zonations in authigenic and hydrothermal quartz characterized by SIMS-, EPMA-, SEM-CL- and SEM-CC-imaging. Mineral Mag 73:633–643

Lehmann K, Pettke T, Ramseyer K (2011) Significance of trace elements in syntaxial quartz cement, Haushi Group sandstones, Sultanate of Oman. Chem Geol 280:47–57

Lüders V (1996) Contribution of infrared microscopy to fluid inclusion studies in some opaque minerals (wolframite, stibnite, bournonite): metallogenic implications. Econ Geol 91:1462–1468

McLaren AC, Cook RF, Hyde ST, Tobin RC (1983) The mechanisms of formation and growth of water bubbles and associated dislocation loops in synthetic quartz. Phys Chem Mineral 9:79–94

Minter WEL, Goedhart M, Knight J, Frimmel HE (1993) Morphology of Witwatersrand gold grains from the Basal Reef: evidence for their detrital origin. Econ Geol 88:237–248

Miyoshi N, Yamaguchi Y, Makino K (2005) Successive zoning of Al and H in hydrothermal vein quartz. Am Mineral 90:310–315

Monecke T, Bombach G, Klemm W, Kempe U, Götze J, Wolf D (2000a) Determination of trace elements in the quartz reference material UNS-SpS and in natural quartz samples by ICP-MS. Geostandard Newlett 24:73–81

Monecke T, Monecke J, Mönch W, Kempe U (2000b) Mathematical analysis of rare earth element patterns of fluorites from the Ehrenfriedersdorf tin deposit, Germany: evidence for a hydrothermal mixing process of lanthanides from two different sources. Mineral Petrol 70:235–256

Monecke T, Kempe U, Götze J (2002) Genetic significance of the trace element content in metamorphic and hydrothermal quartz: a reconnaissance study. Earth Planet Sci Lett 202:709–724

Monecke T, Kempe U, Trinkler M, Thomas R, Dulski P, Wagner T (2011) Unusual rare earth element fractionation in a tin-bearing magmatic-hydrothermal system. Geology 39:295–298

Müller A, Herrington R, Armstrong R, Seltmann R, Kirwin D, Stenina NG, Kronz A (2010) Trace elements and cathodoluminescence of quartz in stockwork veins of Mongolian porphyry-style deposits. Mineral Deposita 45:707–727

Müller A, Wiedenbeck M, Van den Kerkhof AM, Kronz A, Simon K (2003) Trace elements in quartz—a combined electron microprobe, secondary ion mass spectrometry, laser-ablation ICP-MS, and cathodoluminescence study. Eur J Mineral 15:747–763

Novgorodova MI, Veretennikov VM, Boyarskaya RV, Drynkin VI (1984) Geochemistry of trace elements in gold-bearing quartz. Geochem Int 21:101–113

Onasch CM, Vennemann TW (1995) Disequilibrium partitioning of oxygen isotopes associated with sector zoning in quartz. Geology 23:1103–1106

Penniston-Dorland SC (2001) Illumination of vein quartz textures in a porphyry copper ore deposit using scanned cathodoluminescence: Grasberg Igneous Complex, Irian Jaya, Indonesia. Am Mineral 86:652–666

Perny B, Eberhardt P, Ramseyer K, Mullis J, Pankrath R (1992) Microdistribution of Al, Li, and Na in α quartz: possible causes and correlation with short-lived cathodoluminescence. Am Mineral 77:534–544

Phillips GN, Law JDM (2000) Witwatersrand gold fields: geology, genesis, and exploration. Rev Econ Geol 13:439–500

Phillips GN, Myers RE (1989) The Witwatersrand goldfields: Part II. An origin for Witwatersrand gold during metamorphism and associated alteration. Econ Geol Mon 6:598–608

Plötze M (1995) EPR investigations of quartz, scheelite and fluorite from high-thermal trace-metal mineralization (in German). PhD thesis, TU Bergakademie Freiberg, 141 p

Poutivtsev M (2001) Bestimmung der Spurenelementgehalte in Konglomeratquarzen aus vererzten und unvererzten Reefs der Au-U-Lagerstätte Witwtersrand (Südafrikanische Republik) im Vergleich mit Konglomeratquarzen aus Vorkommen in Karelien (Russland) Unpubl Diploma thesis, TU Bergakademie Freiberg, 76 pp

Poutivtsev M, Kempe U, Götze J, Monecke T, Wolf D, Kremenetsky AA (2001) Cathodoluminescece and trace element characteristics of quartz pebbles from the Witwa-tersrand, South Africa. In: Cathodoluminescence in geosciences: new insights from CL in combination with other techniques, Abstracts. TU Bergakademie Freiberg, pp 101–102

Ramseyer K, Mullis J (1990) Factors influencing short-lived blue cathodoluminescence of α-quartz. Am Mineral 75:791–800

Richter DK, Götte T, Götze J, Neuser RD (2003) Progress in application of cathodoluminescence (CL) in sedimentary geology. Mineral Petrol 79:127–166

Robb LJ, Meyer FM (1990) The nature of the Witwatersrand hinterland: conjectures on the source area problem. Econ Geol 85:511–536

Robb LJ, Meyer FM (1991) A contribution to recent debate concerning epigenetic versus syngenetic mineralization processes in the Witwatersrand basin. Econ Geol 86:396–401

Roedder E (1984) Fluid inclusions. Reviews in Mineralogy, vol 12. Mineralogical Society of America, Washington, 646 pp

Rusk BG, Reed MH (2002) Scanning electron microscope–cathodoluminescence analysis of quartz reveals complex growth histories in veins from the Butte porphyry copper deposit, Montana. Geology 30:727–730

Rusk BG, Koenig A, Lowers HA (2011) Visualizing trace element distribution in quartz using cathodoluminescence, electron microprobe, and laser ablation inductively coupled plasma mass spectrometry. Am Mineral 96:703–708

Rusk BG, Lowers H, Reed MH (2008) Trace elements in hydrothermal quartz; relationships to cathodoluminescent textures and insights into hydrothermal processes. Geology 36:547–550

Rusk BG, Reed MH, Dilles JH, Kent AJR (2006) Intensity of quartz cathodoluminescence and trace element content of quartz from the porphyry copper deposit in Butte, Montana. Am Mineral 91:1300–1312

Samson I, Anderson A, Marshall D (2003) Fluid inclusions: analysis and interpretation. Mineral Assoc, Canada, Short Course vol 32, Quebec, 374 pp

Schaefer BF, Pearson DG, Rogers NW, Barnicoat AC (2010) Re–Os isotope and PGE constraints on the timing and origin of gold mineralisation in the Witwatersrand basin. Chem Geol 276:88–94

Seyedolali A, Krinsley DH, Boggs S, O'Hara PF, Dypvik H, Goles GG (1997) Provenance interpretation of quartz by scanning electron microscope-cathodoluminescence fabric analysis. Geology 25:787–790

Shore M, Fowler AD (1996) Oscillatory zoning in minerals: a common phenomenon. Can Mineral 34:1111–1126

Sørensen BE, Larsen RB (2009) Coupled trace element mobilisation and strain softening in quartz during retrograde fluid infiltration in dry granulite protoliths. Contrib Mineral Petrol 157:147–161

Spear FS, Wark DA (2009) Cathodoluminescence imaging and titanium thermometry in metamorphic quartz. J metamorphic Geol 27:187–205

Sprunt ES, Dengler LA, Sloan D (1978) Effects of metamorphism on quartz cathodoluminescence. Geology 6:305–308

Taylor RP (1992) Petrological and geochemical characteristics of the Pleasant Ridge zinnwaldite—topaz granite, Southern New Brunswick, and comparisons with other topaz-bearing felsic rocks. Can Mineral 30:895–921

Thomas R, Förster HJ, Rickers K, Webster JD (2005) Formation of extremely F-rich hydrous melt fractions and hydrothermal fluids during differentiation of highly evolved tin-granite magmas: a melt/fluid-inclusion study. Contrib Mineral Petrol 148:582–601

Van den Kerkhof AM, Hein UF (2001) Fluid inclusion petrography. Lithos 55:27–47

Vennemann TW, Kesler SE, O'Neil JR (1992) Stable isotope compositions of quartz pebbles and their fluid inclusions as tracer of sediment provenance: implications for gold- and uranium-bearing quartz pebble conglomerates. Geology 20:837–840

Vollbrecht A, Oberthür T, Ruedrich J, Weber K (2002) Microfabric analyses applied to the Witwatersrand gold- and uranium-bearing conglomerates: constraints on the provenance and post-depositional modification of rock and ore components. Mineral Deposita 37:433–451

Vollbrecht A, Ruedrich J, Weber K, Oberthür T (1996) Gefügekundliche Untersuchungen an Geröllquarzen der Witwatersrand-Lagerstätte in Südafrika. Z Angew Geol 42:156–161

Wark DA, Watson EB (2006) TitaniQ: a titanium-in-quartz geothermometer. Contrib Mineral Petrol 152:743–754

Wark DA, Hildreth W, Spear FS, Cherniak DJ, Watson EB (2007) Pre-eruption recharge of the Bishop magma system. Geology 35:235–238

Watt GR, Wright P, Galloway S, McLean C (1997) Cathodoluminescence and trace element zoning in quartz phenocrysts and xenocrysts. Geochim Cosmochim Acta 61:4337–4348

Webster JD (2006) Melt inclusions in plutonic rocks. Mineral Ass Can, Short courses vol 36, Quebec, 237 pp

Weil JA (1984) A review of electron spin spectroscopy and its application to the study of paramagnetic defects in crystalline quartz. Phys Chem Mineral 10:149–165

Weil JA (1993) A review of the EPR spectroscopy of the point defects in α-quartz: the decade 1982–1992. In: Helms CR, Deal BE (eds) Physics and Chemistry of $SiO_2$ and the Si-SiO interface 2. Plenum Press, New York, pp 131–144

Whiting KL, Rusk B, Spandler C, Dimond A, Emsbo P (2010) Insights into the origin of Charters Towers Warrior Vein system from fluid inclusions and quartz trace elements. EGRU Newsletter, School of Earth & Environmental Science, Economic Geology Research Unit, James Cook University, Australia, vol 8, pp 12–14

Williams LB, Hervig RL, Bjørlykke K (1997) New evidence for the origin of quartz cements in hydrocarbon reservoirs revealed by oxygen isotope microanalyses. Geochim Cosmochim Acta 61:2529–2538

Zinkernagel U (1978) Cathodoluminescence of quartz and its application to sandstone petrology. Contrib Sedimentol 8:1–96

Zuffa GG (1985) Provenance of arenites. NATO ASI series C 148. Reidel Publ. Co., Boston, p 393

# Index



J. Götze and R. Möckel (eds.), *Quartz: Deposits, Mineralogy and Analytics*, Springer Geology, DOI: 10.1007/978-3-642-22161-3,

MIX
Papier aus verantwortungsvollen Quellen
Paper from responsible sources
FSC® C105338

If you have any concerns about our products,
you can contact us on
**ProductSafety@springernature.com**

In case Publisher is established outside the EU,
the EU authorized representative is:
**Springer Nature Customer Service Center GmbH**
**Europaplatz 3, 69115 Heidelberg, Germany**

Printed by Libri Plureos GmbH
in Hamburg, Germany